A Clinical Lens on Pediatric Engineering

Jamie Leigh Wells
Editor

A Clinical Lens on Pediatric Engineering

Pioneering Science and Technology for Cutting-Edge Patient Care

Springer

Editor
Jamie Leigh Wells
Ventnor, NJ, USA

ISBN 978-3-031-84654-0 ISBN 978-3-031-84655-7 (eBook)
https://doi.org/10.1007/978-3-031-84655-7

This Springer imprint is published by the registered company Springer Nature Switzerland AG
The registered company address is: Gewerbestrasse 11, 6330 Cham, Switzerland

To my dream team for their unconditional love, inspiration and support –

My grandfather, Joseph "Poppop Joe" Friedman, a Renaissance man whom I consider one of the earliest biomedical engineers who instilled in me a lifelong passion for learning fostering limitless curiosity about the world and gratitude for every sunrise.

My father, Dr. Richard N. Wells, the original Dr. Wells, the consummate physician who struck the delicate balance between evidence, humanity and common sense while preserving patient dignity and delivering the highest quality care. Adored by his patients, there was no one better at getting to the bottom line. I will cherish and never forget your exuberance at discovering I was approached to do this book. I wish we could have celebrated its publication together, but I know you are always with me. I hope I made you proud. When you hear hoofbeats, think horses not zebras!

Mollie Joe Wells, the human bulldog, my child of canine persuasion and shadow, a legend, icon, idol and epic love. My better

half of our certified animal-assist therapy team who brought mobile joy and healing to all lucky enough to meet her. There was no greater co-worker, particularly wonderful when preparing this book.

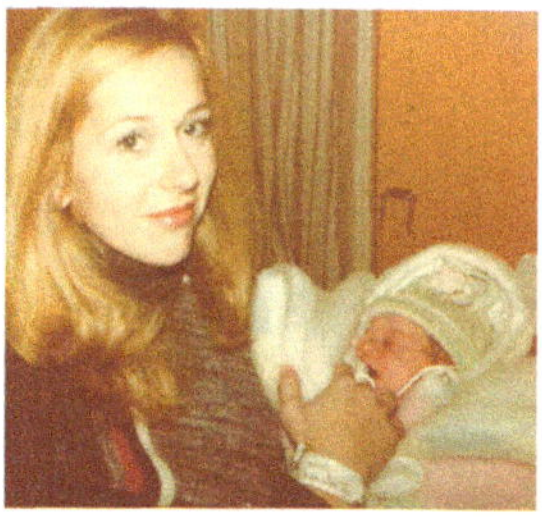

My mother, Estine F. Wells, "honorary physician" and sage, patient and family advocacy tour de force and the ultimate caregiver. You make every person you meet and environment you enter better and more beautiful.

Foreword

Children differ from adults in their physiology, pharmacokinetics, behavior, socialization, development, and psychology. In other words, they aren't little adults. Nonetheless, medical innovations have often treated them that way.

In this compelling series of monographs, Jamie Wells has assembled an impressive group of experts who describe how science and medicine have progressively tailored therapies to children. Prior to those advances, the notion that adult data could simply be extrapolated to children often put them at unnecessary risk. Advances include pediatric pharmacotherapy, surgical interventions, public health, social engineering, gene editing, radiology, medical devices, immunosuppression, solid organ transplantation, and minimally invasive surgery. It is only in understanding the narrative arc of addressing the specific needs of children that we can best understand what the future holds for further advances.

Paul A. Offit

Vaccine Education Center, The Children's Hospital of Philadelphia, Philadelphia, PA USA
Department of Pediatrics, Perelman School of Medicine, University of Pennsylvania, Philadelphia, PA USA

Preface

> A physician is obligated to consider more than a diseased organ, more even than the whole man – he must view the man in his world.
>
> —Harvey W. Cushing, MD

As pioneer Dr. Harvey W. Cushing—considered the *father of modern neurosurgery* having revolutionized the discipline—aptly proclaimed, "A physician is obligated to consider more than a diseased organ, more even than the whole man—he must view the man in his world." Notably, with respect to the field of pediatrics and medical transformation, understanding met and unmet clinical needs along a continuum is a crucial first step to engineering biomedical solutions that not only fix problems and address challenges we actually have, but also prevent unnecessary or reverberating adverse events and beget even more avenues for scientific progress applicable throughout the entire lifespan. Historically, children were viewed as *just little adults*; therefore, clinical treatments and tools for adults were simply modified to tackle this issue of scale while ignoring the varied biopsychosocial, environmental, disease- and age-specific, cognitive or ethical complexities of the developing and growing child. This oft myopic approach in large part led to an existing innovation lag in pediatrics when compared to adults. Fortunately, we have entered the dawning era of pediatric engineering—a newly emerging field and applied science—which is ripe for discovery and well on its way to addressing this health inequity. This landmark book on pediatric engineering from a clinical perspective seeks to open minds, debunk myths, and offer actionable insights to facilitate the success of current and budding pediatric innovators and leaders to advance child health.

Atop the 30,000-foot view, this work will take you on an exciting journey from where we have been to where we are now and where we are headed in the future, detailing the fascinating evolution of the many feats of engineering and game-changing advances in pediatric medicine. It will underscore the importance of nuance in securing optimal diagnostic, therapeutic, research, and overall patient outcomes while resisting the status quo reactive, not proactive default of painting with a broad brush and cultivating silos, whether from policy to protocol, or prototype to digital platform, and regulatory landscape. When fabricating and refining novel and existing treatments or technologies—especially for such a heterogeneous population that warrants interventions in the fetal realm to the premature infant, toddler, school-age child and, adolescent on up to the young adult, it is vital to explore and acquire good evidence (not simply data, albeit bad data, for data's sake) all the while interweaving a consistently agile and pediatric clinical lens. Designing the most elegant technology or digital solution in a vacuum without constant consideration of human factors and surroundings, particularly the distinct often dynamic needs of the end user or primary stakeholders, can relegate a cutting-edge medical device to a shelf, unused or worse, yet, creating new problems that do meaningful harm. Additionally, we will delve into the long held fallacy that investing in pediatric research has *limited* returns, previously prompting companies to direct their resources to *safer* bets while perpetuating myths well overdue for debunking. We identify and address these real and imagined financial disincentives, elucidate further the pediatric engineering renaissance currently underway, and offer the position, providing many examples, that such allocation in pediatric discovery could actually inform on the adult sphere and overall population, not just the other way around which has been the longstanding precedent.

On a more granular 30-foot level, Part I: *Know Thy Patient—Design and Adapt Accordingly* serves to bring readers of all experiences to the same page and examines why it is critical to know who the pediatric patient is, how they are similar and disparate from adults, and what barriers exist to their care. It explores their contrasting physiology, drug metabolism, ethical paradigm, and dedicated healthcare environs. Chapter 1 deeply dives into the history of pediatric medicine, pediatric engineering, and innovation demonstrating the vast scope of breakthroughs (e.g., fetal surgical interventions, artificial placenta, iron lung, cardiac devices, operative tools), occasions of failure, developmental origins of adult disease and need for tailored therapies. Chapter 2 delves into pediatric pharmacokinetics and pharmacodynamics underscoring the multifariousness of drug absorption, distribution, metabolism, and excretion from infancy to adolescence while detailing the regulatory history and possible solutions like acquiring real world evidence (RWE), artificial intelligence (AI), and Bayesian dosing. Chapter 3 provides a holistic approach to the built environment richly exploring evidence-based architectural engineering design of pediatric healthcare facilities and children's hospitals to improve outcomes, enhance patient care and community health, alleviate inequities, ease access, reduce costs, and remain flexible for evolving biomedical equipment and technological innovation. Chapter 4 reviews fundamental bioethical principles and frameworks addressing the diverse challenges specifically applicable to pediatrics, including case examples, discussion of minor status, cognitive differences, hurdles to pediatric clinical trials, experimental therapies, and novel technologies.

In Part II: *Facilitating Progress from Bench to Bedside—Wins, Cautionary Tales, and Lessons in Translation*, Chapter 5—authored by the leadership of the Southwest-Midwest National Pediatric Device Innovation Consortium (SWPDC), a "free and no-strings-attached" virtual accelerator based at Texas Children's Hospital and Baylor College of Medicine and one of five US Food and Drug Administration (FDA) P50 grant-supported pediatric device consortia, provides a comprehensive road map on how to manifest an idea through to clinical implementation integrating all pediatric stakeholder perspectives and instructing pediatric innovators on *how*, *what*, *where*, *when*, *who*, and *why* to get support to achieve their desired success. Punctuated throughout with innumerable examples, this handy guide is a must-read for anyone endeavoring to pioneer pediatric healthcare (and medicine, in general), avoid pitfalls, and amplify the incredible promise for patients now and well into future generations. Chapter 6 pivots to the vanguard of regenerative medicine for our youngest patients detailing and advising on the rapidly evolving science and multifaceted ecosystem required to navigate and coalesce much needed disruption to cure some of our most lethal, debilitating diseases.

Culminating in Chapters 7–13 of Part III: *Pediatric Engineering in Clinical Context—Digital to Device Innovation Past, Present and Future*, you will appreciate the expansive scope of riveting innovation that has already taken place, is still needed and is on the horizon in every facet of care delivery in pediatrics. Topics investigated here and throughout the textbook are vast, widely ranging beyond eye cancer and liquid biopsy, molecular profiling, precision medicine, radiology imaging modalities, artificial intelligence as it applies to multiple fields, machine learning, nanotechnology, gene therapy, liver transplantation, immunosuppression, prognostic biomarkers, orthopedic surgery, oncology, neurodiversity and autism, wearable tech, augmented and virtual reality, fetal surgery, minimally-invasive surgery, cryoanalgesia, 3D printing for tumor resection operative planning, materials science and tissue engineering, and digital health technologies for substance use treatment and prevention from the team at the new Yale Center for Immersive Technologies in Pediatrics and Yale School of Medicine.

My goal is for this book to be a catalyst in this pediatric engineering sea change by seeking to cultivate a growth mindset when it comes to progressing scientific and medical advancement in the arena of pediatrics. To shift toward actionable and successful innovation in this sphere, bridging interdisciplinary knowledge gaps is essential. With the establishment of pediatric device consortia, expanding public-private partnerships, hospital and academic innovation centers stretching across engineering programs, medical schools and clinical settings along with educational, research, and pediatric-directed initiatives, there is no better time than the present or more conducive habitat to effect positive change. This is why contributors to this work possess assorted expertise (e.g., cardiology, critical care, neonatology, ophthalmology, emergency medicine, developmental-behavioral, architecture, urology, bioethics, radiology, pharmacology, engineering, nursing, business, biotech, FDA and NIH advisory, regulatory, addiction), reflect hospital, organization and division leadership, and traverse institutions (e.g., Children's Hospital of Philadelphia, Children's Hospital Los Angeles, NYU Langone Health). When identifying solutions, particularly technical ones, collaboration spanning the spectrum from the patient-physician-caregiver (e.g., parent, legal guardian, family) triad to the research, organizational alliances, regulatory, engineering, nursing and translational teams (and more) coupled with a humanity-centered, multidisciplinary design strategy help course correct discovery pathways in real time. You will gain a complete understanding of why this mitigates suffering, emotional and financial costs and accelerates novel technologies that can authentically revolutionize care. It is this inclusion of end user and primary stakeholder considerations—namely patient and practitioner—early and often that ensures bioethical and unique pediatric considerations remain at the forefront of every phase of development and implementation.

On a personal note, engineering has been a core fabric throughout my life profoundly shaping my outlook. As a child, I was blessed to learn from the best, my grandfather, whom I will always consider an early *biomedical* engineer. He lived with us and we all contributed to his home healthcare. An inventor, philosopher, musician, and artist, this polymath saw opportunity, not obstacle. Faced with intractable intention tremors, he reverted to his engineering training and designed stabilizing contraptions for his upper extremities so he could still draw, write, and play the piano. He saw failure as but one way not to do something and kept on moving. We did calculus problems together and he taught me the Pythagorean theorem when I was four as he believed doing *mental gymnastics* every day keeps one's mind alert. With my early exposure to patients, their difficult plight and the medical province overall, I cemented this mentality by accompanying my physician father on house calls and hospital rounds. I credit the two of them for what I believe is requisite for innovation, a *yes* before *no* perspective replete with a keen eye and curiosity to reassess or refine, continually, the mousetrap. Also, they saw no limits to what I could do as a female in STEMM (**S**cience, **T**echnology, **E**ngineering, and **M**ath, using editor's license with the second "M" for **M**edicine since I believe they are all intrinsically linked), this factor was of no consequence and, undoubtedly, positively influenced my resilience and fortitude.

This way of thinking framed my academic and career pursuits and drive to solve problems and seek improvements. Whether as President of the Yale Science and Engineering Association (YSEA), Director of the Massachusetts Institute of Technology's (MIT) Research Science Institute (RSI) collaboratively sponsored by the Center for Excellence in Education (CEE), a leadership council member of the Wistar Institute—our nation's first independent biomedical research facility, author of over 400 published articles informing the public and policy makers on medtech innovation to genetic engineering, pediatrics, and more, judging the world championships of a robotics competition or delivering a keynote at the Northeast Bioengineering Conference (NEBEC), I will always champion integrating silos that all too frequently impede forward momentum. It is my contention there is a lot of lost innovation on the frontlines of healthcare. When Drexel University's School of Biomedical Engineering, Science and Health Systems (BIOMED) spearheaded the nation's first degree program in pediatric engineering, I was especially thrilled to be asked to help establish and build upon expanding Drexel's global footprint in engineering and innovation for pediatric patients. As an Adjunct Professor, I helped develop the curriculum, teach the inaugural and subsequent classes, co-authored a cover feature in the journal *Artificial Organs,* did extensive outreach and served as an Advisory Leadership Board Member for their U.S. Department of Education Ph.D. Training Grant:

Research and Engineering for Pediatrics by Interdisciplinary Collaboration Leveraging Education and Partnerships for Pediatric Healthcare (R-EPIC LEAP for Pediatric Healthcare).

My hope is for any reader running the gamut from seasoned physician or aspiring one, CEO or executive of a health system or biotech company, lay person, researcher, biomedical engineer or graduate student (e.g., PhD, MD, MBA, MPH), to policy maker, educator or institution leadership to read this work, garner a thorough understanding of the terrain, acquire skills to align incentives and become adept at creating workarounds so as to be motivated to action for this remarkable and vulnerable population. As someone who has had the great privilege of caring for patients, I am honored to present this labor of love and strongly encourage the incorporation of a clinical lens in all phases of biomedical trailblazing to understand the pediatric patient in their world, for their (and technology's!) betterment.

Lastly, I would like to extend my most heartfelt appreciation to an outstanding group of contributing authors who brought this vision to fruition on an impeccable level. Given their extraordinary expertise, my expectations were already high. They exceeded them. Due to their generosity with their time, spirit, and tireless commitment to child health, they delivered a scientifically rigorous, flawless result. I remain forever grateful.

Ventnor, NJ, USA Jamie Leigh Wells

Acknowledgments

It is a career pinnacle to craft and complete this groundbreaking and significant work. In aggregate, it is a lifetime of profound experiences, relationships, and lessons that I believe are foundational to this achievement. It is these collective exposures that have empowered me to be a machine learning vessel ever refining and seeking better solutions to improving healthcare quality, patient care, outcomes, innovation, and contributing substantially to the literary canon. This ongoing knowledge accumulation, the persistent sharing of it and the continuous bridging of interdisciplinary gaps will help revolutionize the state of and future of medicine. It is the recognition that these diverse influences along the way buoy such efforts and make them even possible. The advancement of societal medical and scientific progress is a team sport and I am lucky to have been joined by a windfall of the most valuable players at every stage.

First, I would like to express my deepest gratitude to my legions of patients and their families who entrusted me with their care. I am humbled by their faith in me and will be guided by and carry these cherished encounters with me forever. When it comes to describing the meaningful impact being a physician has had on me, words fall short. Thank you to my department chairs and countless colleagues whose exceptional efforts synergized to benefit our patients, improve healthcare, and cultivate more learning. Whenever called upon, they always generously avail themselves and deliver albeit clinically or share their expertise with the next generation in educational forums. That these associations are sure to last a lifetime speaks to their tremendous influence in my life.

I am blessed to have traveled a unique professional path that has permitted active collaboration or interaction spanning public and private sectors, disciplines, industries, academia, policy makers, researchers, physicians, scientists, engineers, nurses, students, executives, journalists, nonprofits, leadership, program directors, hospital custodial, food service and security personnel, administrators, volunteers, technicians, and educators. Each institution, organization, and individual left their indelible mark to shape my outlook or perspective, for which I am eternally grateful. Often, I was inspired and my intellectual curiosity nourished. Among those I would also like to acknowledge are the wonderful editorial and production teams at

Springer Nature. Thank you for your always gracious assistance and enthusiasm over this project. To renowned innovator Paul A. Offit, MD, who contributed the foreword, I will forever treasure your support and endorsement of this book.

It would literally take another book to identify everyone I would like to champion for their excellence, encouragement, invaluable insights, and generosity with their time. I have been fortunate to be a part of many consortiums filled with extraordinary people working to address healthcare challenges, treat patients, or teach aspiring scholars. To my many students and mentees, future and active clinicians, biomedical engineers and professionals, thank you for the privilege of being your steward. These alliances make me better, more determined and motivated.

Finally, I would like to give special mention to Dr. Lee A. Snyder, MD, a model physician, surgeon, and patient advocate, who is always a fantastic sounding board. Thanks especially for the *high level* meetings. Thank you to Professor Amy Throckmorton, PhD for her dedication to the field of pediatric engineering, training the next generation of biomedical engineers, and the incredible work being done at Drexel Biomed in this newly emerging science. Thank you to my brothers, nieces, family (human and canine), and friends for their unwavering support, love, and wisdom.

Contents

About the Editor

Jamie Leigh Wells, MD, FAAP, the President and Founder of the Yale Alumni Health Network (YAHN), whose mission is to break down silos in the health-related disciplines to spur innovation, is an award-winning Board-certified physician leader with many years of experience caring for patients having served as a clinical instructor and attending physician at NYU Langone, Mount Sinai Beth Israel, and St. Vincent's Medical Centers in Manhattan. An RSI alumna and scholar, she was Director of the Research Science Institute (RSI) at the Massachusetts Institute of Technology (MIT) collaboratively sponsored by the Center for Excellence in Education (CEE). An innovator and thought leader, Dr. Wells has published over 400 articles as Director of Medicine for an educational advocacy nonprofit informing the public and policy makers on evidence-based medicine and science and was a longstanding member of The Leadership Council of the Wistar Institute, the nation's first independent biomedical research facility. When Drexel University's School of Biomedical Engineering, Science and Health Systems (BIOMED) spearheaded the nation's first degree program in pediatric engineering, she was especially thrilled to be asked to help establish and build upon expanding Drexel's global footprint in engineering and innovation for pediatric patients. As an Adjunct Professor, she helped develop the curriculum, teach the inaugural and subsequent classes, co-authored a cover feature in the journal *Artificial Organs,* did extensive outreach, and served as an Advisory Leadership Board Member for their US Department of Education PhD

Training Grant: "Research and Engineering for Pediatrics by Interdisciplinary Collaboration Leveraging Education and Partnerships for Pediatric Healthcare (R-EPIC LEAP for Pediatric Healthcare)." Her book, *A Clinical Lens on Pediatric Engineering: Pioneering Science and Technology for Cutting-Edge Patient Care,* published by Springer Nature is a landmark work in the field.

Dr. Wells writes, speaks, consults, is on panels and routinely interviewed as a medical expert to global audiences. A featured speaker on the future of healthcare and the role of technology and community in addressing the world's most pressing problems, Dr. Wells—a US Ambassador in Healthcare for the Global Blockchain Business Council (GBBC), an official partner of the UN Foundation's Global Goals week—has been among the leaders gathered across technology, government, industry, and academia alongside the United Nations General Assembly (UNGA) to unpack the crucial steps necessary to accelerate progress. Her keynote addresses and distinguished lectures include, but are not limited to the Northeast Bioengineering Conference (NEBEC), Math and AI 4 girls annual award ceremony, Biomedical Research Seminar at ODU's Center for Bioelectronics, and the USA Biology Olympiad (having been one of five Americans selected to compete in the International Biology Olympics as a teenager). Her insights have been showcased in acclaimed medical journals like *Clinical Radiology, Review of Ophthalmology* and *NeoReviews* to print media, podcasts and television, such as *The Washington Post, U.S. News & World Report*, Reuters, BBC, SiriusXM, CBS Evening News, ABC, CNN, TLC, Discovery Health, FNC, PBS, WebMD, *Becker's Hospital Review* and Fox Business Network. Her work is broadly read from the general public and academia to the White House. One such article was named to Doximity's Best of the Year list. Dr. Wells' greatest asset is making complicated material palatable for any audience. She champions increased connectivity between disciplines as the optimal path toward genuine innovation.

Often profiled as a trailblazer, she was featured in HealthVenture's series highlighting awe-inspiring women leading transformation in healthcare, as a physician leader by the Women in Medicine Legacy Foundation and in *Physician Outlook Magazine's* #stellardoc campaign during National Physicians' Week. Dr. Wells is deeply committed to advancing societal medical and scientific progress while ensuring patient safety as well as frequently educating, mentoring, and career advising the next generation. Her experiences are vast from being a grant reviewer for Komen and judge of a world robotics competition to a faculty mentor for the Custom-Designed Major Senior Capstone Project—Ethical Engineering and Biotechnology at Drexel's Pennoni Honors College and coach of CEE's team for the International Public Policy Forum's (IPPF) global debate competition.

A National Merit Scholar, she was also a recipient of *USA Today's* scholarship honoring the top twenty high school students in the country. Dr. Wells graduated with honors from Yale University where she was president of the Yale Science and Engineering Association. She was inducted as a junior, and elected president of the Alpha Omega Alpha National Medical Honor Society at Jefferson Medical College where she obtained her medical degree. She was named a New York Super Doctor, repeatedly, in the *NY Times* magazine supplement listing the top 5% of physicians in over 30 medical specialties as chosen by their peers. She has been awarded America's Top Pediatricians, America's Top Physicians Honors of Distinction and Excellence, America's Most Compassionate Doctors Award, Patients' Choice Award (honors given by patients to <3% of the nation's 720,000 active physicians), and has been recognized for her exemplary care of those with Cystic Fibrosis. Dr. Wells was named a Doctor of Excellence which profiles the world's leading doctors who have demonstrated success and leadership in their profession. Believing she wanted to be a brain surgeon, she began her first residency in neurosurgery, ultimately switching fields to pediatrics. A huge proponent of the health benefits of animals, she was certified with her beloved gifted and adorable English bulldog, Mollie Joe, as a therapy team.

Email: drjamiewells@gmail.com

X (formerly Twitter): @jamiewellsmd

Contributors

Nadeen Alturki, MBChB Department of Surgery, Hackensack University Medical Center, Hackensack, NJ, USA

Kammarauche Aneni, MD, MHS The ACCESS Lab, Yale Child Study Center, Yale School of Medicine, New Haven, CT, USA

Yale Center for Immersive Technologies in Pediatrics, Yale School of Medicine, New Haven, CT, USA

Yale Department of Biomedical Informatics and Data Science, Yale School of Medicine, New Haven, CT, USA

Jesse L. Berry, MD The Vision Center at Children's Hospital Los Angeles, Los Angeles, CA, USA

USC Roski Eye Institute, Keck School of Medicine of the University of Southern California, Los Angeles, CA, USA

Norris Comprehensive Cancer Center, Keck School of Medicine, University of Southern California, Los Angeles, CA, USA

The Saban Research Institute, Children's Hospital Los Angeles, Los Angeles, CA, USA

Steven L. Blumer, MD, MBA, CPE, FAAP UPMC Children's Hospital of Pittsburgh, Pittsburgh, PA, USA

University of Pittsburgh School of Medicine, Pittsburgh, PA, USA

Deepa R. Camenga, MD, MHS Yale Center for Immersive Technologies in Pediatrics, Yale School of Medicine, New Haven, CT, USA

Yale Department of Pediatrics, Yale School of Medicine, New Haven, CT, USA

Yale Department of Emergency Medicine, Yale School of Medicine, New Haven, CT, USA

Yale School of Public Health, New Haven, CT, USA

Andrew Chou, MD Department of Orthopaedic Surgery, KK Women's and Children's Hospital, Singapore, Singapore

Jonathan M. Davis, MD Department of Pediatrics, Tufts Medical Center, Boston, MA, USA

Tufts Clinical and Translational Science Institute, Tufts Medical Center, Boston, MA, USA

Wakako Minamoto Eklund, DNP, APRN, NNP-BC, FAANP, FAAN Pediatrix Medical Group of Tennessee, Nashville, TN, USA

Vanderbilt University School of Nursing, Nashville, TN, USA

Brooke M. Ellison, MPP, PhD, CPHQ Stony Brook University, Stony Brook, NY, USA

Technology and Innovation, United Spinal Association, Kew Gardens, NY, USA

Harvard Kennedy School of Government Carr Center for Human Rights Policy, Cambridge, MA, USA

World Economic Forum, Cologny, Switzerland

Harry Truman National Security Project, Washington, DC, USA

Suffolk County Human Rights Commission, Hauppauge, NY, USA

Adrienne Erdman, LSSGB EwingCole, Philadelphia, PA, USA

Bianca Fischer, MD Department of Surgery, Hackensack University Medical Center, Hackensack, NJ, USA

Adam Griesemer, MD New York University (NYU) Transplant Institute, NYU Langone Health, New York, NY, USA

Kimberly Hieftje, PhD XR Pediatrics at Yale, Yale School of Medicine, New Haven, CT, USA

Yale Center for Immersive Technologies in Pediatrics, Yale School of Medicine, New Haven, CT, USA

Yale Department of Pediatrics, Yale School of Medicine, New Haven, CT, USA

R. Brandon Hunter, MD, FAAP Southwest-Midwest Pediatric Device Innovation Consortium, Houston, TX, USA

Baylor College of Medicine, Houston, TX, USA

Texas Children's Hospital, Houston, TX, USA

Carole Kenner, PhD, RN, FAAN, ANEF, IDFCOINN School of Nursing and Health Sciences, The College of New Jersey, Ewing, NJ, USA

Council of International Neonatal Nurses, Inc. (COINN), Yardley, PA, USA

Leslie Kipps, AIA, LEED AP, EDAC, CLSSBB EwingCole, Philadelphia, PA, USA

Chester J. Koh, MD, MBA, FACS, FAAP Southwest-Midwest Pediatric Device Innovation Consortium, Houston, TX, USA

Baylor College of Medicine, Houston, TX, USA

Texas Children's Hospital, Houston, TX, USA

Keith Kuenzler, MD, FACS, FAAP Department of Surgery, Hackensack University Medical Center, Hackensack, NJ, USA

Joseph M Sanzari Children's Hospital, Hackensack, NJ, USA

New York University Langone Medical Center, New York, NY, USA

Joyce P. Lee, PharmD, BCPS Children's Hospital of Philadelphia, Philadelphia, PA, USA

Christine Luk, MBE Southwest-Midwest Pediatric Device Innovation Consortium, Houston, TX, USA

Proxima Clinical Research, Houston, TX, USA

Karim Masrouha, MD New York Ponseti Clubfoot Center, NYU Langone Health, New York, NY, USA

Division of Pediatric Orthopedic Surgery, NYU Langone Health, New York, NY, USA

Division of Musculoskeletal Oncology, NYU Langone Health, New York, NY, USA

NYU Langone Orthopedics, NYU Langone Health, New York, NY, USA

Jhalak N. Mehta, MBE Southwest-Midwest Pediatric Device Innovation Consortium, Houston, TX, USA

Baylor College of Medicine, Houston, TX, USA

Thomas F. Miller, PhD, MBA Bayer Healthcare, LLC, Whippany, NJ, USA

Nikki Montgomery, MA, MEd, GPAC Family Voices, Concord, MA, USA

Colin K. L. Phoon, MPhil, MD Department of Pediatrics and Division of Pediatric Cardiology, Hassenfeld Children's Hospital at NYU Langone, NYU Langone Health, New York, NY, USA

New York University Grossman School of Medicine, NYU Langone Health, New York, NY, USA

Wendy J. Ross, MD, FAAP Jefferson Center for Autism and Neurodiversity, Jefferson Health, Philadelphia, PA, USA

Kanu Priya Singh, PhD XR Pediatrics at Yale, Yale School of Medicine, New Haven, CT, USA

Yale Center for Immersive Technologies in Pediatrics, Yale School of Medicine, New Haven, CT, USA

Yale Department of Pediatrics, Yale School of Medicine, New Haven, CT, USA

Current Affiliation: Department of Technical Communication & Interactive Design, Norman J. Radow College of Humanities & Social Sciences, Kennesaw State University, Kennesaw, GA, USA

Kanwaljit Singh, MD, MPH, MBA International Neonatal Consortium, The Critical Path Institute, Tucson, AZ, USA

Pediatric Programs, The Critical Path Institute, Tucson, AZ, USA

Shreya Sirivolu, BA The Vision Center at Children's Hospital Los Angeles, Los Angeles, CA, USA

USC Roski Eye Institute, Keck School of Medicine of the University of Southern California, Los Angeles, CA, USA

Jennifer Vittorio, MD, FAASLD New York University (NYU) Transplant Institute, NYU Langone Health, New York, NY, USA

Division of Pediatric Gastroenterology & Hepatology, NYU Langone Health, New York, NY, USA

Jamie Leigh Wells, MD, FAAP Ventnor, NJ, USA

Liya Xu, PhD The Vision Center at Children's Hospital Los Angeles, Los Angeles, CA, USA

USC Roski Eye Institute, Keck School of Medicine of the University of Southern California, Los Angeles, CA, USA

Part I
Know Thy Patient: Design and Adapt Accordingly

A Brief History of Pediatrics and the Beginnings of Pediatric Engineering

Colin K. L. Phoon

Pediatrics: Definition(s)

> **Pediatrics**: "A branch of medicine concerned with the development and care of infants and children. Its specialties are the particular diseases of children and their treatment and prevention."—*Mosby's Medical, Nursing, & Allied Health Dictionary* [1]

While this dictionary definition of pediatrics seems straightforward, we can learn a little more about details from other definitions, such as what appears in one of the standard-bearer textbooks, Oski's Pediatrics [2]: "Pediatrics is concerned with the child's physical, mental, and psychosocial health…. The field of pediatrics is about children's health and the factors that affect their health." This chapter is prefaced by this statement: "The field of pediatrics is about the health care of those individuals who will make the future of the world possible." This suggestion as to the long-term ramifications of optimal child health could be found even as far back as at least ancient Rome, when children were sometimes referred to as "*spes gentis*"—or "the hope of the race" [3]. Inherent to each of these definitions are several special considerations:

1. Diseases, and therefore therapies, may be specific only to the pediatric population.
2. Therapies, including surgical intervention, devices and genetic engineering, will need to sustain the young patient for an entire lifetime.
3. Since children are the "future of the world," designing healthcare interventions in children with a long view in mind is critical.

C. K. L. Phoon (✉)
Division of Pediatric Cardiology, Department of Pediatrics, Hassenfeld Children's Hospital at NYU Langone and New York University Grossman School of Medicine, New York, NY, USA
e-mail: colin.phoon@nyulangone.org

J. L. Wells (ed.), *A Clinical Lens on Pediatric Engineering*,
https://doi.org/10.1007/978-3-031-84655-7_1

The Origins of Pediatrics as a Distinct Medical Specialty

For a long time, the health care of children was not necessarily seen as separate from that of adults [4, 5], but with advances in "real," more scientifically based clinical care emerged the realization that infants and children exhibit and develop their own unique diseases and health care challenges. Eventually, it was recognized that children required different care than adults. For much of history, infant mortality rates were astoundingly high, and the first recorded instances of the care of sick children traced to the fifteenth century suggest childhood malnutrition and neurological disorders (chiefly epilepsy) comprised much of pediatric care and hospitalization [6]. The field we now call "pediatrics" evolved over the next three centuries and could be recognized as its own specialty discipline by the mid-eighteenth century [7]. The first children's hospital in the world, L'Hôpital des Enfants-Malades, was established in Paris in 1802 [8], followed by the Hospital for Sick Children at Great Ormond Street in London in 1852 [4, 5, 9] (Fig. 1).

Other children's hospitals quickly followed through the latter half of the nineteenth century and into the first decades of the twentieth century, although children were also cared for in separate and specialized wards within larger general hospitals [5, 10] (Fig. 2).

As it did in Europe, pediatrics as a specialty unto itself in the United States blossomed in the nineteenth century, and Drs. Abraham Jacobi and Job Lewis Smith are its acknowledged founders [5, 7, 11, 12] (Fig. 2). The term "pediatrics" was first introduced into *The New England Journal of Medicine* only in 1880, by Sir William Osler [13]; and while articles entitled "Progress in Pediatrics" appeared sporadically from 1904 onward, pediatrics was accorded its own section in the *Journal* only in 1954 [14]. Initially, the focus of clinicians was naturally on "diseases of children" [3] but certainly, child health and welfare, stimulated in large part by alarmingly high rates of infant mortality, were causes taken up by many quarters, including social work, nursing, public health, the government, philanthropy, and concerned

Fig. 1 The earliest children's hospitals. (**a**) L'Hôpital des Enfants Malades, Paris (https://la-socialisation-des-enfants-malades-97.webself.net/aspect-historique, accessed 04/14/2024). (**b**) The Hospital for Sick Children, Great Ormond Street, London (Wood engraving by W. E. Hodgkin after D. R. Warry, 1872. Wellcome Collection. Public Domain Mark. Source: Wellcome Collection: https://wellcomecollection.org/works/ajeb7hfb, accessed 04/14/2024)

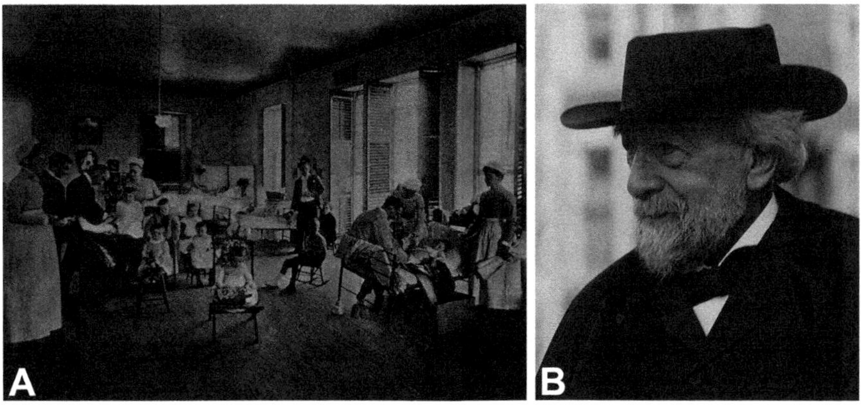

Fig. 2 American pediatrics began flourishing as its own branch of medicine in the mid-late nineteenth century. (**a**) Morning rounds on the children's ward, Bellevue Hospital, New York, ca. 1897 (from the Images from the History of Medicine Collection, National Library of Medicine, http://resource.nlm.nih.gov/101435915, accessed 04/14/2024); (**b**) Dr. Abraham Jacobi (1830–1919), considered the founder of pediatric medicine in the United States, who worked largely in New York City, including at Bellevue Hospital (Adapted from the Images from the History of Medicine Collection, George Grantham Bain Collection, National Library of Medicine, https://www.loc.gov/pictures/item/2014690785/, accessed 04/21/2024)

citizens, in addition to pediatricians [15]. By the mid-twentieth century, there arose a greater emphasis on preventive pediatric medicine, when the phrase "child health" (to describe the field or even in the naming of faculty chairs!) became very popular [3]. These developments in child health and pediatric clinical care paralleled how society viewed children: Starting in the 1830s and especially noticeable after World War II, modern philosophies began to suggest giving the child *agency* modeled to their ability and stage of development[1] [16].

It might be argued that among the earliest pediatric specialties was orthopedics or care for "crippled" and "handicapped" children, beginning in the mid- to late nineteenth century. In contrast to afflictions that arose from accidents in earlier eras (incurred through child labor), most of these children were the victims of infectious diseases such as polio and tuberculosis and nutritional deficiencies like rickets [17, 18]. Homes and hospitals for these children initially did little more than support, educate, and provide (vocational) training; but by the early twentieth century, medical treatments (requiring braces and crutches, among other devices) and surgery became more the focus [17] (Figs. 3 and 4).

Neonatal intensive care was one of the early offshoots of pediatric medicine, with interest and recognized need for such specialized care as early as the 1920s and

[1]The "agentic" child has the support of "loving adults to guide them to achieve their goals; their voices are respected and considered, which empowers the child to act participatory and collaboratively with adults" [16]—and here, the support of healthcare workers, clinicians, and the healthcare system itself (Ed.).

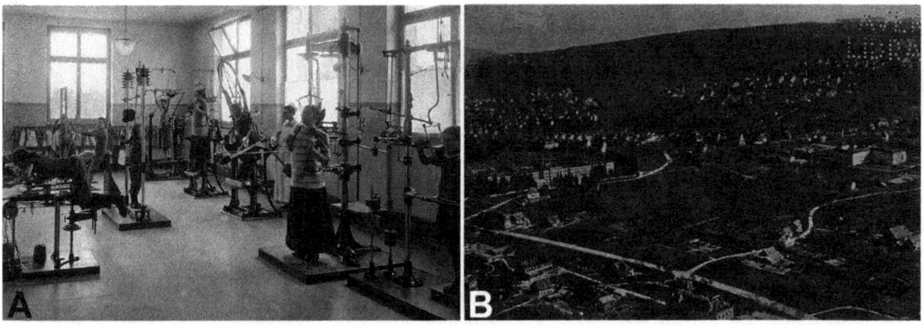

Fig. 3 Balgrist University Hospital developed from the 'Balgrist Sanatorium,' founded in 1912. This institution was dedicated to treating, caring for, educating and training physically disabled children to enable them to join the working world. Balgrist University Hospital was, and still is, operated by the Swiss Balgrist Association, which was established in 1909. (**a**) Children in a well-equipped facility for strength training (https://www.balgrist.ch/en/about-us/clinic/, accessed 04/12/2024. With permission, Universitätsklinik Balgrist Kommunikation); (**b**) Aerial view of Balgrist Hospital for Crippled Children, ca. 1930's (http://resource.nlm.nih.gov/101404783, accessed 04/12/2024)

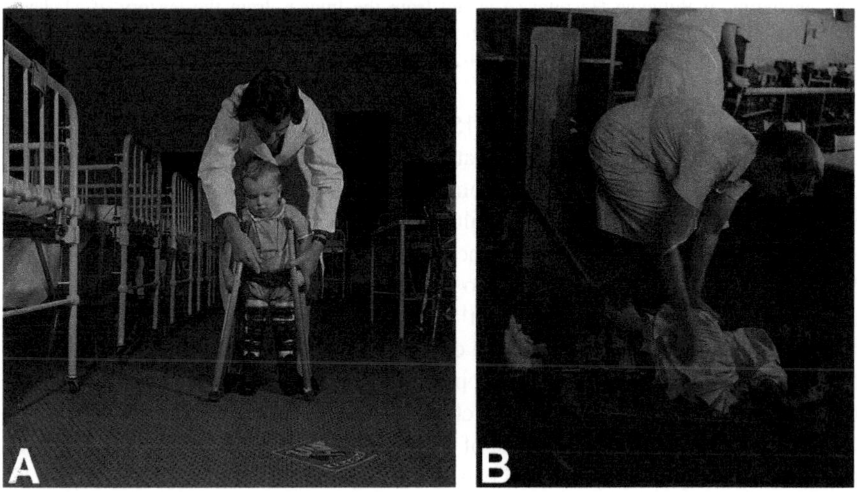

Fig. 4 Many early children's hospitals catered to "crippled children." (**a**) Nurse training: A student nurse, acting as physical therapist, points to the picture book, encouraging a child to learn to use his crutches (from the US Library of Congress Collection, Henle, Fritz, 1909-1993, photographer; https://www.loc.gov/pictures/item/2017694544/, accessed 04/13/2024); (**b**) A woman attends to a small child with leg braces at Crippled Childrens' Clinic, D.C. General Hospital (Images from the History of Medicine Collection, National Library of Medicine, http://resource.nlm.nih.gov/101447430, accessed 04/13/2024)

1930s, although obstetricians had developed some interest in newborn care even earlier [19, 20]. Indeed, "incubator babies" were commonly displayed as side shows around the turn of the twentieth century (Fig. 5), including at Coney Island! [21, 22]. Care of the newborn infant, and especially premature infant, transitioned from

Fig. 5 "Les couveuses d'enfants (intérieur)." Color photo (postcard) of two French nurses observing and assisting newborn infants at an incubator exhibit of Exposition universelle internationale de 1900 in Paris, France (Images from the History of Medicine Collection, National Library of Medicine, http://resource.nlm.nih.gov/101667815, accessed 04/13/2024)

obstetrics to pediatrics (but always with the close involvement of nurses) gradually over the second quarter of the twentieth century [20].

The term "neonatology" was coined in 1960, attributed to Dr. Alexander Schaffer, but it was not until the late 1960s and early 1970s that major changes in clinical care occurred, coinciding with an understanding of newborn (including premature) physiology, advances in technology, miniaturization of blood samples needed for clinical assays, and medications to support the circulation [19]. For older children, pediatric critical care medicine, or pediatric intensive care, emerged during an era of increasingly complex postoperative management, advances in the pediatric medical and surgical subspecialties, and the development of sophisticated life support technology, especially respiratory and ventilatory support [23]. In the early twentieth century, poliomyelitis necessitated the invention of the artificial respirator, the "iron lung," one of the earliest successful attempts at employing technology—made possible by a continuous power source—to treat human disease [24] (Fig. 6).

The first pediatric intensive care unit (PICU) was established in Europe in 1955 at Children's Hospital of Göteburg in Sweden, and starting in the 1970s, many hundreds would be established around the world, with academic training programs to follow [23]. Like other general pediatricians and pediatric subspecialists, pediatric critical care physicians must understand diseases and complications specific to children, but also be facile with new life-supporting and life-saving technology, and judiciously apply sedation and pain management. Facility with technology is especially relevant, including modes of mechanical ventilation, artificial organs (e.g., dialysis, continuous veno-venous hemofiltration; ventricular assist devices;

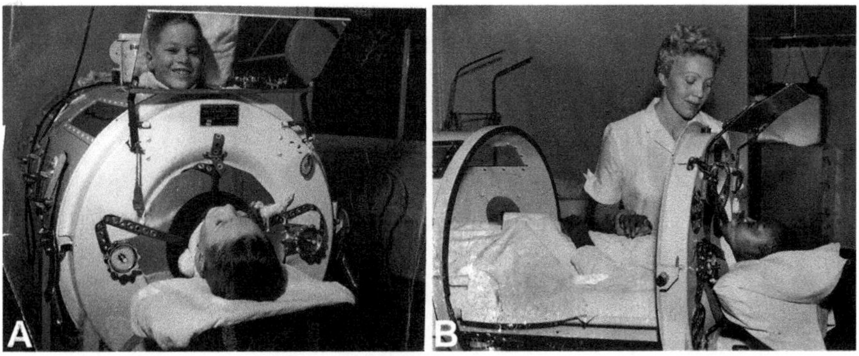

Fig. 6 "Iron lung" respirators heralded the advent of intensive care units. (**a**) "A boy with polio in the Emerson Respirator (Iron Lung) viewing the photographer, Joe Clark, in the machine's mirror, Herman Kiefer Hospital, Detroit, MI, 1955," (NCP 4145); (**b**) "Boy, Floyd Williams, in an iron lung, attended by nurse, Maria Geers, at Herman Kiefer Hospital, Detroit, MI, c. 1955," (NCP 4147). (Source: OHA 250 New Contributed Photographs Collection. Otis Historical Archives, National Museum of Health and Medicine. With permission)

extracorporeal membrane oxygenation), and monitoring (e.g., cardiopulmonary, neurological).

Like clinical care, teaching and research in pediatric medicine lagged behind adult medicine. Academic pediatrics, which includes the formal training of pediatric clinicians and scientific research into child health, emerged in the late nineteenth and early twentieth centuries [5] (Figs. 7 and 8).

Within the U.S. National Institutes of Health (NIH), the National Institute of Child Health and Development (NICHD) was founded only in 1962, whose "mission is to lead research and training to understand human development, improve reproductive health, enhance the lives of children and adolescents, and optimize abilities for all." (https://www.nichd.nih.gov/about, accessed 11/13/2023). In FY2023, the NICHD budget exceeded $1.7 billion, more than double FY2000; this number does not include the research expenditure on other pediatric-related diseases funded by other NIH institutes (https://officeofbudget.od.nih.gov/pdfs/FY25/spending_hist/Mechanism%20Detail%20by%20IC%20FY%202000%20-%20FY%202023%20(V).pdf, accessed 05/15/2024).

Children Are Not Just Little Adults

On their website (https://www.abms.org/about-abms/our-story/, accessed 02/16/2024), the American Board of Medical Specialties states: "The history of the American Board of Medical Specialties (ABMS) began during an extraordinary period in the development of specialty medicine." The Founding Member Boards of the ABMS in 1933 were Dermatology, Obstetrics and Gynecology, Ophthalmology, and Otolaryngology—Head and Neck Surgery. Pediatrics followed quickly in 1935.

Fig. 7 Academic pediatrics—encompassing teaching, research, and clinical care in children's hospitals attached to medical schools—was led in its earliest days by (**a**) Dr. L. Emmett Holt (Babies Hospital, and the College of Physicians and Surgeons at Columbia University) and (**b**) Dr. John Howland (Harriet Lane Home for Invalid Children and the Johns Hopkins Medical School) (Library of Congress, George Grantham Bain Collection https://www.loc.gov/item/2014695071/, accessed 04/13/2024; https://archive.org/details/bookofportraitso00ulma/page/n33/mode/2up)

The American Board of Internal Medicine and American Board of Surgery actually came about a little later, but the internists and (general) surgeons first formed associations ("colleges"), and so the American College of Physicians and the American College of Surgeons were originally formed to serve the same purpose as the specialty boards [25]. (Note: This author wonders if internal medicine and surgery were considered "general practice" prior to that era, and so the earliest Board members—including Pediatrics—in fact distinguished themselves as requiring *additional and specialized training*.)

But, children do not simply exhibit different diseases from adults; they also exhibit different diseases at different ages. This is even evident from the age-specific subspecialties that have developed over the past 50+ years within the field of pediatrics, which include neonatal-perinatal medicine and adolescent medicine (https://www.abp.org/content/subspecialty-certifications-and-admission-requirements, accessed 02/16/2024). We can also count maternal-fetal medicine—indeed, the fetal patient may be cared for by both maternal-fetal medicine specialists and certain pediatric subspecialists such as pediatric cardiologists (fetal cardiologists), neonatologists, pediatric surgeons, and even radiologists (who perform prenatal imaging).

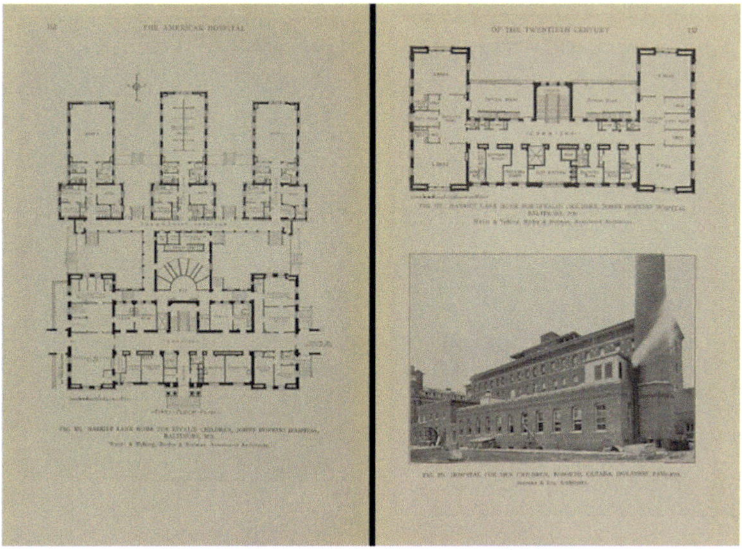

Fig. 8 By the early twentieth century, children's hospitals were designed with pediatric diseases in mind, while also housing research space for academic endeavors. These two pages depict floor plans of the Harriet Lane Home for Invalid Children (the children's hospital of The Johns Hopkins Hospital), as well as the Isolation Pavilion of The Hospital For Sick Children, Toronto, Canada (From *The American hospital of the twentieth century: a treatise on the development of medical institutions, both in Europe and in America, since the beginning of the present century*, pp. 173-174. Images from the History of Medicine Collection, National Library of Medicine, http://resource.nlm.nih.gov/53411000R, accessed 02/14/2024)

Illustrative Case: The Developing Human Being, Up Through Term

The development of specialized approaches for the care of the fetal patient can be illustrated by certain cardiac conditions as well as the significant problem of premature birth. Above all, these required technical—that is, engineering—advances. Advances in clinical ultrasound throughout the 1970s allowed increasingly accurate detection of fetal malformations and abnormalities, including fetal cardiac malformation and rhythm disturbances [26]. With such diagnostic approaches came the ability to monitor the fetal heart, including during interventions such as transplacental (i.e., a pregnant woman takes oral medication, which then passes through the placenta to her fetus) delivery of anti-arrhythmic medications to quell a hemodynamically significant fetal tachyarrhythmia [27]. Mechanically-based interventions to open up stenotic valves or intracardiac channels followed in the 1990s–2000s, based on the observation that certain fetal cardiac conditions progress if left alone [28]. As is often the case with procedural advances, experiments using animal models [29] led the way to human fetal intervention [30–32]. Additional advances in understanding fetal hemodynamics and natural history, diagnostic techniques, and treatment (medications and procedures) are now all part of comprehensive fetal cardiac care [33].

Premature labor and preterm birth, with all of their attendant problems, remain stubbornly common [34, 35]. The first premature infant incubator "station" was established in the United States in the late nineteenth century and the first American newborn intensive care unit (NICU) opened in 1965 [36], but despite significant improvements in survival and outcomes for these children, many long-term morbidities remain, chief among them chronic lung disease and neurological impairment [37–43]. Therefore, the urgent need for a more physiological approach to supporting premature infants has led to the concept of the artificial placenta [44–46] (Fig. 9). Again, animal model systems are leading the way [47, 48], and the U.S. Food and Drug Administration (FDA) has recently met to discuss the future of human artificial placenta trials [49].

Until recently, genetic diagnosis could only be made prenatally via amniocentesis. Amniocentesis was first developed in the early to mid-twentieth century and by the mid-twentieth century, it allowed for rudimentary genetic analysis (karyotyping) as well as biochemical analyses [50, 51]. Most recently, advances in genomics have permitted non-invasive genetic testing of fetuses with the analysis of fetal cell-free DNA in the maternal blood [52].

Illustrative Case: The Newborn Infant

Every pediatrician recognizes that the newborn infant is special: One must be ever-watchful for a host of diseases unique to this age group, including congenital malformations, metabolic disorders, and vulnerability to infections. And yet, the newborn infant may harbor a life-threatening condition without any overt warning signs until it is too late—until the infant becomes critically ill or develops permanent sequelae. Unlike many adult diseases, the time window for effective treatment and intervention may be in the order of days to weeks. Hence the development of newborn screening programs, which had their origins in the 1960s with the development of effective blood testing for phenylketonuria (PKU) (https://www.nichd.nih.gov/health/topics/newborn/conditioninfo/history, accessed 03/30/2024).

Most targets of newborn screening are genetic diseases and inborn errors of metabolism. In New York State, this program screens for more than 50 different disorders using a few small blood spots (https://www.wadsworth.org/programs/newborn/screening, accessed 03/30/2024). In addition, universal newborn hearing screening is a component of the New York Early Hearing Detection and Intervention Program (EHDI). Moreover, all 50 U.S. states plus the District of Columbia now have newborn screening programs that test for critical congenital heart defects (https://www.cdc.gov/ncbddd/heartdefects/features/all-states-supporting-cchd-screening.html/, accessed 03/30/2024).

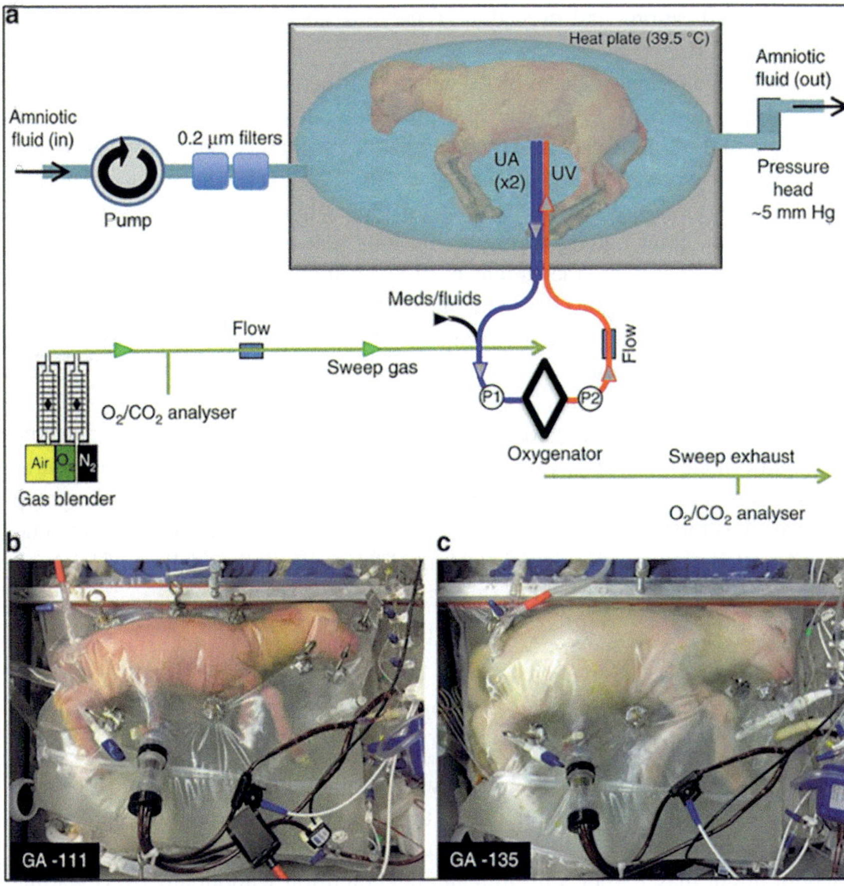

Fig. 9 Artificial placenta: the UA/UV Biobag system design. (**a**) Circuit and system components consisting of a pumpless, low-resistance oxygenator circuit, a closed fluid environment with continuous fluid exchange and an umbilical vascular interface. (**b**) Representative lamb cannulated at 107 days of gestation and on day 4 of support. (**c**) The same lamb on day 28 of support illustrating somatic growth and maturation. (Reprinted with permission, Partridge EA et al.,… Flake AW. *Nat Commun*. 2017)

Illustrative Case: Pediatric Surgery

Pediatric surgery can be traced at least as far back as Egypt's Islamic Era (ca. 1000 CE), and textbooks on pediatric surgery were published in Europe at least as early as the eighteenth century [53]. But, even as of the mid-twentieth century, the question was posed to a young surgical resident—now one of the field's most esteemed surgeons—"Why pediatric surgery?" [54] Few operations were available for children, especially the newborn infant; medical care of the fragile newborn in particular was still quite tenuous; and the distinct surgical challenges of congenital malformations and other diseases of young children posed seemingly insurmountable hurdles. But with advances in medical care (including the development of neonatal [NICUs] and pediatric intensive care units [PICUs]), the development of robust clinical and research training, and the emergence of surgical expertise in broad areas (including trauma—the recognition that pediatric trauma is an important "disease"), pediatric surgery as its own specialty has come of age in the twenty-first century [23, 54].

We have already touched upon fetal cardiac intervention. A special branch of pediatric surgery is fetal surgery. It took millennia for the fetus to be "demystified" and understood in terms of anatomy, physiology, developmental phases, and—importantly—diseases [55]. Ultrasonography was the key step in allowing safe, non-invasive visualization of the fetus, not only to make diagnoses but to guide procedures (such as amniocentesis). Only then could the fetus be considered a patient [55]. The impetus for fetal intervention was the recognition that for certain pediatric diseases, a diagnosis shortly after birth was already too late. Decades of experimental (i.e., on animal models) fetal studies and interventions had to be coupled with advances in anesthesia and surgical tools. In the 1980s, human fetal surgery took its first steps toward the correction of anatomic defects, including obstructive uropathy, hydrocephalus, congenital cystic adenomatoid malformation, and congenital diaphragmatic hernia [55].

Pharmacokinetics and Pharmacodynamics

The study of pharmacokinetics (what a body does to a drug) dates back to the early 1900s, and the term "pharmacokinetics" was first introduced in 1953 by the German pediatrician Dr. FH Dost, in his book "*Der Blütspiegelkinetik der Konzentrationsablaüffe in der Krieslaufflüssigkeit*" ("The Blood Level Kinetics of Concentration Decreases in the Circulatory Fluid") [56–58]. Spurred in part by dramatic clinical events (such as recognition of the chloramphenicol-gray baby syndrome and thalidomide-phocomelia teratogenicity) as well as by new assays and computer technology, clinical pharmacokinetics for age-related drug and dosage recommendations progressed significantly in the 1970s [57]. Our understanding of pharmacodynamics, on the other hand—what a drug does to a body—has lagged behind that of pharmacokinetics [57]. Children may be similar or dissimilar to adults with regard to physiology, drug metabolism, immunology, and response to disease states and therapy; moreover, there may be therapies that are uniquely applicable to children [59]. Effective, personalized pharmacotherapeutic approaches in the pediatric population will require a comprehensive understanding of the physiological and genetic complexities of the pediatric population—this will include pharmacokinetics, pharmacodynamics, and, also, pharmacogenomics [60].

Allometry

One could be forgiven to think that simplistically, "all" that needs to be done to treat children is to scale down drug doses and devices; or, in the case of growing structures, indexing (normalizing) to a simple growth metric. However, how such scaling is done—and assuming nothing else changes in children as they grow and develop—has been an ongoing research question for decades. Accurate and appropriate scaling is also necessary when extrapolating (drug treatment) data from animal studies to humans, especially children. While intuitively one might scale according to body weight, this is not in fact the case: More than 75 years ago, Kleiber reviewed how metabolic rates correlate better with body surface area than with weight, for example [61]. Allometry—defined as the measurement and study of the changes in proportions of the various parts of an organism in relation to the growth of the whole or within a series of related organisms—has been part of pediatric drug and small molecule modeling [see 62–69]; even with common drugs such as acetaminophen, questions still persist today [70]. Allometry is also invoked in understanding normal developmental [71–73] and disease phenotypes [74, 75] and practical issues for device-based monitoring and treatment [76, 77]. The increasing prevalence of obesity in children has prompted additional allometric, as well as pharmacokinetic and pharmacodynamic considerations [78, 79]. Pediatric rare diseases pose a special challenge: There are very few pediatric patients available for study (a problem compounded by phenotypic diversity); there may be no or few valid endpoints or biomarkers; and disease progression is poorly understood. Thus, allometric scaling from adult or animal studies may be even more important in rare pediatric diseases than in dose selection for more common pediatric diseases [80].

The Need for Tailored Therapy in the Pediatric Population

As outlined in another standard-bearer pediatrics textbook, *Nelson's Textbook of Pediatrics*, "The field of pediatrics" summarizes different conditions at different ages, which illustrate the biological, environmental, and social underpinnings of pediatric diseases [81]. All of these conditions will require different "engineering" for both diagnosis and treatment. Below, we provide specific examples of child-specific diseases and child-specific treatments.

Medical Treatments: Medications, Drugs, and Miscellanea

Medications (pharmaceuticals) require special consideration in the pediatric population, particularly in the very young (i.e., fetuses, infants, young children). In addition to allometry, physiological changes throughout infancy and childhood alter drug metabolism and clearance; disease conditions may alter them in still other ways. Pharmacokinetics thus remains an active area of clinical research [82].

A specific example of surfactant and antenatal steroid use in infants born prematurely required an understanding first of the perinatal events occurring in and the physiology of the infant lung, and then development and delivery of treatment [83, 84]. Dr. Mary Ellen Avery paved the way in the mid-twentieth century in the understanding of surfactant deficiency in the prematurely delivered infant [85], culminating in the first clinical use of artificial surfactant in human infants in 1980 [86]. The benefits of antenatal corticosteroid administration to mothers (as opposed to postnatal administration to the babies themselves) were recognized and first reported in 1972 [87]. Today, antenatal corticosteroids and surfactant therapy, along with advances in positive-pressure ventilatory management, form the foundation of care for respiratory distress syndrome in the newborn infant [88].

Surgical Intervention, Surgical Tools, Cardiac Catheterization, Devices

The history of surgery is long and illustrious, spanning centuries… except in children. Even as recently as the mid-twentieth century:

> "[C]hildren did not get a fair shake in surgery…. Surgeons in general were frightened of children and distrusted anesthetists to be able to wake them up after putting them to sleep… The younger and smaller the patient, the more advanced the hazard." [18]

As surgeons, Koop and others recognized in children "their physiological differences, their limited reserve, and their special pharmacological needs" [18]. Surgical intervention may also require specialized tools for very small bodies and/or unique diseases.

Cardiac care in children has led the way in many such technical advances; some of the earliest cardiac operations were performed in children with congenital heart defects [89–91]. For example, Vivien Thomas (with William Longmire) had to develop a new clamp for use in the "Blue Baby operation" reported by Drs. Alfred Blalock and Helen Taussig [90] in the 1940s [92]; also https://exhibits.library.jhu.edu/exhibits/show/the-blue-baby-operation/vivien-thomas, accessed 12/10/2023; (Fig. 10).

The era of open-heart surgery, allowing direct visual repair, began in the early 1950s, and children were very often the beneficiaries. Years of experimental work ultimately led to the first open repairs of congenital heart defects, made possible by hypothermia to slow the body's metabolism [93, 94] and the development of extracorporeal circulation, such as cross-circulation [95] and the heart-lung machine [96, 97]. Many of the patients who had their congenital heart defects repaired developed postoperative complete heart block, but there were also children who were born with congenital complete heart block. These children required a means to increase their heart rates, and the first wearable pacemaker was first implanted in a child by

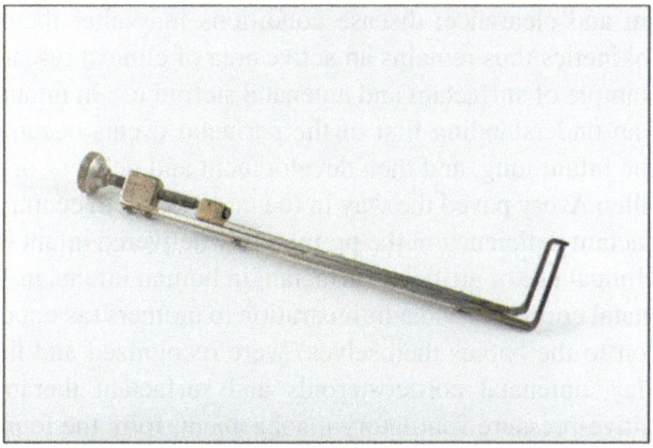

Fig. 10 Clamp for the temporary occlusion of the pulmonary artery, devised for Dr. Alfred Blalock's use by Vivien Thomas and Dr. William Longmire, working with the Baltimore surgical supply house Murray Baumgartner & Co. (From the online exhibition, The Blue Baby Operation: Vivien Thomas, Johns Hopkins Medical Institutions, https://exhibits.library.jhu.edu/exhibits/show/the-blue-baby-operation/vivien-Thomas, accessed 12/10/2023. With permission)

Dr. C. Walton Lillehei in 1957 [98], although other types of pacemakers had been used for a couple of decades at least [99].

Most everyone is familiar with Dr. Christiaan Barnard's milestone in performing the first successful human heart transplant in Cape Town, South Africa, on December 3, 1967 [100, 101]. Fewer are aware that just days later, the second transplant was performed in an infant with severe congenital heart disease in the U.S. by Dr. Adrian Kantrowitz in Brooklyn, New York, on December 6, 1967 [102].

As surgical approaches continued to benefit from both advances in technology and technique, non-surgical alternative procedures began to develop as well. In 1966, Rashkind and Miller reported their initial experience with balloon atrial septostomy to increase the mixing of blood between the atria in cyanotic infants with d-transposition of the great arteries, thereby increasing arterial oxygenation [103]. In the late 1970s and early 1980s, further technical advances in balloon catheters combined with experiments in the pre-clinical research laboratory led to the first trans-catheter, percutaneous balloon pulmonary valvuloplasty to relieve stenosis (obstruction) of the pulmonary valve, performed by Dr. Jean Kan and her team, ushering in the era of interventional pediatric cardiology [104, 105]. Working with the company that manufactured the balloon catheters and with a surgical parallel already in place (the Brock procedure), Dr. Kan quickly was able to get this new therapeutic device cleared to market, allowing widespread adoption of the procedure [105]. We note that both the balloon atrial septostomy and balloon valvuloplasty, as well as many related "spin-off" procedures, continue to be in widespread use today [106–109].

In 2007, Congress passed the Food and Drug Administration (FDA) Amendments Act of 2007 (FDAAA), which contained several sections addressing pediatric devices and medications specifically (https://www.govinfo.gov/content/pkg/

PLAW-110publ85/pdf/PLAW-110publ85.pdf, accessed 04/20/2024). Section 515A and other provisions in the FDAAA were intended to encourage the development of devices for use in pediatric patients, and indeed, the need for the legislation included that "[s]ome products are designed specifically for children, while others are borrowed from adult applications or produced for more general use… Children have specific medical needs that must be considered when medical and surgical devices are prescribed." An examination of recent history may prove illuminating (see FDA Pre-Market Approval Reports to Congress, FY2021; https://www.fda.gov/media/177481/download?attachment, accessed 04/20/2024): In FY2021, 59 devices were approved, 15 (25%) of which were devices indicated for use in a pediatric population or subpopulation, and of the remaining 44 devices that were indicated for use in adults, all 44 (100%) demonstrated use (treat, diagnose, cure) in a disease or condition for which there is a pediatric subpopulation that also suffers from such a disease or condition. More recently, the FDA even has its own program to fund the development of pediatric devices (https://www.fda.gov/industry/medical-products-rare-diseases-and-conditions/pediatric-device-consortia-grants-program). In the introduction to the FDA's Pediatric Device Consortia Grants Program, it is stated: "Pediatric medical device development faces unique challenges as children often differ from adults in size, growth, development, body chemistry, and disease propensity. Due in part to these issues, only a small number of medical devices approved are labeled for pediatric use."

Engineering for Pediatric Public Health: Illustrative Example of Accidents or Unintentional Injuries

Injuries have been a leading killer—and in recent decades, the leading killer—of children for centuries, in addition to causing long-term disabilities [110]. Until recently, injuries were commonly referred to as "accidents" but in fact, unlike accidents, injuries are not necessarily random, can be predicted, and causes are mostly known. Moreover, we now understand the significant inequities in the field of childhood injuries and injury prevention, wherein the underprivileged and poor are disproportionately impacted [110, 111]. Only about 50 years ago was the scientific method applied to injury prevention [111]. Illustrative examples briefly reviewed here will be child labor, automobile passenger safety, and access to bottled medications. We will see that starting in the late nineteenth century, accelerating in the early twentieth century, and up to the present time, solutions to non-intentional injuries are found mostly in the public health and policy (social engineering) arenas.

In all the millennia of human existence, child labor has been part of the fabric of society. This included agricultural work—working on farms [112–114]—and with the Industrial Revolution, working in factories [112, 113]. It seems children were employed for both moral purposes (i.e., Puritanical ideals of hard work) and economic purposes. Data on the extent of child labor are scant prior to the mid-late nineteenth century, when the origins of reform took root. The conditions, of course, were difficult and, until after the Civil War, unregulated in the U.S [112]. In recent decades, the specific dangers to the health of children have been more carefully documented and understood, including in agriculture, which worldwide is the most hazardous occupation—and the one most likely to involve children [113, 114]. In a Congressional report, Mayer summed up the early growing reform movement:

> "Regulation of child labor has been motivated by diverse concerns: economic, humane, and more broadly social. In the 19th and early 20th centuries, child workers were often viewed as an alternative source of low-wage labor who vied with their parents and other adults for employment—*even at the cost of their own health* and education." [113] (emphasis mine)

Legislation to regulate child labor and child labor initiatives began to take hold in the early twentieth century [113]. To this day, "education, engineering, and enforcement" form the "3 E's of prevention" in agricultural injuries in children, necessitating a multipronged approach to reducing death and disability [114]; these principles, of course, apply to other areas of high injury risk to children.

Road traffic injuries are the leading cause of injury deaths in children, as well as the leading cause of traumatic brain injury and extremity injuries with long-term disability [110, 115]. Among the many engineering strategies applied to reduce road traffic-associated injuries, car seats are perhaps the strategy most associated with children. For decades, car seats served only to boost children for improved visibility of their surroundings. The invention of the (backward-facing) child car safety seat is credited to the British inventor Jean Ames in 1962, while Leonard Rivkin patented a forward-facing seat in a metal frame with a 5-point harness system (U.S. Patent Office, Patent #1,307,942; https://ppubs.uspto.gov/dirsearch-public/print/downloadPdf/3107942, accessed 05/01/2024) [116]. Only in 1978 did Tennessee become the first U.S. state to mandate child safety seat use [111].

Injuries and accidents also include poisoning from unintentional ingestions, which were estimated to be the third leading cause of unintentional injury death in the European Region in 2008 by the World Health Organization (WHO) [115]. Again, the path to reducing mortality and morbidity is multipronged, but includes packaging—that is, child-resistant packaging of medications and toxic substances [115]. "Safety closure caps" to prevent accidental drug poisoning in children were first developed in the mid-twentieth century [117].

Over the last century, the incidence of injury deaths has decreased significantly, attributed to three factors: First, a general reduction in exposure to dangerous jobs and safety improvements related to transportation and housing; second, improvements in the medical care of the injured, including the pediatric patient; and third, the development of a broad array of evidence-based injury control strategies, including engineering strategies [111]. The examples above illustrate the critical roles of public health and social policy in improving the health of children.

Genetic Engineering: Gene Therapy

The history of gene therapy is tied to the history of pediatrics, since many genetic or heritable diseases manifest in infants and children. Roth and Marson have chronicled that the first known genetic disorder was alkaptonuria (described at the turn of the twentieth century), giving rise to the recognition of inborn errors of metabolism; the first hereditable alteration in a protein linked to disease was identified in sickle cell anemia in the late 1940s—indeed, the first genetic diseases to be molecularly characterized were hemoglobinopathies; and the first human genetic alteration found to be associated with disease was Down syndrome, identified in 1959 as the chromosomal abnormality trisomy 21 [118].

Gene therapy was first conceptualized and attempted in vitro in the 1960s and 1970s, and in 1972, Friedmann and Roblin published their seminal concept paper, "Gene therapy for human genetic disease?" in *Science* [119]. This paper outlined the necessary steps for successful gene therapy, including genetic modification by DNA, preliminary criteria (including ethical criteria) for such modification and therapy, and control of gene therapy (to prevent misuse).

Using a retroviral-mediated transfer approach, the first approved gene therapy clinical research in the U.S. was initiated in 1990 at the NIH to treat severe combined immunodeficiency due to a genetic defect in adenosine deaminase (ADA-SCID) in two children [120]. This was the culmination of years of basic science investigation and clinical work (for a reflection by one of the pioneers, Dr. Michael Blaese, see [121]). However, the promise of gene therapy progressed haltingly for a long time, due to issues with safety and efficacy, including disputed results, treatment failures, and highly publicized deaths in experimental trials [122]. The gene therapy community was especially rocked by the death of 18-year-old Jesse Gelsinger in 1999, during a gene therapy trial to treat inborn ornithine transcarbamylase deficiency at the University of Pennsylvania [123, 124].

The first gene therapy was approved in 2003 in China; it took almost 15 years for the U.S. to approve its first gene therapy in 2017. Since then, the approval of gene therapies has accelerated, benefitting both pediatric and adult patients [122, 125]. We note that there are several different approaches to gene targeting, gene delivery, et cetera, which seek to overcome hurdles to both short- and long-term efficacy [118, 126, 127]. Most recently, gene therapy based on gene editing has allowed us to tackle such important childhood-onset diseases such as sickle cell disease, beta-thalassemia, and congenital deafness [128–132]; also https://www.bbc.co.uk/news/health-68921561.amp, (accessed 05/12/2024).

Developmental Origins of Adult Disease: When (Supposedly) Healthy Children May Grow Up to Be Unhealthy Adults

After World War II, remarkable clinical observations tied the nutritional status of pregnant mothers and their unborn children and young infants to adult-onset cardiovascular disease [133–135]. To put this in the simplest possible terms, *In the absence of overt disease, underweight newborn infants are more likely to develop adult-onset cardiovascular and metabolic diseases.* Therefore, health in the fetus and young infant—not just disease—may carry repercussions for lifelong health. This concept was most ardently espoused and researched by Dr. David Barker and became known as the "Barker Hypothesis" [134–136], later renamed the fetal origins of adult disease (FOAD), developmental origins of adult disease (DOAD), developmental origins of health and disease (DOHaD), or simply, the developmental origins theory [137–140]. The basic idea is that certain insults to the developing fetus (such as undernutrition) that occur during critical points in gestation permanently change the body's structure, function, and metabolism long after birth, and into adulthood, leading to a host of adult non-communicable diseases [137]. Epigenetics likely plays a mechanistic role, but there may also be environmental factors at play [140, 141]. Thus, we now recognize that optimizing wellness very early in life, not just treating or even curing disease, is necessary for optimal health throughout one's life course.

When Children Were (Are) Failed

Much of written history tends to prop achievements and luminaries on a pedestal, highlighting milestones and advances in care. History may also provide valuable lessons when medical care or public health has failed to protect children.

Harm from a Natural Substance

Oxygen must be considered one of the most important therapeutic substances discovered in all of medicine. Although Joseph Priestley is usually credited with its discovery (naming it "dephlogisticated air") [142, 143], Severinghaus' deeper examination of history suggests that "[e]ight sages over five centuries share oxygen's discovery": Ibn al-Nafis (ca. 1250 CE), Michael Servetus, Michael Sendivogius, John Mayow, Carl Wilhelm Steele, Priestley, Antoine Lavoisier, and in the 1760s–1770s, Henry Cavendish [143]. Oxygen has been used to help newborn infants with respiratory difficulties and cyanosis as early as 1780 [142], but "modern" (liberal, prolonged) exposure of newborn infants to oxygen, including with specially constructed devices and tubes, began in the early-mid-1900s [144–146].

Shortly thereafter, a new disease afflicting the eyes of premature infants was described [147]—first dubbed retrolental fibroplasia and more recently renamed retinopathy of prematurity (ROP)—and ascribed to excessive exposure to oxygen by Drs. Mary Crosse in Birmingham and Kate Campbell in Melbourne [142, 148, 149]. Since then, iatrogenic damage from supplemental oxygen use has decreased due to several factors, including the recognition of underlying mechanisms; technological advances to regulate oxygen, monitor blood oxygenation, and detect early ROP; and clinical trials to study optimal levels of supplemental oxygen to minimize damage to various organ systems [150, 151]. However, the optimal range of "normoxia" in this highly vulnerable population of infants remains debated, despite several large clinical trials [152, 153].

Harm from Medication

Thalidomide was introduced in the mid-late 1950s as a sedative that was also particularly effective for the hyperemesis occurring in pregnancy [154–157]. At the time, medications were not routinely tested for effects on the unborn child. Soon thereafter, letters and reports began to appear associating its use with congenital abnormalities, especially phocomelia [158–162]. Thalidomide was pulled off the market, but not before many thousands of children worldwide were affected. Because of the anatomical limb deformities (especially of the upper limbs), the focus of treatment was on enabling function, including through surgical interventions, prosthetic devices, and engineering (reviewed in [157]). It should be noted that thalidomide was never approved for public use in the U.S., due to the efforts of Dr. Frances Oldham Kelsey at the FDA [163]; and her insistence on safety data prevented the teratogenic outbreak that Europe witnessed [164, 165]. She, in fact, drew from the observations of none other than Dr. Helen Taussig [162, 166]. For her far-reaching work, President John F. Kennedy awarded Dr. Kelsey the Gold Medal for Distinguished Federal Civilian Service in 1962, among many other accolades (Fig. 11).

Nevertheless, thalidomide has enjoyed a resurgence—albeit with tight restrictions on and monitoring of its use—in part due to the hypothesis that the limb deformities may have arisen from the drug's interference with vasculogenesis, leading to the discovery of its anti-angiogenic, anti-inflammatory, and immunomodulatory effects. It is now used in a variety of inflammatory (including dermatological, rheumatological, gastrointestinal, and cardiac), infectious, and malignant conditions [156].

Fig. 11 Dr. Frances Oldham Kelsey and President John Kennedy, standing at a microphone during the awards ceremony at which she received the Gold Medal for Distinguished Federal Civilian Service. (1962) (Images from the History of Medicine Collection, National Library of Medicine, http://resource.nlm.nih.gov/101447160, accessed 04/13/2024)

Harm from Technology

Since the introduction of electronic health records and computerized physician order entry (CPOE) in the late twentieth century, medication errors have decreased overall but introduced new sources of error [167–170], including, in some earlier instances, increased mortality [171]. Even in today's electronic era, children remain at risk for medical errors, especially medication errors. In a 2008 New York Times article on medical errors in children, "Medical errors pose a greater threat to children than to adults for a number of reasons. They are physically small, and their kidneys, liver and immune system are still developing. Even a tiny increase in the dose of medication can have serious effects, especially in babies born prematurely. And if children take a turn for the worse, they can deteriorate more rapidly than adults. Children also are less able to communicate what they are feeling, making it difficult to diagnose their problem or know when a symptom or complication develops." [172]. The nature of actual or potential medication errors is somewhat different from that in the adult population [173]. To this day, work is ongoing to improve pediatric medication safety [170, 174], and actual improvements in outcomes remain something of an open question [175].

Conclusion

We have recognized for centuries that children are not just "little adults" with respect to their unique developmental trajectories, disease conditions, and drug-/small molecule-based and device-based therapies. With advances in our understanding of diseases and pathophysiology, medical treatments, surgical approaches, device technology, molecular targeting, and social engineering, child mortality has declined steeply, but only in the past two centuries (Fig. 12) [176].

And yet, much work remains to be done to optimize the long-term health and well-being of all individuals within the pediatric-age population. Only in this century did the U.S. government formalize legislation as one step towards advancing research and drug discovery specifically for the pediatric population: The Best Pharmaceuticals for Children Act (enacted into law on January 4, 2002) and the Pediatric Research Equity Act (enacted into law, December 3, 2003) were designed

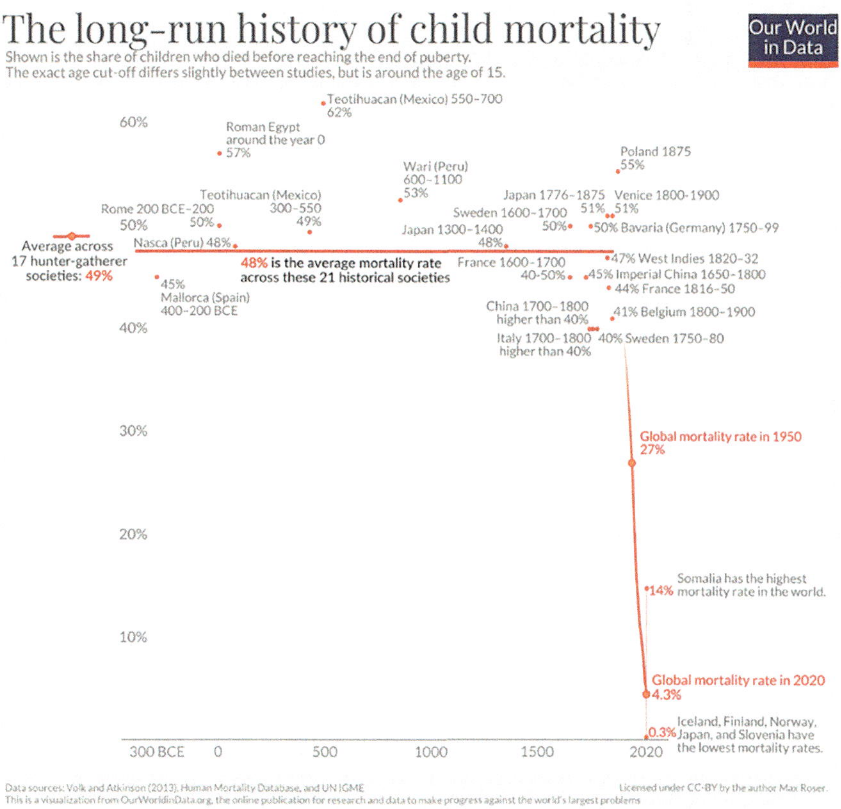

Fig. 12 "Mortality in the past: every second child died" Published online at OurWorldInData.org. Retrieved from: 'https://ourworldindata.org/child-mortality-in-the-past' [Online Resource; from Max Roser (2023)] (accessed 04/10/2024)

to encourage more pediatric studies of drugs used for children [177]. Stakeholder input, especially from patients and families, can also help prioritize research [178]. Even with such advances in pediatric treatments, however, durability over an entire lifetime remains an open question, especially where devices are concerned [179]. A few examples include cochlear implants [180], surgery for spinal deformities [181, 182], and cardiac valve replacement/repair [183–187]. It is this author's hope that this introduction to the history of pediatrics has revealed the field's breadth and richness. Pediatrics has evolved and progressed dramatically in the past 150 years, and we are now at an exciting point in its history to start seeing truly tailored and individually engineered therapies for children of all ages, even starting before birth.

Competing Interests Dr. Phoon has no conflicts of interest to declare that are relevant to the content of this chapter.

References

1. *Mosby's Medical, Nursing, & Allied Health Dictionary*, Elsevier Inc. (Mosby); 2001. Gale OneFile: Health and Medicine, link.gale.com/apps/doc/A173832451/HRCA?u=nysl_me_nyuniv&sid=bookmark-HRCA&xid=dda5fbd6. Accessed 14 Nov 2023.
2. Alden ER, Mulvey HJ, Hartline JV, Perelman R. The field of pediatrics. In: McMillan JA, Feigin RD, DeAngelis C, Jones MD, editors. Oski's Pediatrics, Principles and Practice. 4th ed. Philadelphia, PA: Lippincott Williams & Wilkins; 2006. p. 8–13.
3. Pearn J. Paediatrics: the etymology of a name. Arch Dis Child. 2011;96:759–63.
4. Cone TE Jr. History of American Pediatrics. Boston: Little, Brown and Company; 1979.
5. Phoon CKL. The origins of pediatrics as a clinical and academic specialty in the United States. *Hektoen International: A Journal of Medical Humanities*. Winter, 2018. http://hekint.org/2018/03/15/origins-pediatrics-clinical-academic-specialty-united-states/
6. Williams AN, Sharma RM. Children in hospitals before there were children's hospitals. Pediatrics. 2014;134:425–7.
7. Viner R. Abraham Jacobi and the origins of scientific pediatrics in America. In: Markel H, Stern AM, editors. Formative Years: Children's Health in the United States, 1880-2000. Ann Arbor: University of Michigan Press; 2009. p. 23–46.
8. Cone TE Jr. L'Hôpital Des Enfants-Malades, the world's first children's hospital, founded in Paris in 1802. Pediatrics. 1981;67:670.
9. Manke CB. The growth and development of a specialty: The history of pediatrics. Clin Pediatr. 2000;39:705–14.
10. Young DG, Carachi R. Britain. In: Carachi R, Young DG, Buyukunal C, editors. A History of Surgical Paediatrics. Singapore: World Scientific Publishing Co. Pte. Ltd; 2009. p. 55–66.
11. Faber HK. Job Lewis Smith, forgotten pioneer. J Pediatr. 1963;63(Suppl):794–802. https://doi.org/10.1016/s0022-3476(63)80265-6.
12. Viner R. Abraham Jacobi and German medical radicalism in antebellum New York. Bull Hist Med. 1998 Fall; 72:434-463. https://doi.org/10.1353/bhm.1998.0162
13. Osler W. On the systolic brain murmur of children. Boston Med Surg J. 1880;103:29–30.
14. Hostetter MK. What we don't see. N Engl J Med. 2012;366:1328–34. https://doi.org/10.1056/nejmra1111421.
15. Markel H. For the welfare of children: The origins of the relationship between U.S. public health workers and pediatricians. In: Markel H, Stern AM, editors. Formative Years:

Children's Health in the United States, 1880-2000. Ann Arbor: University of Michigan Press; 2009. p. 47–65.
16. Haring U, Sorin R, Caltabiano NJ. Reflecting on childhood and child agency in history. Palgrave Communications. 2019;5:1–9.
17. Cooter R. The cause of the crippled child. In: Surgery and Society in Peace and War. Science, Technology and Medicine in Modern History. London: Palgrave Macmillan; 1993. p. 53–78. https://doi.org/10.1007/978-1-137-10235-5_4.
18. Koop CE. A perspective on the early days of pediatric surgery. J Pediatr Surg. 1998;33:953–60. https://doi.org/10.1016/s0022-3468(98)90514-0.
19. Philip AG. The evolution of neonatology. Pediatr Res. 2005;58:799–815. https://doi.org/10.1203/01.pdr.0000151693.46655.66.
20. Baker JP. Technology in the nursery: Incubators, ventilators, and the rescue of premature infants. In: Markel H, Stern AM, editors. Formative Years: Children's Health in the United States, 1880-2000. Ann Arbor: University of Michigan Press; 2009. p. 66–90.
21. Smith MD. Incubator babies. Am J Nurs. 1911;11:791–5.
22. Silverman WA. Incubator-baby side shows (Dr. Martin A. Couney). Pediatrics. 1979;64:127–41.
23. Epstein D, Brill JE. A history of pediatric critical care medicine. Pediatr Res. 2005;58:987–96. https://doi.org/10.1203/01.pdr.0000182822.16263.3d.
24. Markel H. The genesis of the iron lung. Early attempts at administering artificial respiration to patients with poliomyelitis. Arch Pediatr Adolesc Med. 1994;148:1174–80. https://doi.org/10.1001/archpedi.1994.02170110060009.
25. Brownlee RC. The American Board of Pediatrics: its origin and early history. Pediatrics. 1994;94:732–5.
26. Kleinman CS, Hobbins JC, Jaffe CC, Lynch DC, Talner NS. Echocardiographic studies of the human fetus: prenatal diagnosis of congenital heart disease and cardiac dysrhythmias. Pediatrics. 1980;65:1059–67.
27. Kleinman CS, Donnerstein RL, Jaffe CC, DeVore GR, Weinstein EM, Lynch DC, et al. Fetal echocardiography: A tool for evaluation of in utero cardiac arrhythmias and monitoring of in utero therapy: analysis of 71 patients. Am J Cardiol. 1983;51:237–43. https://doi.org/10.1016/s0002-9149(83)80042-3.
28. Gardiner HM. Progression of fetal heart disease and rationale for fetal intracardiac interventions. Semin Fetal Neonatal Med. 2005;10:578–85. https://doi.org/10.1016/j.siny.2005.08.001.
29. Kohl T, Szabo Z, Suda K, Petrossian E, Ko E, Kececioglu D, et al. Fetoscopic and open transumbilical fetal cardiac catheterization in sheep. Potential approaches for human fetal cardiac intervention. Circulation. 1997;95:1048–53. https://doi.org/10.1161/01.cir.95.4.1048.
30. Kohl T, Sharland G, Allan LD, Gembruch U, Chaoui R, Lopes LM, et al. World experience of percutaneous ultrasound-guided balloon valvuloplasty in human fetuses with severe aortic valve obstruction. Am J Cardiol. 2000;85:1230–3. https://doi.org/10.1016/s0002-9149(00)00733-5.
31. Marshall AC, van der Velde ME, Tworetzky W, Gomez CA, Wilkins-Haug L, Benson CB, et al. Creation of an atrial septal defect in utero for fetuses with hypoplastic left heart syndrome and intact or highly restrictive atrial septum. Circulation. 2004;110:253–8. https://doi.org/10.1161/01.cir.0000135471.17922.17.
32. Tworetzky W, Wilkins-Haug L, Jennings RW, van der Velde ME, Marshall AC, Marx GR, et al. Balloon dilation of severe aortic stenosis in the fetus: potential for prevention of hypoplastic left heart syndrome: candidate selection, technique, and results of successful intervention. Circulation. 2004;110:2125–31. https://doi.org/10.1161/01.cir.0000144357.29279.54.
33. Haxel CS, Johnson JN, Hintz S, Renno MS, Ruano R, Zyblewski SC, et al. Care of the fetus with congenital cardiovascular disease: From diagnosis to delivery. Pediatrics. 2022;150(Suppl 2):e2022056415C. https://doi.org/10.1542/peds.2022-056415c.
34. Romero R, Dey SK, Fisher SJ. Preterm labor: one syndrome, many causes. Science. 2014;345:760–5. https://doi.org/10.1126/science.1251816.
35. Dudley DJ, Ennen CS. The vexing problem of preterm birth prevention. JAMA. 2023;330:323–5. https://doi.org/10.1001/jama.2023.7244.

36. Pearson HA, Anunziato D, Baker JP, Gartner LM, Howell DA, Strain JE, et al; Historical Archives Advisory Committee. Committee Report. American Pediatrics: Milestones at the millennium. Pediatrics. 2001; 107:1482-1491. https://doi.org/10.1542/peds.107.6.1482
37. Thygesen SK, Olsen M, Østergaard JR, Sørensen HT. Respiratory distress syndrome in moderately late and late preterm infants and risk of cerebral palsy: a population-based cohort study. BMJ Open. 2016;6:e011643. https://doi.org/10.1136/bmjopen-2016-011643.
38. Cheong JL, Doyle LW, Burnett AC, Lee KJ, Walsh JM, Potter CR, et al. Association between moderate and late preterm birth and neurodevelopment and social-emotional development at age 2 years. JAMA Pediatr. 2017;171:e164805. https://doi.org/10.1001/jamapediatrics.2016.4805.
39. Heikkilä K, Pulakka A, Metsälä J, Alenius S, Hovi P, Gissler M, et al. Preterm birth and the risk of chronic disease multimorbidity in adolescence and early adulthood: A population-based cohort study. PLoS One. 2021;16:e0261952. https://doi.org/10.1371/journal.pone.0261952.
40. Boyd SM, Tapawan SJ, Badawi N, Popat H. Protecting the brain of the micropreemie. Semin Fetal Neonatal Med. 2022;27:101370. https://doi.org/10.1016/j.siny.2022.101370.
41. Haffner DN, Sankovic A. A neurologist's guide to neonatal neurodevelopmental assessments and preterm brain injury. Semin Pediatr Neurol. 2022;42:100974. https://doi.org/10.1016/j.spen.2022.100974.
42. Jin JH, Lee SH, Youk TM, Yoon SW. Long-term outcomes of preterm infants in the first 6 years of life: a nationwide population-based study in Korea. Eur J Pediatr. 2023;182:641–50. https://doi.org/10.1007/s00431-022-04728-w.
43. Collaco JM, Eldredge LC, McGrath-Morrow SA. Long-term pulmonary outcomes in BPD throughout the life-course. J Perinatol. 2024. https://doi.org/10.1038/s41372-024-01957-9
44. Westin B, Nyberg R, Enhorning G. A technique for perfusion of the previable human fetus. *Acta Paediatr* (Stockh). 1958; 47:339-349. https://doi.org/10.1111/j.1651-2227.1958.tb07643.x
45. Callaghan JC, Angeles J, Boracchia B, Fisk L, Hallgren R. Studies of the first successful delivery of an unborn lamb after 40 minutes in the artificial placenta. Can J Surg. 1963;6:199–206.
46. Davis RP, Bryner B, Mychaliska GB. A paradigm shift in the treatment of extreme prematurity: the artificial placenta. Curr Opin Pediatr. 2014;26:370–6. https://doi.org/10.1097/mop.0000000000000083.
47. Partridge EA, Davey MG, Hornick MA, McGovern PE, Mejaddam AY, Vrecenak JD, et al. An extra-uterine system to physiologically support the extreme premature lamb. Nat Commun. 2017;8:15112. https://doi.org/10.1038/ncomms15112.
48. Flake AW. A supportive physiologic environment for the extreme premature infant: Improving life outside the womb. J Pediatr Surg. 2022;57:167–71. https://doi.org/10.1016/j.jpedsurg.2021.10.025.
49. Kozlov M. Human trials of artificial wombs could start soon. Here's what you need to know. (News Feature). Nature. 2023;621:458–60. https://doi.org/10.1038/d41586-023-02901-1.
50. Sachs L, Serr DM, Danon M. Prenatal diagnosis of sex using cells from the amniotic fluid. Science. 1956;123:548. https://doi.org/10.1126/science.123.3196.548.
51. Nadler HL. Prenatal detection of genetic defects. J Pediatr. 1969;74:132–43. https://doi.org/10.1016/s0022-3476(69)80023-5.
52. Harris S, Reed D, Vora NL. Screening for fetal chromosomal and subchromosomal disorders. Semin Fetal Neonatal Med. 2018;23:85–93. https://doi.org/10.1016/j.siny.2017.10.006.
53. Holschneider A, Gdanietz K. Germany. In: Carachi R, Young DG, Buyukunal C, editors. A History of Surgical Paediatrics. Singapore: World Scientific Publishing Co. Pte. Ltd; 2009. p. 167–99.
54. Haller JA Jr. Why pediatric surgery? A personal journey through the first 50 years. Ann Surg. 2003;237:597–606. https://doi.org/10.1097/01.sla.0000064563.41539.b2.
55. Harrison MR. Fetal Surgery. In: Carachi R, Young DG, Buyukunal C, editors. A History of Surgical Paediatrics. Singapore: World Scientific Publishing Co. Pte. Ltd; 2009. p. 746–88.
56. Wagner JG. The history of pharmacokinetics. Drug Intelligence Clin Pharm. 1977;11:747–8.
57. Bartels H. Drug therapy in childhood: what has been done and what has to be done? *Pediatr Pharmacol* (New York). 1983; 3:131-143.

58. Rescigno A, Bocchialini BM. Pharmacokinetics: Unfolding of a concept. In: Rescigno A, Thakur AK, editors. *New Trends in Pharmacokinetics*. NATO ASI Series, Series A: Life Sciences, Vol 221, (Proceedings of a NATO Advanced Study Institute on New Trends in Pharmacokinetics, held September 4-15, 1990, Erice, Sicily, Italy). New York & London: Plenum Press; 1991. p.1-25.
59. Institute of Medicine (US) Roundtable on Research and Development of Drugs, Biologics, and Medical Devices. In: Yaffe S, editor. Rational Therapeutics for Infants and Children: Workshop Summary. Washington (DC): National Academies Press (US); 2000. 3, *Pharmacokinetics and Pharmacodynamics in Children versus Adults*. Available from: https://www.ncbi.nlm.nih.gov/books/NBK225507/
60. Watanabe H, Nagano N, Tsuji Y, Noto N, Ayusawa M, Morioka I. Challenges of pediatric pharmacotherapy: A narrative review of pharmacokinetics, pharmacodynamics, and pharmacogenetics. Eur J Clin Pharmacol. 2024;80:203–21. https://doi.org/10.1007/s00228-023-03598-x.
61. Kleiber M. Body size and metabolic rate. Physiol Rev. 1947;27:511–41. https://doi.org/10.1152/physrev.1947.27.4.511.
62. Anderson BJ, Holford NH. Understanding dosing: children are small adults, neonates are immature children. Arch Dis Child. 2013;98:737–44. https://doi.org/10.1136/archdischild-2013-303720.
63. Barbour AM, Fossler MJ, Barrett J. Practical considerations for dose selection in pediatric patients to ensure target exposure requirements. AAPS J. 2014;16:749–55. https://doi.org/10.1208/s12248-014-9603-x.
64. Mahmood I. Prediction of drug clearance in premature and mature neonates, infants, and children ≤2 years of age: A comparison of the predictive performance of 4 allometric models. J Clin Pharmacol. 2016;56:733–9. https://doi.org/10.1002/jcph.652.
65. Mahmood I. Misconceptions and issues regarding allometric scaling during the drug development process. Expert Opin Drug Metab Toxicol. 2018;14:843–54. https://doi.org/10.1080/17425255.2018.1499725.
66. Mahmood I, Tegenge MA. A comparative study between allometric scaling and physiologically based pharmacokinetic modeling for the prediction of drug clearance from neonates to adolescents. J Clin Pharmacol. 2019;59:189–97. https://doi.org/10.1128/aac.30.1.46.
67. Liu T, Ghafoori P, Gobburu JV. Allometry is a reasonable choice in pediatric drug development. J Clin Pharmacol. 2017;57:469–75. https://doi.org/10.1002/jcph.831.
68. Malik PRV, Temrikar ZH, Chelle P, Edginton AN, Meibohm B. Pediatric dose selection for therapeutic proteins. J Clin Pharmacol. 2021;61(Suppl 1):S193–206. https://doi.org/10.1002/jcph.1829.
69. Bouazza N, Dokoumetzidis A, Knibbe CAJ, de Wildt SN, Ambery C, De Cock PA, et al; c4c Expert Group on Pharmacometrics. General clinical and methodological considerations on the extrapolation of pharmacokinetics and optimization of study protocols for small molecules and monoclonal antibodies in children. Br J Clin Pharmacol. 2022; 88:4985-4996. https://doi.org/10.1111/bcp.15571
70. Holladay J, Winch P, Morse J, Anderson BJ, McKee CT, Rice-Weimer J, et al. Acetaminophen pharmacokinetics in infants and children with congenital heart disease. Paediatr Anaesth. 2023;33:46–51. https://doi.org/10.1111/pan.14579.
71. Sluymans T, Colan SD. Theoretical and empirical derivation of cardiovascular allometric relationships in children. J Appl Physiol (1985). 2005;99:445–57. https://doi.org/10.1152/japplphysiol.01144.2004.
72. Dallaire F, Sarkola T. Growth of cardiovascular structures from the fetus to the young adult. Adv Exp Med Biol. 2018;1065:347–60. https://doi.org/10.1007/978-3-319-77932-4_22.
73. Werneck AO, Conde J, Coelho-E-Silva MJ, Pereira A, Costa DC, Martinho D, et al. Allometric scaling of aerobic fitness outputs in school-aged pubertal girls. BMC Pediatr. 2019;19:96. https://doi.org/10.1186/s12887-019-1462-2.
74. Billig EM, O'Meara WP, Riley EM, McKenzie FE. Developmental allometry and paediatric malaria. Malar J. 2012;11:64. https://doi.org/10.1186/1475-2875-11-64.

75. Tamura R, Watanabe N, Nakamura S, Yoshimura N, Ozawa S, Hirono K, et al. Evaluation of the effects of ontogenetic or maturation functions and chronic heart failure on the model analysis for the dose-response relationship of warfarin in Japanese children. Eur J Clin Pharmacol. 2019;75:913–20. https://doi.org/10.1007/s00228-019-02652-x.
76. Borland LM. Allometric determination of the distance from the central venous pressure port to wedge position of balloon-tip catheters in pediatric patients. Crit Care Med. 1986;14:974–6. https://doi.org/10.1097/00003246-198611000-00013.
77. Wolstencroft P, Arnold P, Anderson BJ. Dose estimation for bivalirudin during pediatric cardiopulmonary bypass. Paediatr Anaesth. 2021;31:637–43. https://doi.org/10.1111/pan.14125.
78. Xiong Y, Fukuda T, Knibbe CAJ, Vinks AA. Drug dosing in obese children: Challenges and evidence-based strategies. Pediatr Clin North Am. 2017;64:1417–38. https://doi.org/10.1016/j.pcl.2017.08.011.
79. Morse JD, Cortinez LI, Anderson BJ. Considerations for intravenous anesthesia dose in obese children: Understanding PKPD. J Clin Med. 2023;12:1642. https://doi.org/10.3390/jcm12041642.
80. Durairaj C, Bhattacharya I. Challenges, approaches and enablers: effectively triangulating towards dose selection in pediatric rare diseases. J Pharmacokinet Pharmacodyn. 2023;50:445–59. https://doi.org/10.1007/s10928-023-09868-6.
81. Stanton BF, Behrman RE. The field of pediatrics. In: Kleigman RM, Stanton BF, St. Geme III JW, Schor NF, Behrman RE, editors. Nelson Textbook of Pediatrics. 20th ed. Philadelphia: Elsevier; 2016. p. 1–18.
82. Wang C, Allegaert K, Tibboel D, Danhof M, van der Marel CD, Mathot RA, et al. Population pharmacokinetics of paracetamol across the human age-range from (pre)term neonates, infants, children to adults. J Clin Pharmacol. 2014;54:619–29. https://doi.org/10.1002/jcph.259.
83. Clements JA, Avery ME. Lung surfactant and neonatal respiratory distress syndrome. Am J Respir Crit Care Med. 1998;157(4 Pt 2):S59–66. https://doi.org/10.1164/ajrccm.157.4.nhlb1-1.
84. Hallman M, Herting E. Historical perspective on surfactant therapy: Transforming hyaline membrane disease to respiratory distress syndrome. Semin Fetal Neonatal Med. 2023;28:101493. https://doi.org/10.1016/j.siny.2023.101493.
85. Avery ME, Mead J. Surface properties in relation to atelectasis and hyaline membrane disease. AMA J Dis Child. 1959;97(5 Pt 1):517–23. https://doi.org/10.1001/archpedi.1959.02070010519001.
86. Fujiwara T, Maeta H, Chida S, Morita T, Watabe Y, Abe T. Artificial surfactant therapy in hyaline-membrane disease. Lancet. 1980;1(8159):55–9. https://doi.org/10.1016/s0140-6736(80)90489-4.
87. Liggins GC, Howie RN. A controlled trial of antepartum glucocorticoid treatment for prevention of the respiratory distress syndrome in premature infants. Pediatrics. 1972;50:515–25.
88. Sweet DG, Carnielli VP, Greisen G, Hallman M, Klebermass-Schrehof K, Ozek E, et al. European consensus guidelines on the management of respiratory distress syndrome: 2022 update. Neonatology. 2023;120:3–23. https://doi.org/10.1159/000528914.
89. Gross RE, Hubbard JP. Surgical ligation of a patent ductus arteriosus: report of first successful case. JAMA. 1939;112:729–31.
90. Blalock A, Taussig HB. The surgical treatment of malformations of the heart in which there is pulmonary stenosis or pulmonary atresia. JAMA. 1945;128:189–202.
91. Crafoord C, Nylin KGV. Congenital coarctation of the aorta and its surgical treatment. J Thorac Surg. 1945;14:347–61.
92. Thomas VT. Partners of the Heart: Vivien Thomas and His Work with Alfred Blalock. Philadelphia: University of Pennsylvania Press; 1985. p. 96–7.
93. Lewis FJ, Taufic M. Closure of atrial septal defects with the aid of hypothermia; experimental accomplishments and the report of one successful case. Surgery. 1953;33:52–9.
94. Swan H, Zeavin I, Blount SG Jr, Virtue RW. Surgery by direct vision in the open heart during hypothermia. J Am Med Assoc. 1953;153:1081–5. https://doi.org/10.1001/jama.1953.02940290013005.
95. Lillehei CW, Cohen M, Warden HE, Varco RL. The direct-vision intracardiac correction of congenital anomalies by controlled cross circulation; results in thirty-two patients with ven-

tricular septal defects, tetralogy of Fallot, and atrioventricularis communis defects. Surgery. 1955;38:11–29.
96. Miller BJ, Gibbon JH Jr, Greco VF, Smith BA, Cohn CH, Allbritten FF Jr. The production and repair of interatrial septal defects under direct vision with the assistance of an extracorporeal pump-oxygenator circuit. J Thorac Surg. 1953;26:598–616; discussion 631-2.
97. Mazurak M, Kusa J. Milestone in congenital cardiac surgery: 65 years of the heart-lung machine. Arch Dis Child. 2020;105:92–4. https://doi.org/10.1136/archdischild-2018-315684.
98. Weirich WL, Gott VL, Lillehei CW. The treatment of complete heart block by the combined use of a myocardial electrode and an artificial pacemaker. Surg Forum. 1957;8:360–3.
99. Cohen MI, Thurber C. The history of cardiac pacing in the young and a look to the future. Curr Opin Pediatr. 2022;34:476–83. https://doi.org/10.1097/mop.0000000000001160.
100. Barnard CN. The operation. A human cardiac transplant: an interim report of a successful operation performed at Groote Schuur Hospital, Cape Town. S Afr Med J. 1967;41:1271–4.
101. Barnard CN. Human cardiac transplantation. An evaluation of the first two operations performed at the Groote Schuur Hospital, Cape Town. Am J Cardiol. 1968;22:584–96. https://doi.org/10.1016/0002-9149(68)90166-5.
102. Kantrowitz A, Haller JD, Joos H, Cerruti MM, Carstensen HE. Transplantation of the heart in an infant and an adult. Am J Cardiol. 1968;22:782–90.
103. Rashkind WJ, Miller WW. Creation of an atrial septal defect without thoracotomy. A palliative approach to complete transposition of the great arteries. JAMA. 1966;196:991–2.
104. Kan JS, White RI Jr, Mitchell SE, Gardner TJ. Percutaneous balloon valvuloplasty: a new method for treating congenital pulmonary-valve stenosis. N Engl J Med. 1982;307:540–2. https://doi.org/10.1056/nejm198208263070907.
105. Kan JS. Jean S. Kan, MD: a conversation with Colin K.L. Phoon, MPhil, MD. Am J Cardiol. 2008;101:129–38. https://doi.org/10.1016/j.amjcard.2007.07.074.
106. Kang SL, Benson L. Recent advances in cardiac catheterization for congenital heart disease. F1000Res. 2018;7:370. https://doi.org/10.12688/f1000research.13021.1.
107. Goldstein BH, Kreutzer J. Transcatheter intervention for congenital defects involving the great vessels: JACC Review Topic of the Week. J Am Coll Cardiol. 2021;77:80–96. https://doi.org/10.1016/j.jacc.2020.11.019.
108. Barry OM, Bouhout I, Turner ME, Petit CJ, Kalfa DM. Transcatheter cardiac interventions in the newborn: JACC Focus Seminar. J Am Coll Cardiol. 2022;79:2270–83. https://doi.org/10.1016/j.jacc.2022.03.374.
109. Turner ME, Bouhout I, Petit CJ, Kalfa D. Transcatheter closure of atrial and ventricular septal defects: JACC Focus Seminar. J Am Coll Cardiol. 2022;79:2247–58. https://doi.org/10.1016/j.jacc.2021.08.082.
110. Grossman DC. The history of injury control and the epidemiology of child and adolescent injuries. Future Child. 2000 Spring-Summer; 10:23-52.
111. Dukleska K, Borrup K, Campbell BT. Childhood injury prevention: Where we've been and where we need to be. Semin Pediatr Surg. 2022;31:151220. https://doi.org/10.1016/j.sempedsurg.2022.151220.
112. Abbott E. A study of the early history of child labor in America. Am J Sociol. 1908;14:15–37.
113. Mayer G. Child labor in America: History, policy, and legislative issues. Washington, DC: Congressional Research Service; 2013. p. 7–5700.
114. Wright S, Marlenga B, Lee BC. Childhood agricultural injuries: an update for clinicians. Curr Probl Pediatr Adolesc Health Care. 2013;43:20–44. https://doi.org/10.1016/j.cppeds.2012.08.002.
115. Jullien S. Prevention of unintentional injuries in children under five years. BMC Pediatr. 2021;21(Suppl 1):311. https://doi.org/10.1186/s12887-021-02517-2.
116. Nuraresya I, Nirmal U, Ng PK. A comprehensive review on the development of car booster seats for children. Curr J Appl Sci Tech. 2019;38:1–21.

117. Arena JM. Safety closure caps; safety measure for prevention of accidental drug poisoning in children. J Am Med Assoc. 1959;169:1187–8. https://doi.org/10.1001/jama.1959.73000280001010.
118. Roth TL, Marson A. Genetic disease and therapy. Annu Rev Pathol. 2021;16:145–66. https://doi.org/10.1146/annurev-pathmechdis-012419-032626.
119. Friedmann T, Roblin R. Gene therapy for human genetic disease? Science. 1972;175:949–55. https://doi.org/10.1126/science.175.4025.949.
120. Blaese RM, Culver KW, Miller AD, Carter CS, Fleisher T, Clerici M, et al. T lymphocyte-directed gene therapy for ADA- SCID: initial trial results after 4 years. Science. 1995;270:475–80. https://doi.org/10.1126/science.270.5235.475.
121. Wilson JM. Recollections from a pioneer who provided the foundation for the success of gene therapy in treating severe combined immune deficiencies. Hum Gene Ther Clin Dev. 2016;27:53–6. https://doi.org/10.1089/humc.2016.29013.int.
122. Daley, J. Gene therapy arrives. Sci Am. 2020; 322(1) (https://www.scientificamerican.com/article/gene-therapy-arrives/, accessed 04/30/2024)
123. Raper SE, Chirmule N, Lee FS, Wivel NA, Bagg A, Gao GP, et al. Fatal systemic inflammatory response syndrome in a ornithine transcarbamylase deficient patient following adenoviral gene transfer. Mol Genet Metab. 2003;80:148–58. https://doi.org/10.1016/j.ymgme.2003.08.016.
124. Branca MA. Gene therapy: cursed or inching towards credibility? Nat Biotechnol. 2005;23:519–21. https://doi.org/10.1038/nbt0505-519.
125. Senior M. Fresh from the biotech pipeline: record-breaking FDA approvals. [News Feature] Nat Biotechnol. 2024; 42:355-361. https://doi.org/10.1038/s41587-024-02166-7
126. Byrne BJ, Corti M, Muntoni F. Considerations for systemic use of gene therapy. Mol Ther. 2021;29:422–3. https://doi.org/10.1016/j.ymthe.2021.01.016.
127. Trivedi PD, Byrne BJ, Corti M. Evolving horizons: Adenovirus vectors' timeless influence on cancer, gene therapy and vaccines. Viruses. 2023;15:2378. https://doi.org/10.3390/v15122378.
128. Daley GQ. Welcoming the era of gene editing in medicine. N Engl J Med. 2024;390:1642–5. https://doi.org/10.1056/nejmp2314279.
129. Frangoul H, Locatelli F, Sharma A, Bhatia M, Mapara M, Molinari L, et al.; CLIMB SCD-121 Study Group. Exagamglogene Autotemcel for severe sickle cell disease. N Engl J Med. 2024; 390:1649-1662. https://doi.org/10.1056/nejmoa2309676
130. Locatelli F, Lang P, Wall D, Meisel R, Corbacioglu S, Li AM, et al.; CLIMB THAL-111 Study Group. Exagamglogene Autotemcel for transfusion-dependent beta-thalassemia. N Engl J Med. 2024; 390:1663-1676. https://doi.org/10.1056/nejmoa2309673
131. Lv J, Wang H, Cheng X, Chen Y, Wang D, Zhang L, et al. AAV1-hOTOF gene therapy for autosomal recessive deafness 9: a single-arm trial. Lancet. 2024;403(10441):2317–25. https://doi.org/10.1016/s0140-6736(23)02874-x.
132. Spyridakou C, Chan J, Tan J, Clement E, Nash R, Hall A, et al. Preparing for Otoferlin gene therapy trials: A survey of NHS Paediatric Audiology and Cochlear Implant services on diagnosis and management of Auditory Neuropathy Spectrum Disorder. Int J Pediatr Otorhinolaryngol. 2024;177:111870. https://doi.org/10.1016/j.ijporl.2024.111870.
133. Ravelli GP, Stein ZA, Susser MW. Obesity in young men after famine exposure in utero and early infancy. N Engl J Med. 1976;295:349–53.
134. Barker DJ, Osmond C, Golding J, Kuh D, Wadsworth ME. Growth in utero, blood pressure in childhood and adult life, and mortality from cardiovascular disease. BMJ. 1989;298:564–7. https://doi.org/10.1136/bmj.298.6673.564.
135. Barker DJ, Winter PD, Osmond C, Margetts B, Simmonds SJ. Weight in infancy and death from ischaemic heart disease. Lancet. 1989;2(8663):577–80. https://doi.org/10.1016/s0140-6736(89)90710-1.
136. Barker DJ. The fetal and infant origins of adult disease. BMJ. 1990;301:1111. https://doi.org/10.1136/bmj.301.6761.1111.

137. Barker DJ. The origins of the developmental origins theory. J Intern Med. 2007;261:412–7. https://doi.org/10.1111/j.1365-2796.2007.01809.x.
138. Calkins K, Devaskar SU. Fetal origins of adult disease. Curr Probl Pediatr Adolesc Health Care. 2011;41:158–76. https://doi.org/10.1016/j.cppeds.2011.01.001.
139. Oken E, Bastain TM, Bornkamp N, Breton CV, Fry RC, Gold DR, et al.; program collaborators for Environmental influences on Child Health Outcomes. When a birth cohort grows up: challenges and opportunities in longitudinal developmental origins of health and disease (DOHaD) research. J Dev Orig Health Dis. 2023; 14:175-181. https://doi.org/10.1017/s2040174422000629
140. Wells JCK, Desoye G, Leon DA. Reconsidering the developmental origins of adult disease paradigm: The 'metabolic coordination of childbirth' hypothesis. Evol Med Public Health. 2024;12:50–66. https://doi.org/10.1093/emph/eoae002.
141. Faa G, Fanos V, Manchia M, Van Eyken P, Suri JS, et al. The fascinating theory of fetal programming of adult diseases: A review of the fundamentals of the Barker hypothesis. J Public Health Res. 2024;13:22799036241226817. https://doi.org/10.1177/22799036241226817.
142. Silverman WA. A cautionary tale about supplemental oxygen: the albatross of neonatal medicine. Pediatrics. 2004;113:394–6. https://doi.org/10.1542/peds.113.2.394.
143. Severinghaus JW. Eight sages over five centuries share oxygen's discovery. Adv Physiol Educ. 2016;40:370–6. https://doi.org/10.1152/advan.00076.2016.
144. Flagg PJ. Treatment of asphyxia in the new-born: Preliminary report of the practical application of modern scientific methods. JAMA. 1928;91:788–91.
145. Henderson Y. The prevention and treatment of asphyxia in the new-born. JAMA. 1928;90:583–6.
146. Smith CA, Kaplan E. Adjustment of blood oxygen levels in neonatal life. Am J Dis Child. 1942;64:843–59.
147. Terry TL. Extreme prematurity and fibroblastic overgrowth of persistent vascular sheath behind each crystalline lens:I. Preliminary report. Am J Ophthalmol. 1942;25:203–4.
148. Campbell K. Intensive oxygen therapy as a possible cause of retrolental fibroplasia; a clinical approach. Med J Aust. 1951;2:48–50.
149. Crosse VM. The problem of retrolental fibroplasia in the city of Birmingham. Trans Ophthalmol Soc UK. 1951;71:609–12.
150. Hartnett ME, Lane RH. Effects of oxygen on the development and severity of retinopathy of prematurity. J AAPOS. 2013;17:229–34. https://doi.org/10.1016/j.jaapos.2012.12.155.
151. Mathias M, Chang J, Perez M, Saugstad O. Supplemental oxygen in the newborn: Historical perspective and current trends. *Antioxidants* (Basel). 2021; 10:1879. https://doi.org/10.3390/antiox10121879
152. BOOST-II Australia and United Kingdom Collaborative Groups, Tarnow-Mordi W, Stenson B, Kirby A, Juszczak E, Donoghoe M, Deshpande S, et al. Outcomes of two trials of oxygen-saturation targets in preterm infants. N Engl J Med. 2016;374:749–60. https://doi.org/10.1056/nejmoa1514212.
153. Askie LM, Darlow BA, Finer N, Schmidt B, Stenson B, Tarnow-Mordi W, et al.; Neonatal Oxygenation Prospective Meta-analysis (NeOProM) Collaboration. Association between oxygen saturation targeting and death or disability in extremely preterm infants in the neonatal oxygenation prospective meta-analysis collaboration. JAMA. 2018; 319:2190-2201. https://doi.org/10.1001/jama.2018.5725
154. Loos H. Clinical experience with the sedative contergan-forte & the sedative contergan which is administered daily. Medizinische. 1958;12:482–3.
155. Somers GF. Pharmacological properties of thalidomide (alpha-phthalimido glutarimide), a new sedative hypnotic drug. Br J Pharmacol Chemother. 1960;15:111–6. https://doi.org/10.1111/j.1476-5381.1960.tb01217.x.
156. Franks ME, Macpherson GR, Figg WD. Thalidomide. Lancet. 2004;363(9423):1802–11. https://doi.org/10.1016/s0140-6736(04)16308-3.

157. Vargesson N, Hooper G, Giddins G, Hunter A, Stirling P, Lam W. Thalidomide upper limb embryopathy - pathogenesis, past and present management and future considerations. J Hand Surg Eur *Vol.* 2023; 48:699-709. https://doi.org/10.1177/17531934231177425
158. McBride WG. Thalidomide and congenital abnormalities. Lancet. 1961;278(1358):90927–8.
159. Lenz W. Thalidomide and congenital abnormalities. Lancet. 1962;279(7219):45.
160. Somers GF. Thalidomide and congenital abnormalities. Lancet. 1962;1(7235):912–3. https://doi.org/10.1016/s0140-6736(62)91943-8.
161. Speirs AL. Thalidomide and congenital abnormalities. Lancet. 1962;1(7224):303–5. https://doi.org/10.1016/s0140-6736(62)91248-5.
162. Taussig HB. A study of the German outbreak of phocomelia. The thalidomide syndrome. JAMA. 1962;180:1106–14.
163. Kelsey FO. Problems raised for the FDA by the occurrence of thalidomide embryopathy in Germany, 1960-1961. Am J Public Health Nations Health. 1965;55:703–7. https://doi.org/10.2105/ajph.55.5.703.
164. Kuehn BM. Frances Kelsey honored for FDA legacy: award notes her work on thalidomide, clinical trials. JAMA. 2010; 304:2109-2110, 2112. https://doi.org/10.1001/jama.2010.1652
165. Watts G. Frances Oldham Kelsey. Lancet. 2015;386(10001):1334. https://doi.org/10.1016/s0140-6736(15)00339-6.
166. McNamara DG, Manning JA, Engle MA, Whittemore R, Neill CA, Ferencz C. Helen Brooke Taussig: 1898 to 1986. J Am Coll Cardiol. 1987;10:662–71. https://doi.org/10.1016/s0735-1097(87)80211-5.
167. Walsh KE, Adams WG, Bauchner H, Vinci RJ, Chessare JB, Cooper MR, et al. Medication errors related to computerized order entry for children. Pediatrics. 2006;118:1872–9. https://doi.org/10.1542/peds.2006-0810.
168. Liang MQ, Boudjellab A, Kwon H, Jouvet P, Lebel D, Thibault M, et al. Implementation of a computerized provider order entry system in a pediatric hospital in Canada. Stud Health Technol Inform. 2021;281:590–4. https://doi.org/10.3233/shti210239.
169. Liu KW, Shih YF, Chiang YJ, Chen LJ, Lee CH, Chen HN, et al. Reducing medication errors in children's hospitals. J Patient Saf. 2023;19:151–7. https://doi.org/10.1097/pts.0000000000001087.
170. Satir AN, Pfiffner M, Meier CR, Caduff GA. Prescribing errors in children: what is the impact of a computerized physician order entry? Eur J Pediatr. 2023;182:2567–75. https://doi.org/10.1007/s00431-023-04894-5.
171. Han YY, Carcillo JA, Venkataraman ST, Clark RS, Watson RS, Nguyen TC, et al. Unexpected increased mortality after implementation of a commercially sold computerized physician order entry system. Pediatrics. 2005;116:1506–12. https://doi.org/10.1542/peds.2005-1287.
172. Tarkan L. Small patients, big consequences in medical errors. The New York Times, 09/15/2008. https://www.nytimes.com/2008/09/15/health/healthspecial2/15mistakes.html, accessed 04/12/2024.
173. Kaushal R, Bates DW, Landrigan C, McKenna KJ, Clapp MD, Federico F, et al. Medication errors and adverse drug events in pediatric inpatients. JAMA. 2001;285:2114–20. https://doi.org/10.1001/jama.285.16.2114.
174. Wimmer S, Toni I, Botzenhardt S, Trollmann R, Rascher W, Neubert A. Impact of a computerized physician order entry system on medication safety in pediatrics-The AVOID study. Pharmacol Res Perspect. 2023;11:e01092. https://doi.org/10.1002/prp2.1092.
175. Frederick SA. Advanced technology in pediatric intensive care units: Have they improved outcomes? Pediatr Clin North Am. 2016;63:293–301. https://doi.org/10.1016/j.pcl.2015.12.003.
176. Roser, M. Mortality in the past: every second child died. In: OurWorldInData.org, 2023. Retrieved from: 'https://ourworldindata.org/child-mortality-in-the-past' [Online Resource] (accessed 04/10/2024)
177. Committee on Pediatric Studies Conducted Under the Best Pharmaceuticals for Children Act (BPCA) and the Pediatric Research Equity Act (PREA); Board on Health Sciences Policy; Institute of Medicine; Field MJ, Boat TF, editors. *Safe and Effective Medicines for*

Children: Pediatric Studies Conducted Under the Best Pharmaceuticals for Children Act and the Pediatric Research Equity Act. Washington (DC): National Academies Press (US); 2012 Feb 29. Available from: https://www.ncbi.nlm.nih.gov/books/NBK202036/. https://doi.org/10.17226/13311.

178. Burns J, Basken A, Acosta R, Garnier-Villarreal M, Kulkarni A, Hayes DA. Identification of research priorities in CHD: empowering patients and families through participation in the development of formal research agendas. Cardiol Young. 2023;33:288–93. https://doi.org/10.1017/s104795112200066x.
179. Hollister SJ. Paediatric devices that grow up. Nat Biomed Eng. 2017;1:777–8. https://doi.org/10.1038/s41551-017-0147-0.
180. Conboy PJ, Gibbin KP. Paediatric cochlear implant durability: the Nottingham experience. Cochlear Implants Int. 2004;5:131–7. https://doi.org/10.1179/cim.2004.5.4.131.
181. Hammad AM, Balsano M, Ahmad AA. Vertebral body tethering: An alternative to posterior spinal fusion in idiopathic scoliosis? Front Pediatr. 2023;11:1133049. https://doi.org/10.3389/fped.2023.1133049.
182. Sullivan MH, Jackson TJ, Milbrandt TA, Larson AN, Kepler CK, Sebastian AS. Evidence-based indications for vertebral body tethering in spine deformity. Clin Spine Surg. 2024;37:82–91. https://doi.org/10.1097/bsd.0000000000001521.
183. Etnel JR, Elmont LC, Ertekin E, Mokhles MM, Heuvelman HJ, Roos-Hesselink JW, et al. Outcome after aortic valve replacement in children: A systematic review and meta-analysis. J Thorac Cardiovasc Surg. 2016; 151:143-152.e1-3. https://doi.org/10.1016/j.jtcvs.2015.09.083
184. Smood B, Hara H, Cleveland DC, Cooper DKC. In search of the ideal valve: optimizing genetic modifications to prevent bioprosthetic degeneration. Ann Thorac Surg. 2019;108:624–35. https://doi.org/10.1016/j.athoracsur.2019.01.054.
185. Li RL, Russ J, Paschalides C, Ferrari G, Waisman H, Kysar JW, Kalfa D. Mechanical considerations for polymeric heart valve development: Biomechanics, materials, design and manufacturing. Biomaterials. 2019;225:119493. https://doi.org/10.1016/j.biomaterials.2019.119493.
186. Singh SK, Faridmoayer E, Vitale N, Woodard E, Xue Y, Abramov A, et al. Valved conduits for right ventricular outflow tract reconstruction: A review of current technologies and future directions. Pediatr Cardiol. 2023. https://doi.org/10.1007/s00246-023-03346-z
187. Zhu MZL, Buratto E, Wu DM, Ishigami S, Schulz A, Brizard CP, et al. Long-term outcomes of mechanical aortic valve replacement in children. Semin Thorac Cardiovasc Surg Pediatr Card Surg Annu. 2024;27:52–60. https://doi.org/10.1053/j.pcsu.2023.12.003.

Pediatric Pharmacotherapy: Beyond Scaling—Understanding Developmental Dynamics and Therapeutic Adaptations

Joyce P. Lee

Introduction

It is common for those familiar with pediatric medicine to say "children are not small adults." Historically, most therapeutics and medical devices or equipment in healthcare are designed for adults and modified for children predicated on the overly simplified notion that issues are exclusive to scale. The landscape is much more complex; size is but one consideration of many systemic changes that occur in the transitions from infancy through adolescence. Medical advances have come a long way to recognize and highlight differences in the physical, psychological, social, and emotional development of children at various stages. The majority of this chapter will focus on the reasons and methods for approaching medication use in children differently than adults. However, it is not adequate only to identify the differences. Understanding the similarities first in basic cellular and physiological systems of children compared to adults and other mammals can facilitate the extrapolation of readily available information. For instance, children's kidneys, like adults, have a loop of Henle and proximal and distal renal tubules. Therefore, it is not surprising that diuretics, widely used in adults to increase water and sodium excretion and alleviate volume overload in conditions such as acute kidney injury, hypertension, diabetes insipidus, and heart failure, are also effective in children [1].

Extrapolation of adult data for pediatrics, when used appropriately, may be beneficial in increasing the availability of medications and medical devices for use in children. Maximizing the use of available adult data can potentially minimize the number of children needed, if at all, for clinical trials [2]. However, it needs to be cautioned that differences in development as well as variations of disease progression and pathogenesis should be carefully considered when extrapolating data. It is

J. P. Lee (✉)
Children's Hospital of Philadelphia, Philadelphia, PA, USA
e-mail: LeeJ32@chop.edu

J. L. Wells (ed.), *A Clinical Lens on Pediatric Engineering*,
https://doi.org/10.1007/978-3-031-84655-7_2

important to take the next step in understanding the developmental stage of the pediatric patient as well as differences in disease presentation between adults and children to find the most effective dose and avoid adverse events. To illustrate this important concept further using the previous diuretic example: premature neonates require less frequent dosing of furosemide compared to adults due to longer elimination half-life, which helps to avoid drug accumulation and risk for toxicities. Additionally, bumetanide can displace bilirubin and should be avoided in patients at risk for kernicterus, a rare but serious neurological disorder caused by excess bilirubin that more commonly presents in infants. Ongoing research and use of age-specific pharmacokinetic and pharmacodynamic data have empowered healthcare providers to tailor medication regimens more precisely to each patient's developmental stage. By incorporating nuanced insights into how growth impacts drug absorption, distribution, metabolism, and elimination, therapeutic outcomes can be optimized while minimizing the risk of adverse events. This continuously evolving understanding enhances safe and efficacious use of medications in pediatric and adolescent medicine.

Pharmacokinetic and Pharmacodynamic Differences

Pharmacokinetics (PK) is the study of what the body does to a drug and how the drug moves through the body, specifically looking at absorption, distribution, metabolism, and excretion (known as ADME). Conversely, pharmacodynamics (PD) is the study of what the drug does to the body and the relationship between drug concentration and the resulting effect [3]. PK/PD can be described as an exposure-response relationship and its understanding is fundamental to the development of drugs. Physiologic differences in children can have a profound effect on basic PK/PD principles. Considering absorption of orally administered drugs as an example, gastric emptying is slower in neonates and infants, thereby leading to prolonged or increased time to maximal effect. Gastric pH is initially neutral at birth and decreases within the first 24 h, so the absorption of acid-labile drugs, such as amoxicillin, is greater in neonates and infants. Drug dosing and frequency may need to be adjusted to account for these differences. Administration of medications via other routes are also affected by the developmental stages. Intramuscular absorption may be compromised in newborns who have poor peripheral perfusion or low muscle mass. On the contrary, topical absorption may be increased in neonates and infants due to thinner stratum corneum, greater perfusion of the epidermis, and increased body surface area.

Age-dependent body composition changes in extracellular fluid volume, total body water, and water-to-body fat ratio can affect drug distribution after the drug enters the bloodstream. Contrary to what one may expect, younger infants may require larger doses of medications that are water-soluble because their volume of distribution is larger. Protein binding, cardiac output, tissue perfusion, and membrane permeability are a few other factors that impact drug distribution and need to be considered when optimizing medication dosing regimens in pediatric patients. The development of cytochrome P450 (CYP) monooxygenases matures at different ages and can affect the half-life, dosing frequency, and toxicity of medications [4].

Pharmacokinetic factors often need to be considered in relation to one another because they may have counterintuitive influences on medication dosing. Gentamicin is an aminoglycoside antibiotic used to treat severe bacterial infections. It has been associated with nephrotoxicity and ototoxicity and has a narrow therapeutic window, necessitating individualized drug monitoring to adjust the dose and monitor for toxicity. Administration of gentamicin at higher doses with longer dosing intervals, known as extended interval dosing, allows for better efficacy and reduces toxicity [5]. Infants and younger children require larger conventional doses, with the same dosing frequency, compared to older children and adults given the drug's water solubility and larger volume of distribution in younger individuals. Conversely, neonates require lower doses, but dosing frequency remains similarly extended due to gentamicin's significantly longer half-life in this population. This discrepancy could lead to confusion when comparing neonatal once-daily dosing to extended interval dosing in children and adults. Understanding the nuanced differences in how pharmacokinetic parameters influence dosage, frequency, and terminology of drug regimens is crucial to avoid inadvertent dosing errors in infants and children.

Pharmacokinetics and considerations related to medication use and dosing may also be influenced by the development of coordination and maturity. For example, effective use of an inhaler for asthma requires hand-mouth coordination as well as deep and timely inhalations. While spacers and nebulizers have helped children overcome these barriers, even the study of objective results such as peak expiratory flow rate is dependent on the patient's ability to perform it.

Research into the treatment of emerging issues in pediatric patients often lags behind the diagnosis of these conditions. Childhood mental health and neurodevelopmental disorders such as attention-deficit/hyperactivity disorder (ADHD), anxiety, and depression have increased in school-age children and adolescents. However, the use of psychotropic medications in treating pediatric mental health conditions does not have adequate considerations for the quality of research in diverse younger populations and understanding of long-term outcomes and adverse events. ADHD is one of the most common neurobehavioral health disorders of childhood and treatment guidelines recommend behavioral therapy in combination with medication management. In school-age children, stimulants, which increase neurotransmitters dopamine and norepinephrine, are often prescribed. Side effects such as loss of appetite and insomnia along with risk for abuse, misuse, and addiction may limit their use. Moreover, challenges with dosing administration during school have led to the development of different extended-release formulations that are not interchangeable and can increase the risk for treatment or administration errors. Alternatively, non-stimulants lack adequate FDA-approved labeling and are often prescribed off-label with limited evidence to support appropriate use and dosing [6].

Adolescents also present a unique challenge even though organ function is similar to adults and pharmacokinetic considerations may not be as complex as with smaller children. The developmental stages surrounding puberty can impact prescribing decisions such as with prolonged steroid use and the potential to affect growth. Considerations also need to be taken for medications that are contraindicated in pregnancy and should be avoided in sexually active adolescents. The complexity surrounding adolescent social interactions, risk-taking, independence, and behaviors can all impact adherence to medications used to treat chronic illnesses such as diabetes. Extended-release formulations, long-acting implantable devices, and integration with electronics such as watches and phones are necessary to ensure adherence to and self-management of important treatments [7].

Regulatory History

In the United States, new human drugs and biological products must be approved by the U.S. Food and Drug Administration (FDA). Traditional pharmaceutical drugs used in humans are chemically synthesized with a known structure while biologics are complex mixtures manufactured using biological sources such as blood components, proteins (antibodies), cells, or genetic material. The pharmaceutical company is required to demonstrate through an extensive application and clinical trial process that its drug is safe and effective for the intended indication and quality standards are followed for manufacturing. Historical lack of pediatric information on approved drug labeling can be attributed to reasons including small financial incentives for pharmaceutical companies and challenges in conducting studies, such as not having pediatric equipment, expertise, and informed consent. Low profitability and limited patient population are reasons children as well as patients of all ages with rare diseases, one that affects fewer than 200,000 people in the United States, were considered "therapeutic orphans" in which there was a lack of approved products [8].

In response to sentinel events like the sulfanilamide elixir tragedy where over 100 people, including many children, died from diethylene glycol poisoning and thalidomide, leading to birth defects in Europe, the United States passed several pieces of legislation to protect the public (Fig. 1). The Food, Drug and Cosmetic Act

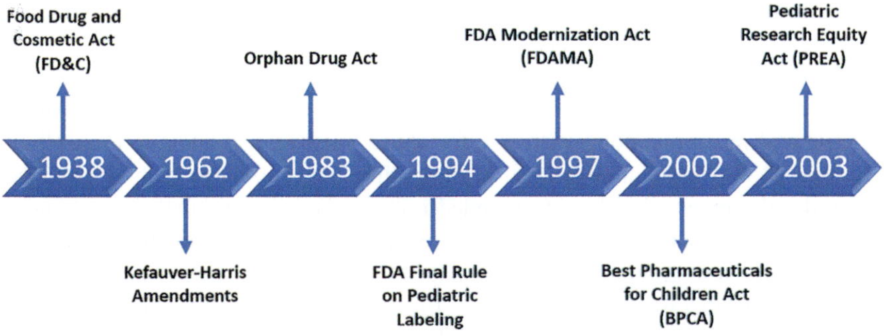

Fig. 1 Timeline of historical regulatory events related to pediatric drug development and safety. *FDA* U.S. Food and Drug Administration

(FD&C) of 1938 required companies to prove their drugs are safe before marketing. The Kefauver-Harris Amendments in 1962 further mandated evidence for efficacy. As a result, many companies avoided clinical pediatric studies and instead labeled the medications as not for use in children. In 1983, the Orphan Drug Act was passed to encourage drug development and provide financial incentives for research in drugs for rare diseases, of which approximately half of the patient population includes children. In 1994, the FDA Final Rule on Pediatric Labeling was passed to increase *Pediatric Use* information on drug labeling. It required manufacturers to survey existing data to determine if pediatric use information could be extrapolated. Then in 1997, the FDA Modernization Act (FDAMA) provided additional patent exclusivity incentives for companies who conducted pediatric studies on their products.

Significant pediatric research did not occur until the passage of the Best Pharmaceuticals for Children Act (BPCA) in 2002 and Pediatric Research Equity Act (PREA) in 2003. BPCA identified drugs in need of study, sponsored trials via funding from the National Institutes of Health (NIH) and provided financial incentives for pediatric exclusivity. PREA required new drug and biologic applications to have a pediatric assessment, unless waived or deferred by the FDA. The two acts together provided incentives and requirements for pediatric studies and added pediatric information to over 400 drugs in the first 5 years of implementation.

Although regulatory and clinical advances have expanded pediatric labeling and safety, there are still a large number of drugs that lack pediatric information and clinical trials in children often lag behind that of adults. "Off-label" is the terminology used to describe the use of a medication for an indication or dose that is not included in the FDA-approved labeling. Off-label use is common in pediatrics, especially in subpopulations such as neonates, infants and children younger than 2 years old, and children with rare diseases, where even less research exists. It is important to note that off-label use is not illegal or intrinsically dangerous. There are, of course, risks associated with off-label use and clinical expertise coupled with published literature, such as case reports, are necessary to help guide appropriate use [9].

"Key Potentially Inappropriate Drugs in Pediatrics: The KIDs List" is the first published list of medications that are potentially inappropriate for use in children, almost 30 years after Beers and colleagues published a similar list of medications for older adults. This review of published literature identified medications that have a higher risk of serious harm in children, such as risk of respiratory depression and death with codeine due to pharmacogenomic effects, increased rate of propofol-related infusion syndrome in children due to higher relative doses, and tooth discoloration and delay in skeletal development with tetracyclines [10]. A study looking at the incidence of adverse events in different age groups using data from large clinical trials found that younger children, between ages 0 and 9 years, had the highest incident rate of adverse events [11]. It highlights the importance of having well-designed studies that include pediatric patients so practitioners can be better informed about the safety and risks of medication use.

Extrapolation of data from adults is challenging because similar diseases may have differences in pathophysiology or disease progression. Additionally, as mentioned above, pharmacology of a drug may also vary among different populations. A systematic approach to extrapolation of data with careful consideration of disease similarity, drug pharmacology, and response to treatment of a reference population to a target population can help guide safe usage of a drug and balance unnecessary exposure of children to investigational trials [12].

Drug Design

As drug design is primarily focused on adults, children and their caregivers often encounter another unique challenge when dealing with medication administration. A young child may not be able to swallow a whole tablet or capsule or palatability may be an issue. Liquid medications may help in most cases; however, smaller neonates or preterm babies may not be able to tolerate the volume, or alternatively, the concentration may not allow for a large enough volume to be prepared for the dose. Some liquid formulations such as elixirs may also contain alcohol or other excipients that may be a risk for certain subpopulations with under-developed or impaired organ function. This risk is not exclusive to oral medications as exemplified by benzyl alcohol preservative in intravenous formulations that have caused fatal illness in neonates and preterm infants.

Children's medications are commonly dosed by weight and commercially available products designed for adults often cannot accommodate the exact dose needed. Manipulation of products such as opening capsules, splitting tablets in half or cutting patches can be inaccurate or can alter an intended drug release design. Even in children's hospitals, commercial products, which are typically formulated for the standard adult dose, require manipulation prior to use in children. Not only does this increase the workload for pharmacy and nursing staff, but it also introduces the potential for medication preparation errors as well as stability and sterility concerns. To meet patient needs, compounding of extemporaneous preparations may be required when a commercial product is not available. Like off-label use, extemporaneous preparations are not FDA-approved and can have risks associated with stability, contamination, and uniformity. Additionally, the availability of different concentrations can lead to dose conversion or calculation errors.

Caregiver education on how to measure the correct dose and store the medication is also important to prevent adverse events with unintended administration errors. Inaccuracies with measuring volume using kitchen spoons, dosage cups or droppers and understanding which line to measure on various types of oral and injectable syringes also pose challenges to the lay caregiver. Add in potential frequent dose changes or titrations in chronic illnesses with many medications, illiteracy or language barriers, and multiple caregivers in the home or school; it is not a surprise that medication errors are common. The American Academy of Pediatrics (AAP) underscored these issues in a policy statement on preventing home medication administration errors. It was noted more than half of American children take at least one medication each week and medication administration errors commonly occur, with studies showing almost half of caregivers have administered doses that deviated more than 20% from the prescription and a quarter of caregivers administered doses that deviated by over 40% [13].

The development of medications that are more palatable or visually appealing has resulted in formulations containing increased sweeteners, flavors, and dyes, potentially posing issues for children with specific allergies or dietary restrictions. Furthermore, certain medication formulations, such as gummy dietary supplements or fentanyl lozenge on a handle, bear a resemblance to candy, which can be a concerning risk for inadvertent ingestion by children. Balancing the need for accessible products while prioritizing safety in children continues to be a multifaceted challenge for drug developers, regulatory agencies, and healthcare providers.

Advancements

It cannot be denied that pediatric advancements in drug development and drug use have made tremendous progress over the last few decades. Children are still a minority population and much of what was discussed in this chapter, like extrapolation and off-label use, has been reactive, not proactive. Physicians not specialized in pediatrics and other healthcare professionals such as pharmacists, respiratory therapists, and dieticians, learn little about pediatrics in school and need to supplement with additional, specialized education and training to care for children. Multidisciplinary teamwork among healthcare providers as well as caregivers is essential to address the complex needs of children. Even healthcare systems and electronic health records are designed mainly with the adult population in mind. Pediatric solutions often require special modules, customization, or workarounds to adult default settings.

The future looks promising though, thanks to remarkable advancements in technology. With nearly universal use of computers and digital health records, there is potential to use massive amounts of data gathered, termed Big Data, to transform traditional research. Real-world data (RWD) from electronic medical records, billing databases, registries, as well as home devices such as smartphones and smartwatches can be analyzed to provide real-world evidence (RWE). Data extracted from technology may provide more robust, relevant, and valuable information compared to controlled population studies [14]. For example, electronic health record data from institutions across the country can be used to compare safety rates of hospital-acquired conditions such as falls or surgical site infections and medication infusion pump data across hospitals could be used to standardize alerts and medication concentrations. Bayesian dosing software applying pediatric population pharmacokinetic models can be utilized to calculate dose and frequency of medications with narrow therapeutic windows. Internationally, many other countries are also using electronic data and registries for clinical research, adverse event reporting, and longitudinal health data collection [15, 16].

The integration of artificial intelligence (AI) technology, or the use of computer systems to simulate human intelligence, has been experiencing a notable surge within healthcare. AI processes include machine learning, decision-making and reasoning, natural language processing, robotics, and more. In healthcare, it can do more than automated administrative tasks and virtual assistants for patient communication. AI can be used to analyze large quantities of data and text for predictive modeling. It can analyze medical images, detecting subtle changes that humans may or may not catch. Furthermore, it can screen large databases to aid in drug discovery, development, and research [17].

AI technology has significant potential to impact the quality of clinical research, practice, and education throughout all aspects of healthcare, including that of pediatric medicine. However, caution and advocacy need to be exercised to prevent children from being excluded in AI applications. Assessments of pediatric inclusion in clinical trials integrating the use of AI have shown only about 20% include children in the study population and only 4% were pediatric-specific studies [18]. Reasons children may be excluded are reminiscent of historical reasons children were not included in clinical trials: lack of financial return, limited available pediatric data, and ethical concerns surrounding issues such as informed consent and privacy. Potential bias may also be introduced into the AI algorithms that exclude children, perpetuating inequity. It is important to learn from history to prevent children from becoming therapeutic orphans again. Similar to basic pharmacology principles where children cannot be treated simply as little adults, AI data also should not be extrapolated or generalized from adults to children. Proactive pediatric advocacy and inclusion in AI technology development is vital to prevent further health disparity and potential harm to children.

Conclusion

Children are not small adults and pharmacotherapy cannot simply be scaled down to fit pediatric needs. The nuanced complexities of developmental stages and growth must be taken into consideration to manage medication use safely and effectively in children. The future of pediatric research and practice relies upon learning from the past while advocating for progress and innovation.

Competing Interests The author has no conflicts of interest to declare that are relevant to the content of this chapter.

References

1. Stephenson T. How children's responses to drugs differ from adults. Br J Clin Pharmacol. 2005 Jun;59(6):670-3. https://doi.org/10.1111/j.1365-2125.2005.02445.x. PMID: 15948930; PMCID: PMC1884865.
2. Conklin LS, Hoffman EP, van den Anker J. Developmental Pharmacodynamics and Modeling in Pediatric Drug Development. J Clin Pharmacol. 2019 Sep;59 Suppl 1(Suppl 1):S87-S94. https://doi.org/10.1002/jcph.1482. PMID: 31502687; PMCID: PMC6741426.
3. System Pharmacists, A. S. (2023). 2017-ASHP-Intro to Pharmacokinetics and Pharmacodynamics. SIMIODE, QUBES Educational Resources. https://doi.org/10.25334/6QHC-YT04
4. Batchelor HK, Marriott JF. Paediatric pharmacokinetics: key considerations. Br J Clin Pharmacol. 2015 Mar;79(3):395-404. https://doi.org/10.1111/bcp.12267. PMID: 25855821; PMCID: PMC4345950.
5. Darmstadt GL, Miller-Bell M, Batra M, Law P, Law K. Extended-interval dosing of gentamicin for treatment of neonatal sepsis in developed and developing countries. J Health Popul Nutr. 2008 Jun;26(2):163-82. PMID: 18686550; PMCID: PMC2740664.
6. Shahidullah JD, Roberts H, Parkhurst J, Ballard R, Mautone JA, Carlson JS. State of the Evidence for Use of Psychotropic Medications in School-Age Youth. Children (Basel). 2023 Aug 26;10(9):1454. https://doi.org/10.3390/children10091454. PMID: 37761415; PMCID: PMC10528957.
7. Kang M, Kim K. Prescribing for adolescents. Aust Prescr. 2019 Feb;42(1):20-23. https://doi.org/10.18773/austprescr.2019.004. Epub 2019 Feb 1. PMID: 30765905; PMCID: PMC6370610.
8. Burckart GJ, Kim C. The Revolution in Pediatric Drug Development and Drug Use: Therapeutic Orphans No More. J Pediatr Pharmacol Ther. 2020;25(7):565–73. https://doi.org/10.5863/1551-6776-25.7.565. PMID: 33041711; PMCID: PMC7541025.
9. American Academy of Pediatrics. Policy statement. Off-label use of drugs in children. Pediatrics. 2014;133(3):563–7. https://doi.org/10.1542/peds.2013-4060.
10. Meyers RS, Thackray J, Matson KL, McPherson C, Lubsch L, Hellinga RC, Hoff DS. Key Potentially Inappropriate Drugs in Pediatrics: The KIDs List. J Pediatr Pharmacol Ther. 2020;25(3):175–91. https://doi.org/10.5863/1551-6776-25.3.175. PMID: 32265601; PMCID: PMC7134587.
11. Luo J, Eldredge C, Cho CC, Cisler RA. Population Analysis of Adverse Events in Different Age Groups Using Big Clinical Trials Data. JMIR Med Inform. 2016 Oct 17;4(4):e30. https://doi.org/10.2196/medinform.6437. PMID: 27751983; PMCID: PMC5088342.
12. ICH Harmonized Guideline: E11A Pediatric Extrapolation (2022). Retrieved from https://www.fda.gov/media/161190/download
13. Yin HS, Neuspiel DR, Paul IM, Franklin W, Tieder JS, Adirim T, Alvarez F, Brown JM, Bundy DG, Ferguson LE, Gleeson SP, Leu M, Mueller BU, Connor Phillips S, Quinonez RA, Rea C, Rinke ML, Shaikh U, Shiffman RN, Vickers Saarel E, Spencer Cockerham SP, Mack Walsh K, Jones B, Adler AC, Foster JH, Green TP, Houck CS, Laughon MM, Neville K, Reigart JR, Shenoi R, Sullivan JE, Van Den Anker JN, Verhoef PA. Preventing Home Medication Administration Errors. Pediatrics. 2021 Dec 1;148(6):e2021054666. https://doi.org/10.1542/peds.2021-054666. PMID: 34851406.
14. Magalhães T, Dinis-Oliveira RJ, Taveira-Gomes T. Digital Health and Big Data Analytics: Implications of Real-World Evidence for Clinicians and Policymakers. Int J Environ Res Public Health. 2022 Jul 8;19(14):8364. https://doi.org/10.3390/ijerph19148364. PMID: 35886214; PMCID: PMC9325235.

15. McMahon AW, Cooper WO, Brown JS, Carleton B, Doshi-Velez F, Kohane I, Goldman JL, Hoffman MA, Kamaleswaran R, Sakiyama M, Sekine S, Sturkenboom MCJM, Turner MA, Califf RM. Big Data in the Assessment of Pediatric Medication Safety. Pediatrics. 2020 Feb;145(2):e20190562. https://doi.org/10.1542/peds.2019-0562. Epub 2020 Jan 14. PMID: 31937606.
16. Subbiah V. The next generation of evidence-based medicine. Nat Med. 2023 Jan;29(1):49-58. https://doi.org/10.1038/s41591-022-02160-z. Epub 2023 Jan 16. PMID: 36646803.
17. Corsello A, Santangelo A. May Artificial Intelligence Influence Future Pediatric Research?-The Case of ChatGPT. Children (Basel). 2023 Apr 21;10(4):757. https://doi.org/10.3390/children10040757. PMID: 37190006; PMCID: PMC10136583.
18. Nagy M, Sisk B, Lai A, Kodish E. Will artificial intelligence widen the therapeutic gap between children and adults? Pediatr Investig. 2023 Dec 1;8(1):1-6. https://doi.org/10.1002/ped4.12407. PMID: 38516139; PMCID: PMC10951493.

Beyond the Clinical: How a Holistic Approach to Design Builds a Strong Foundation for Modern Pediatric Healthcare

Leslie Kipps and Adrienne Erdman

Evolution of Pediatric Healthcare

Healthcare Design Shift

Over the past several decades, healthcare design has undergone a significant transformation, shifting towards creating spaces tailored to specific patient populations and emphasizing individualized care. This shift is evident in the gradual transition from traditional hospital wards with multiple beds (Fig. 1) to general care wards with multi-occupancy patient rooms to specialized units with private patient rooms and designated areas for family members [3, 12]. Additionally, the location of care delivery has evolved from tertiary hospital locations, with a growing emphasis on outpatient specialty care centers in suburban areas, reflecting changes in healthcare delivery models driven by market demands and innovations such as patient- and family-centered care [7]. The distinction between inpatient and outpatient care has also become more prevalent in hospitals themselves, with popular structural approaches incorporating outpatient services on lower floors for ease of access and inpatient services on higher floors to reduce disruptions for overnight patients [3].

In the 1980s and 1990s, hospitals started to include non-medical amenities such as cafeterias and shops [3, 12]. Now, they are competing for patients by offering amenities like private rooms and massage therapy, which can influence patient recommendations and choices more than clinical quality alone [32, 53]. Patients may choose hospitals based on amenities because they are easier to understand and more visible from the onset of the medical experience compared to complex and access-restricted clinical spaces and quality data [32, 53]. In fact, patients report that the

L. Kipps (✉) · A. Erdman
EwingCole, Philadelphia, PA, USA
e-mail: lkipps@ewingcole.com; aerdman@ewingcole.com

J. L. Wells (ed.), *A Clinical Lens on Pediatric Engineering*,
https://doi.org/10.1007/978-3-031-84655-7_3

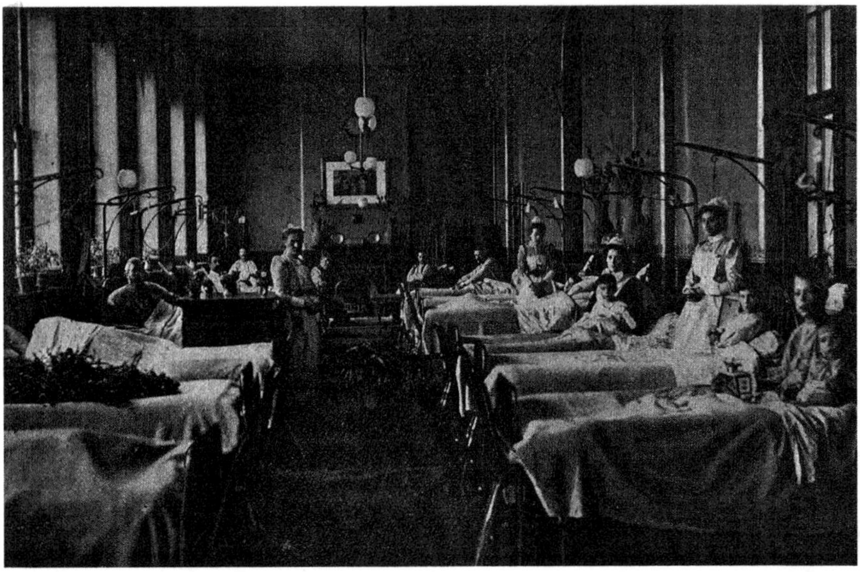

Fig. 1 The Joseph Storrs Fry ward at Bristol General Hospital circa 1890, showing open ward layout and mixing of adult and pediatric patients. (Source: Historical Images Archive/Alamy Stock Photo)

nonclinical experience at a hospital is twice as important as the hospital's clinical reputation when making their choice [32].

Implications for Pediatric Care

This evolution in healthcare design has shifted from simply designating healthcare spaces for children and their families to creating environments that reduce fear and minimize barriers to experiencing childhood as expected, ultimately setting the stage for healthy child-to-adult development. For example, as a result of global healthcare shifts, a child can attend a sleep study at a regional outpatient specialty center and make it to school on time after they are discharged in the morning.

Historically, pediatric hospitals emerged in the mid- to late nineteenth century, initially serving as places for "foundlings"[1] before evolving into dedicated facilities for pediatric care [6]. The development of pediatric hospitals was influenced by societal changes, including increased sickness and mortality among children during

[1] "Foundling" is a term historically used to describe an infant or young child who has been abandoned or deserted by their parents or guardians and found by someone else. Today, the term is less commonly used, as social and legal frameworks have evolved to better address issues related to child abandonment and welfare.

the Industrial Revolution [3], and the establishment of the American Pediatric Society's mission in 1888: "the advancement of physiology, pathology, and therapeutics of infancy and childhood" [2]. Over time, there was a gradual transition from facilities that mixed children and adult populations (Fig. 1) to dedicated pediatric care centers, with a notable increase in specialized facilities in the mid- to late twentieth century [3, 12].

Family-centered care gained prominence in response to the growing recognition of children's developmental needs for safety, security, and familial support. Initiatives such as the UK's publication of "The Welfare of Children in Hospital" in 1959 emphasized the importance of including parents and guardians in healthcare. Similarly, research findings highlighted the benefits of allowing children to stay with their families during hospitalization, leading to a shift towards family-centered care models [3, 12].

In Europe, numerous countries have adopted the European Association for Children in Hospital (EACH) charter, which outlines the rights of children regardless of their age, illness, social or cultural background, disability, or inpatient or outpatient status. [8] The articles of the EACH charter (Table 1) advocate for integrating healthcare with children's social, developmental, and physical needs while respecting their rights and the rights of their families. While not formally recognized outside of Europe, the charter details foundational principles for patient- and family-centered care that are increasingly prevalent in global markets, including the United States.

Design Aesthetic

As the focus has been directed to how a diverse range of children and their families experience an episode of care, the physical designs have evolved from simply decorating rooms with cartoonish appliqué to more sophisticated environments comfortable for all ages, while maintaining whimsy and engaging imagination (Fig. 2).

Pediatric facilities started as a duplication of adult setups, making them susceptible to general trends in healthcare, such as increased emphasis on technology, efficiency, and expandability and reduced focus on overall comfort and experience that occurred in the mid-twentieth century [3, 9]. As a result, some facilities developed a sterile feeling with consistently bright lights and high noise levels. However, with research demonstrating the harmful effect of these environments on infant vitality and the increased recognition of children's unique physiological, physical, cognitive, and social needs, healthcare facilities evolved into pediatric-specific designs [3].

Early pediatric-specific designs incorporated bright primary colors, animated characters, and animals (Fig. 2, top) [3, 9, 13]. In the 1990s and 2000s, this cartoonish style additionally incorporated hospital branding, custom mascots, and distinct shapes into various physical elements, including custom millwork and sculptures (Fig. 2, middle) [30, 31]. However, these designs were later recognized as appealing

Table 1 Articles of the EACH charter

Article number	Article description
Article 1	Children shall be admitted to hospital only if the care they require cannot be equally well provided at home or on a day basis
Article 2	Children in hospital shall have the right to have their parents or parent substitute with them at all times
Article 3	Accommodation should be offered to all parents, and they should be helped and encouraged to stay Parents should not need to incur additional costs or suffer loss of income In order to share in the care of their child, parents should be kept informed about ward routine and their active participation encouraged
Article 4	Children and parents shall have the right to be informed in a manner appropriate to age and understanding Steps should be taken to mitigate physical and emotional stress
Article 5	Children and parents have the right to informed participation in all decisions involving their health care Every child shall be protected from unnecessary medical treatment and investigation
Article 6	Children shall be cared for together with children who have the same developmental needs and shall not be admitted to adult wards There should be no age restrictions for visitors to children in hospital
Article 7	Children shall have full opportunity for play, recreation and education suited to their age and condition and shall be in an environment designed, furnished, staffed, and equipped to meet their needs
Article 8	Children shall be cared for by staff whose training and skills enable them to respond to the physical, emotional, and developmental needs of children and families
Article 9	Continuity of care should be ensured by the team caring for children
Article 10	Children shall be treated with tact and understanding, and their privacy shall be respected at all times

only to young children and inappropriate for older pediatric populations [9, 13, 28]. Given the potential for children aged 0–18 (or sometimes even early 20s), design trends shifted in the 2010s and 2020s to provide variable spaces and aesthetics according to different age groups [11]. To accommodate variations in patient ages, designs have begun to feature more abstract art and softer color palettes incorporated in furniture and a subset of finishes; more costly environmental features use simple forms and neutral colors to maximize flexibility (Fig. 2, bottom) [11]. Family-centered care has influenced healthcare design towards more homey, socially supportive approaches that provide patients with developmentally appropriate care, services, and access to education, social interaction, and entertainment [3]. Additionally, pediatric facilities have begun incorporating positive distractions into the environment, particularly in procedural settings, to reduce patient stress and manage pain without medications. Examples of positive distractions include art, interactive media, wall and ceiling murals, and multisensory experiences [1, 4, 10].

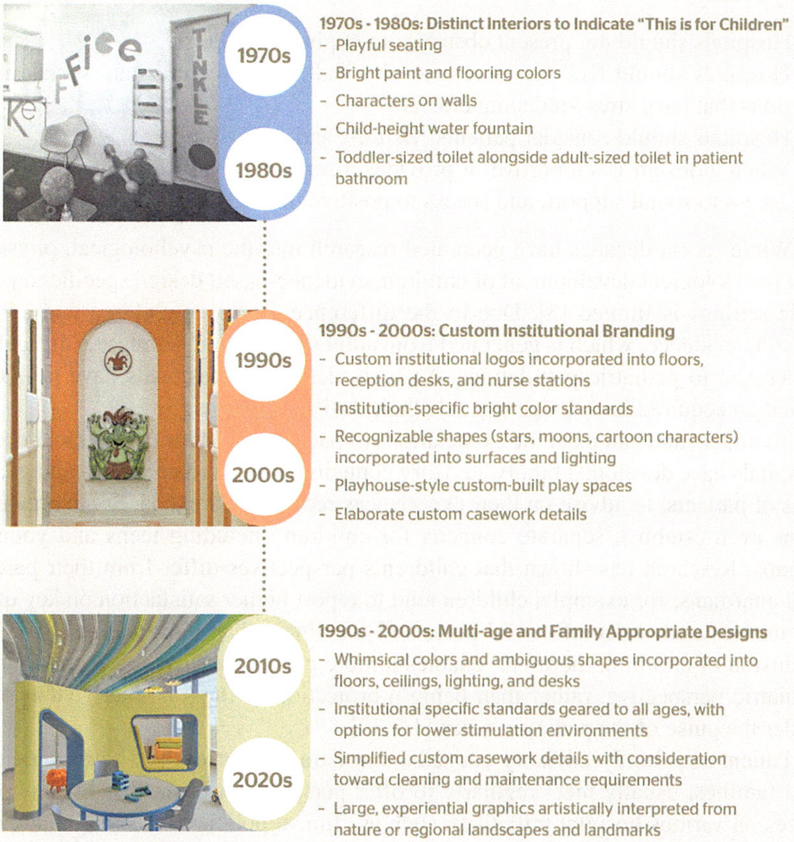

Fig. 2 Evolution of pediatric healthcare design over the last 50 years. (Photo Sources: (top) reproduced with permission from Malkin, 1990, Figure 3-60, Medical and Dental Space Planning for the 1990s, Von Nostrand Reinhold; (middle) Jeff Veire; (bottom) Halkin/Mason)

Shift Toward Informed Decision-Making

Since the late twentieth century, evidence-based design has grown in popularity, becoming a predominant approach in healthcare design, likely due to the prevalence of evidence-based medicine [28]. Roger Ulrich's landmark research on the potential for views of nature to improve patient recovery time sparked a movement in associating physical designs with measurable clinical outcomes [15, 28]. In 1991, Ulrich published his Theory of Supportive Design [4], which remains critical in healthcare design today:

1. Hospitals should not present obstacles to coping with stress.
2. Hospitals should facilitate access to physical design features and social situations that have stress-reducing effects.
3. Hospitals should consider patients, visitors, and staff.
4. When a design is supportive, it provides a sense of control over surroundings, access to social support, and access to positive distractions.

While recent decades have generated research into the psychological, physical, and physiological development of children, evidence-based design specific to pediatric settings is limited [3]. Due to the differences between children and adults, existing evidence, which is generated from adult samples and settings, is difficult to generalize to pediatric populations. As such, design professionals have primarily relied on acquired knowledge to guide their design decisions [3].

To understand and meet the expectations of patients and their families, children's hospitals have developed family advisory councils, comprised of parents and guardians of patients, to advise on their experiences receiving care [11]. Some organizations even establish separate councils for children, including teens and younger groups. Research has shown that children's perspectives differ from their parents and guardians; for example, children tend to report higher satisfaction on key quality metrics such as comfort and perceived timeliness than their parents [14]. The inclusion of children, therefore, enables a more comprehensive and representative pediatric perspective, rather than being a projection of the guardians' viewpoints under the guise of the pediatric voice [14, 30].

Patient and family advisory councils, consisting of current and former patients and families, usually meet regularly to offer perspectives from their lived experiences on various hospital initiatives, such as clinical policies, research, and operational guidelines [10]. During design and construction projects, councils are often invited to contribute to the design and provide feedback on proposed ideas. By incorporating the perspectives of patients and families, designers and organizational leaders can tailor project designs to meet their needs and preferences more comprehensively.

Design trends have also evolved to recognize unique characteristics on an individual level, leading to designs considering neurodiversity, disabilities, cultural and social background, and other impacts of health conditions [3, 10, 11]. Just as hospital experiences are personal, facility design must consider opportunities for personalization and flexibility [9]. As such, design professionals are moving away from one-size-fits-all approaches, recognizing the critical nature of including the target patient population in the design process and incorporating design elements reflective of important cultural and social symbols from the surrounding community [10, 12, 13].

Accompanying this movement is the increased use of human-centered design methodologies, such as participatory design, journey mapping, ethnographic study, qualitative surveys, interviews, and focus groups [10, 30]. Human-centered design methods have been employed in various industries, such as product development, operational workflows, and organizational strategy, to develop an understanding of intended users and their needs, behaviors, and motivations [16]. Iteration, foundational to human-centered design, enables design professionals to evaluate in-progress designs under conditions representative of the intended real-world use, ensuring they are headed in the right direction or otherwise course-correcting before committing to the design [18, 19]. Given the often-lengthy timeline of design and construction projects, this iterative approach aligns well with the variable nature of pediatrics, both across age groups and as individual children grow. As pediatric healthcare design continues to evolve, designers are expected to integrate human-centered design with evidence-based design to provide a holistic approach to informing design decisions [10, 30].

Design Inputs for Pediatric Settings

Human-Centered Approach

The Institute of Patient- and Family-Centered Care defines patient- and family-centered care as "an approach to the planning, delivery, and evaluation of healthcare that is grounded in mutually beneficial partnerships among healthcare providers, patients, and families. It redefines the relationships in healthcare by placing an emphasis on collaborating with people of all ages, at all levels of care, and in all healthcare settings" [52]. Human-centered design provides a harmonious framework for creating child-friendly healthcare facilities and associated operations. Human-centered design is a systems-based approach to understanding how humans think, perceive, and behave within the social, environmental, and organizational context that surrounds them [17, 18]. Core concepts include:

- **Empathy**: The ability to understand and share the feelings, thoughts, and perspectives of others. It involves recognizing and acknowledging someone else's experiences, imagining oneself in their situation, and responding with compassion and sensitivity [19].
- **Stakeholder engagement**: The process of involving individuals, groups, or organizations who have a vested interest in a particular project, decision, or issue to ensure alignment between their values, concerns, and perspectives and a designed solution. Stakeholders can include a wide range of parties such as end users, customers, staff, investors, community members, government agencies, and non-profit organizations [10, 18].
- **Divergent thinking**: A thought process or method used to generate creative ideas and solutions by exploring multiple possible answers or perspectives. Unlike convergent thinking, which aims to find the single best solution to a problem, divergent thinking encourages exploring various possibilities, considering

different angles, and generating numerous ideas, even if they seem unrelated or unconventional [18, 19].

- **Iteration**: The process of repeating a sequence of steps or actions with the aim of gradually refining and improving a design, idea, or solution. It involves revisiting and revising previous versions, incorporating feedback and insights gained from each repetition to make incremental enhancements and ultimately achieve greater effectiveness, efficiency, and quality [19].

Human-centered methods facilitate a deeper understanding of user needs, empower users, and provide contextual evidence of the effectiveness of design solutions as designers develop them [10, 16–19]. By generating evidence throughout the project, designers can innovate without existing literature [19]. Specific methods can vary based on project context and research objectives but are typically qualitative in nature [10]. The approach is called mixed methods when combined with quantitative methods and evidence-based design. Qualitative methods typically involve gathering non-numerical data, such as interviews, observations, or textual analysis, to understand individuals' or groups' underlying meanings, motivations, and experiences. Quantitative methods involve collecting numerical data through closed-question surveys, experiments, or statistical analysis to quantify relationships, patterns, or trends within a population. By combining qualitative and quantitative data collection and analysis techniques, a mixed methods approach can provide richer insights, enhance the validity and reliability of findings, and offer a more nuanced understanding of complex phenomena [17].

Varying Needs Across Ages

Pediatric healthcare design is at risk of being perceived as the same as adult healthcare design, just smaller [11]. Children, however, are not smaller adults; they differ substantially with respect to physiology, physical characteristics, cognition, and social needs [1, 3]. Compared to adults, children vary significantly across age groups, prioritize play and exploration over other aspects of life, and are dependent on parents, guardians, or other family members [3, 20, 25, 26, 33]. Children also have a reduced ability to cope with change, which can result in regression, anger, fear, depression, and anxiety [3, 27, 30]. As a result, children are more susceptible to potential negative impacts from the change in routine and setting associated with hospitalization [3, 5, 10, 27].

Healthcare designers have a unique opportunity to reduce stress and uncertainty in hospitalized children, potentially preventing long-term psychological trauma [5, 10, 27]. Place identity, the personal meaning attributed to a location, influences how individuals perceive themselves and others within that environment, shaping their sense of belonging and identification [3, 26].

Throughout childhood, sensory abilities, cognition, physical capabilities, emotional regulation, and social skills develop according to genetics, physiology, and interactions with surroundings and other people [20]. As such, children have variable characteristics and preferences based on their age and corresponding developmental stage, communities, and health conditions [3, 20, 33]. Designing according to the specific characteristics and goals of the target pediatric population better aligns the environment with their needs, thereby providing a sense of belonging and environmental empowerment [30, 36, 41]. Table 2 defines distinct pediatric age groups and briefly describes each age group's characteristics [3, 11, 20, 25, 41].

Table 2 Characteristics of pediatric age groups

Age group	Developmental needs	Physical characteristics	Cognitive characteristics	Sensory characteristics	Psychosocial characteristics
Infants (0–18 months)	Basic care Feeding Comfort Safety	Rapid physical growth and development. Reflexive actions turn into intentional movement. Gross motor milestones. Require 12–15 h of sleep	Develop an understanding of cause and effect and object permanence. Start to develop memory	Heightened sensitivity to touch, taste, smell, and sound. Vision gradually develops	Form attachments and trust through caregiver interactions
Toddlers and Preschoolers (1–6 years)	Safety Sensory, cognitive, and physical stimulation Opportunities for exploration and learning Social interaction	Gradual refinement of motor skills. Fine motor milestones. Toilet training. Develop strength through physical activities. Require 10–14 h of sleep	Develop language and basic reasoning skills. Develop memory	Fully developed vision. Explore through sensory experiences	Express emotions but struggle to emotionally regulate. Egocentric thinking. Navigate peer interactions. Self-care milestones
School Age (6–12 years)	Structured learning Socialization Extracurricular activities	Coordination of motor skills. Develop strength through physical activities. Require 9–11 h of sleep	Improved memory. Develop abstract thinking	Refine senses and develop preferences	Develop friendships, empathy, and a sense of identity

(continued)

Table 2 (continued)

Age group	Developmental needs	Physical characteristics	Cognitive characteristics	Sensory characteristics	Psychosocial characteristics
Adolescent (12–18 years)	Autonomy Peer relationships Identity exploration Guidance for future plans	Significant physical changes associated with puberty. May participate in strength-building activities. Require 8–10 h	Engage in complex decision-making, logical thinking, and hypothetico-deductive reasoning. Considers future planning and ideological problems	Seek sensory stimulation through activities such as music, sports, and social interaction	Seek identity formation, peer relationships, and autonomy while exploring their roles in society

Built environment designers can consider pediatric age-specific needs by integrating features that accommodate various developmental stages:

- **Safety measures**: Implementing childproofing features to prevent accidents for infants and toddlers, such as safety gates, cushioned edges, and secure window locks [1, 3].
- **Sensory stimulation**: Incorporating colorful, interactive elements for toddlers and preschoolers to engage their senses and promote cognitive development. Catering to sensory needs, such as quiet zones for relaxation, interactive displays for stimulation, and calming environments for children who may be overwhelmed by sensory input [3].
- **Play spaces**: Designing age-appropriate playgrounds and recreational areas with equipment and toys suitable for different age groups, from swings and slides for younger children to sports facilities for older ones [1, 3, 11, 33].
- **Educational environments**: Creating stimulating learning environments with flexible spaces for group activities and quiet zones for focused learning, catering to the educational needs of toddlers and preschoolers versus school-aged children versus adolescents. Enable educational participation from wheelchairs or stretchers [3, 11].
- **Social spaces**: Including communal areas that encourage peer-to-peer interactions, such as hangout spots or study areas with comfortable seating for adolescents and parallel play opportunities for younger children [1, 3, 11].
- **Accessibility**: Ensuring accessibility for children with disabilities or special needs, such as ramps, elevators, sensory-friendly features, and communication aids [1].

- **Separate zones**: Designate distinct areas for different age groups within the hospital, ensuring that infants, toddlers, school-aged children, and adolescents have environments suited to their developmental needs [1, 11].
- **Autonomy and self-expression**: Provide developmentally appropriate opportunities for children and adolescents to personalize their experiences, control their physical surroundings, or make choices on how to spend their time. Examples include room controls for lighting, window shades, and temperature as well as the ability to display personal items from home [1, 3, 11, 27, 28, 30].

Social Determinants of Health

Social determinants of health (SDOH) significantly impact individuals' health, well-being, and quality of life. These determinants also contribute to health disparities and inequities, where individuals facing barriers to resources and opportunities experience worse health outcomes [21, 45]. SDOH comprises five categories: economic stability, education access and quality, healthcare access and quality, neighborhood and built environment, and social and community context [21].

Economic stability plays a pivotal role in shaping health outcomes, encompassing factors like employment, housing stability, and poverty. Poor job conditions (for guardians and/or adolescents), housing insecurity, and transportation limitations exacerbate health disparities, leading to issues such as depression, stress-related illnesses, and food insecurity [21]. Additionally, childhood poverty can result in developmental delays and long-term health issues. Education access and quality, including early childhood development and higher education enrollment, significantly impact lifelong outcomes. Adverse events like poverty hinder development and exacerbate health and academic disparities, while limited language skills and health literacy pose barriers to healthcare access. Higher education enhances employment opportunities and income potential, contributing to better health outcomes [21].

Healthcare access and quality are essential for optimal health outcomes but are hindered by obstacles like lack of insurance, transportation challenges, and provider shortages. Delays in receiving preventive care and treatment worsen overall health, disproportionately affecting marginalized communities [21, 45]. Moreover, the neighborhood and built environment significantly influence health, with barriers like access to healthy food, exposure to crime and violence, and housing quality exacerbating health disparities [21, 34]. Social and community contexts, including social support and discrimination, also impact health outcomes, with strong social ties positively affecting health and social isolation posing risks [34, 36].

Design professionals should consider community-specific data on SDOH and leverage strategies to mitigate risks, address barriers, and proactively promote healthy choices and behaviors [22, 36]. Sources for community data include the Spatial Equity Analysis Toolkit from ArcGIS [23] and the PRECEDE tool from Perkins & Will [24]. For example, design professionals could include outdoor spaces with diverse plants, options for physical activity, and pathways that accommodate individuals with mobility challenges. To encourage spending time outdoors, designers can provide consistent and unhindered visibility to the outdoor space from the inside [36]. Other considerations include the following:

- **Accessibility**: Ensure that hospitals are easily accessible, including by way of public transportation options and proximity to areas of need [1, 3, 21, 35, 36].
- **Cultural competence**: Design spaces that respect and accommodate diverse cultural practices and beliefs, including signage in multiple languages, culturally sensitive décor, and accommodations for religious practices [35, 36].
- **Community engagement**: Involve community members in the design process to ensure that the hospital meets the specific needs and preferences of the local population, fostering trust and collaboration [21, 29, 36, 37].
- **Equitable care delivery:** To reduce disparities in healthcare outcomes, design care pathways and services that prioritize equity and access for all patients, regardless of socioeconomic status, race, or ethnicity [21, 45, 54].
- **Supportive services**: Integrate resources and services within the hospital, such as social work, financial counseling, and transportation assistance, to address social needs that impact health outcomes [21, 30].
- **Health promotion and education**: Design spaces for health education and promotion activities, including community outreach programs, wellness classes, and resources for preventive care, to empower patients and communities to take control of their health [21, 45].
- **Collaboration with community partners**: Establish partnerships with local organizations and agencies to address broader social determinants of health, such as housing insecurity, food access, and education, through coordinated efforts and referrals [37].

Patient Support System

The ultimate goal of pediatric care in hospitals is to provide comprehensive care that integrates healthcare with children's social, developmental, and physical needs while respecting their rights and the rights of their families [3, 8]. According to the principles of patient- and family-centered care, effective pediatric care involves children receiving care in hospitals only when necessary, ensuring guardians' presence and involvement in their child's care, providing appropriate

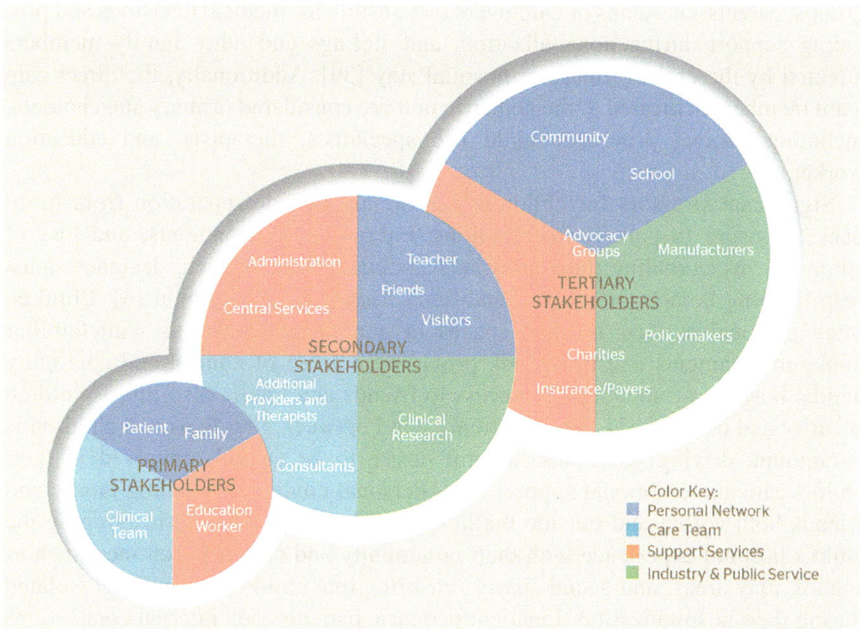

Fig. 3 Patient Stakeholder Map. (Source: Authors)

information and support to children and guardians, promoting informed participation in healthcare decisions, creating environments conducive to recreation and education, employing staff with the necessary skills to meet children's needs, ensuring continuity of care, and treating children with dignity and respect while respecting their privacy [8].

To facilitate healthcare systems in achieving these aims, designers must recognize the support system surrounding the pediatric patient [16, 39]. Their support system includes various stakeholders in their lives and healthcare, including family, care teams, support staff, and organizations who collectively influence the patient's overall care experience [18, 39]. One study identified up to 25 individual stakeholders for a pediatric oncology patient [39]. Figure 3 shows an example stakeholder map, demonstrating the complexity of the patient's social and care support system. By defining a stakeholder map for pediatric patients in a hospital, healthcare providers can better understand the relationships, needs, and priorities of the various stakeholders involved in pediatric care. This facilitates effective communication, collaboration, and decision-making to improve outcomes for children and their families [15, 18, 39, 41].

Primary stakeholders encompass individuals directly affected by the hospital's design, services and policies. These include pediatric patients of various age groups, parents/guardians or caregivers responsible for medical decisions and providing support during hospitalization, and siblings and other family members affected by the child's illness or hospital stay [39]. Additionally, the direct care team members dedicated to the pediatric unit are considered primary stakeholders, including clinical providers, child life specialists, therapists, and education workers.

Significant stressors for children in hospitals include separation from loved ones, exposure to painful interventions, unfamiliar environments, and loss of autonomy, exacerbating their fears and anxieties [30]. As such, designers must help to bring home, community, and daily routine to the hospital [3]. Children often personalize their hospital spaces to maintain connections with familiar home environments and personnel, providing a sense of control and continuity amidst hospital protocols [30]. Access to friends and family also offers children security and comfort [3]. The emotional bond between guardians and children is paramount, driving guardians' natural desire to be actively involved in their child's care and emotional support [27]. Personal contacts, including family and friends both within and outside the hospital, play a crucial role in bridging the child's hospital experience with their community and daily life routines, such as school, play areas, and social spaces, ensuring that children do not feel isolated during their hospitalization. Inpatient pediatric patients seek internal connections within the healthcare community and external connections with friends, school, and normalcy, with play as a means to foster bonding among patients, family, and caregivers [38].

Balancing hospital life with home responsibilities poses additional challenges for families, impacting partnerships, family dynamics, employment, and sibling relationships [33]. The involvement of the entire family in a child's hospitalization can lead to psychological and social distress for parents, guardians, and siblings, which can impact their ability to participate effectively in their child's care and lead to long-term adjustment difficulties for both parents/guardians and children [27, 33]. As such, designing with the patient's family in mind facilitates the child's acceptance of hospital routines, reduces stress levels, and improves overall health outcomes.

Case Study 1 demonstrates how Nemours Children's Hospital, a Delaware-based pediatric hospital surrounded by a nature setting, designed their central Child Life Clubhouse to promote all ages, physical abilities and family involvement.

Case Study 1 Child Life Clubhouse at Nemours Children's Hospital in Wilmington, DE (Photo Source: Halkin/Mason)

Secondary stakeholders consist of individuals and groups indirectly affected by hospital design, operations, and outcomes, encompassing extended hospital staff members involved in pediatric care, hospital maintenance, and hospital administrators responsible for resource allocation and policy-setting [39]. Given the complex nature of pediatric healthcare, no single entity can fully meet all the needs of a child, making children a vulnerable demographic whose well-being relies on support from adults and various structures [40]. Effective support systems must cater to both the child and family needs within their home and community context, adapting dynamically to changes during their growth and development. Traditional healthcare models often fail to address the multifaceted issues affecting child health, necessitating coordination among caregivers and stakeholders involved in the child's care [40].

The care team may only include primary care physicians or on-unit direct care teams. Other children's care teams may include a wide array of stakeholders such as medical specialists, teachers, social workers, therapists, and community agencies [39, 40]. Pediatric medical subspecialists play a crucial role in the care team for children with complex conditions, often serving as team leaders or co-managing with primary care clinicians [40]. Team-based care for children differs from that for adults due to the profound influence of early life experiences on lifelong health outcomes, the societal benefits of investing in health promotion during childhood, and the dynamic nature of childhood itself. Adverse childhood experiences (ACEs) and toxic stress can lead to lasting physical and mental health challenges, underscoring the importance of identifying and supporting affected children and families [40]. As children's needs evolve during development, the care team must adapt accordingly, providing tailored support and addressing health consequences stemming from adversity, social determinants, and complex conditions, including mental health issues [40].

Effective communication and collaboration among healthcare professionals are essential for ensuring patient safety, efficiency, and quality of care [39, 55]. Patients who experience good continuity of care report higher satisfaction and better health outcomes, emphasizing the importance of enabling and supporting teamwork in healthcare settings. However, achieving effective teamwork can be challenging due to varying working routines and goals among staff members [39]. Thus, fostering teamwork requires efforts to align protocols, processes, and goals across different specialties and disciplines within healthcare organizations [39].

Designers can positively influence care coordination and communication by carefully considering social interaction spaces for staff, opportunities for continuing education, and enabling as much care as possible to be performed in the patient room [55]. Specific design solutions may include a larger room with additional data and power infrastructure to support various medical equipment and large care teams.

Tertiary stakeholders comprise external entities and organizations that can exert influence on or be impacted by pediatric healthcare services. These include:

- Insurance companies or third-party payers that offer coverage for pediatric healthcare services
- Regulatory agencies tasked with establishing healthcare standards and guidelines affecting the delivery of pediatric care

- Community organizations, advocacy groups, and charitable foundations that offer support for pediatric healthcare initiatives or furnish resources for patients and families

Defining the Project Vision and Associated Care Delivery Model

Due to the variability of pediatric patients, their support systems, and approaches to care across different communities, no two healthcare facilities are the same. Care variability can even extend to distinct divisions of the same hospital. In pursuing a new facility design, designers and other project stakeholders must define their vision of the future, their values, and the associated measurable objectives the project must achieve [28, 44].

Underpinning this vision is the desired care delivery model(s) that a healthcare facility wants to provide. Care models are defined as the organizational factors and clinical protocols that establish clinical care provided by the hospital [42, 43, 45]. Care models include details such as staff roles, staff-to-patient ratios, shift schedules, patient cohorts, and contingency plans [42, 43]. Supplemental-to-care models are the ancillary support services that enable the delivery of care, such as materials management, environmental services, central sterile processing, pharmacy, labs, and scheduling/registration. Healthcare systems are intricate networks where many individuals interact in ways that are not always foreseeable. These systems are highly interconnected, so one participant's actions can affect others [42].

Considering the design of healthcare buildings and services, simultaneously, is crucial to prevent issues in service delivery due to poor connections. Healthcare facility design is complex due to the involvement of multiple stakeholders and the interaction of various outcomes and flows. Hospitals often operate at high utilization levels, making layout and flow critical [44]. Mapping the seven healthcare flows can facilitate an in-depth understanding of how healthcare is delivered: patients, clinicians, visitors, medication, supplies, equipment, and information [42]. The flow details help designers understand which rooms or spaces need to exist, how many, and how they spatially relate. Furthermore, understanding the equipment, materials, and staff required for different activities enables an understanding of what size of rooms or spaces are needed. Knowledge of what technologies or care models are on the horizon is critical to space planning through monitoring healthcare policy and scientific and technological advancements, specifically those in medical devices and pharmaceuticals.

Ultimately, by knowing their stakeholders and vital operational parameters in detail and how they interrelate, designers can better ensure a healthcare environment that functions and supports its users and the surrounding community [42].

Translating Inputs into Design

The inputs gathered on the healthcare organization, patient population, patient support system, and community become the foundation for each following step. The design process is flexible based on the organizational culture, budget, and schedule. Phases are sometimes combined or have less-defined edges if the design does not need to have cost estimating done between each phase before proceeding to the next. In general, most projects have some representation of the following:

- **Programming**: This phase's purpose is to develop the project's vision, including service lines, size and requirements.
- **Schematic Design**: Site configuration, departmental adjacencies, initial floor plans and building massing[2] are developed during this phase.
- **Design Development**: Technical aspects of the project in concert with the functional and aesthetic goals are combined to reach a common understanding and affirmation of the complete project design.
- **Construction Documentation**: Final documentation as a set of instructions for the building's assembly is delivered in this phase. Most design decisions have already been made, and this involves coordinating the materials and infrastructure systems to meet the project's design requirements.

Design Layers

During the early design phases, designers must synthesize the project's needs into a physical design that maximizes meeting project goals within the overall project size, timeline, and budget parameters. One problem could have multiple operational and physical solutions. Evaluating the options against the key project criteria and assembling a combination of "design layers" to support the desired outcomes advances the design in a way that supports patients, families, and clinicians instead of providing space then creating operations around the physical environment. These layers comprise the entirety of the built environment design, reciprocally influencing each other to create an overall fit-for-purpose design. Figure 4 shows each of these layers, centered around operations and processes and defined iteratively.

Operations & Processes are the series of organizational activities and protocols that define the overall facility functions. They are the foundations around which the design layers are added for support. For example, protocols such as allowing patients to bring items from home including blankets, stuffed animals, and toys will impact the equipment and furniture layer to consider laundering capability and open storage for the items [10].

[2] "Building Massing" is an architectural design term describing the shape, size, and assembly arrangement of building components, as well as its scale in relationship to other buildings and elements within the surrounding context.

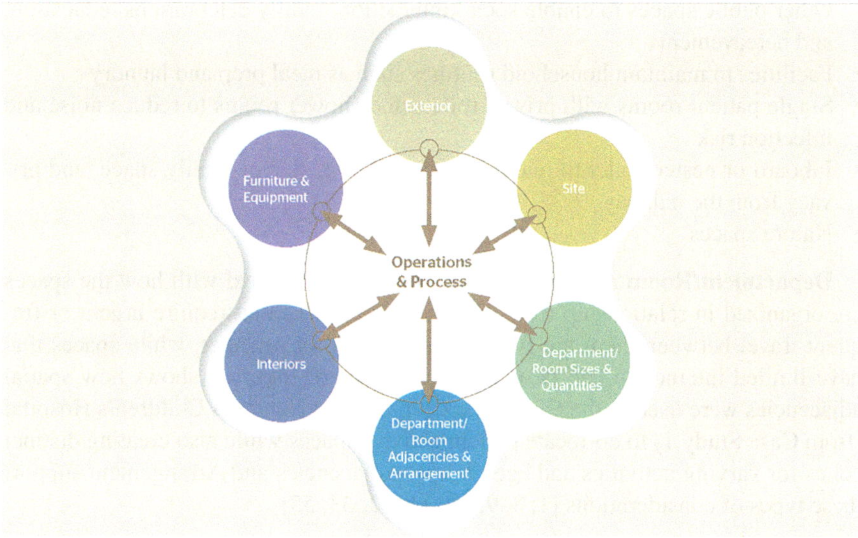

Fig. 4 Design Layers, each developed iteratively in coordination with operations and process. (Source: Authors)

Site addresses the specific location where a building or structure is situated, incorporating land topography, existing infrastructure, and the surrounding community. Site location influences arrival, parking, and entrances as well as views in and out of the building. Key considerations include [1, 3, 9, 11, 12, 44]:

- Proximity of the site to other community facilities such as schools, museums, hotels, and parks
- Positioning of the building to maximize daylighting
- Complementing the surrounding community's identity and infrastructure
- Accessibility by a range of means including walking, car, public transportation, or even domestic and international travel
- Integration of gardens or other outdoor amenity spaces

Exterior pertains to the outer surface of the building, which can evoke a branding expression of the building function and communicate what to expect inside based on the size and placement of windows. An example of an exterior recommendation is to design the façade to maximize daylighting while regulating the thermal impact of sunlight [1, 9].

Department/Room Sizes and Quantities refer to the distinct spaces within a building, typically incorporating unique operations and processes. The characteristics thereof are influenced by how the design layers are combined and include considerations such as [1, 3, 9, 11, 12, 33]:

- Activity spaces distributed throughout the unit and facility, varied by activity type, age group, and capacity

- Other public spaces to enable social interaction, family celebrations, education, and bereavement
- Facilities to maintain household routines such as meal prep and laundry
- Single patient rooms with private toilets and shower rooms to reduce noise and infection risk
- Inboard or nested toilet to maximize daylighting, size of family space, and privacy from the hallway
- Nature spaces

Department/Room Adjacencies and Arrangement deal with how the spaces are organized in relationship to each other. Two spaces that require urgent or frequent travel between them are typically placed closer together, while spaces that have limited interaction can be placed further apart. Figure 5 shows how spatial adjacencies were used in the Child Life Clubhouse at Nemours Children's Hospital (from Case Study 1) to co-locate patient activity spaces while also creating distinct zones for varying activities and age groups. Adjacencies and Arrangement support these types of considerations [1, 3, 9, 10, 11, 12, 33, 55]:

- Separation of "on-stage" (activities with direct patient or public interaction) and "off-stage" (support activities not visible to patients or public) to reduce noise in public-facing spaces and improve operational efficiency
- Proximity between clinical spaces and storage to reduce travel distances for healthcare providers and increase time in direct patient care
- Neighborhoods that co-locate patients, providers, and resources based on acuity level, specialty, or age
- Separation of adolescents and babies
- Visibility and accessibility of outdoor and activity spaces

Fig. 5 Floor plan for Nemours Children's Hospital Child Life Clubhouse, showing spatial proximity of activity spaces varied for activity types and age groups. (Source: EwingCole)

- Visibility of patients from nursing stations and unit entrances
- Single-loaded corridors (with patient rooms on only one side of the corridors) with view and access to activity/play areas
- Distance between patient rooms and entrances/exits, stairways, and elevators to minimize abduction risk
- Limited number of entrances/exits to minimize abduction risk

Interiors are the design of the inside of the building, including material finishes, fixtures, architectural features, and lighting. Considerations include [1, 3, 5, 9, 10, 11, 28, 33, 55]:

- Reduced scale for kids, such as lower ceilings, horizontal surfaces, and windows
- Clearly defined clinical, patient, and family zones in patient rooms
- Barriers between rooms and sound-attenuating surfaces to facilitate noise control and privacy, particularly in patient rooms
- Hotel-like finishes, furniture, fixtures, art, and experiential technology
- Signage considering infographics, visual aesthetics, color, form, composition, rhythm, and overall consistency for branding expression and wayfinding
- Distinction and diversity among spaces to facilitate patient engagement, place-making, and wayfinding
- Unique art and architectural fixtures to support wayfinding, provide a sense of place, and connect to familiar cultural or community motifs
- Flooring with impact and slip resistance that is additionally suitable for crawling

Furniture & Equipment are the purchased items that enable spaces to fulfill their intended purposes. Details that fit in this layer include [1, 3, 9, 10, 11, 12, 33, 54]:

- Child-sized furniture and medical devices, such as shorter seating, neonatal isolettes, and smaller vitals equipment
- Careful placement of objects to prevent visual obstruction for clinical monitoring, especially in spaces with small children or crawlers
- Horizontal surfaces within patient rooms that can be used for remote working, homework, and family meals
- Art, interactive media, aquariums, and speakers to provide positive distractions
- Comfortable bench, bed, or chair for families to sleep in or near the patient room
- Wall-mounted computer or workstation-on-wheels that enables simultaneous clinician charting and patient/family discussions within care spaces
- Furniture or equipment that accommodates a guardian sitting with or immediately next to the pediatric patient during an exam or procedure, as appropriate
- Opportunities for personalization or accommodation of items from home

Figure 6 demonstrates how these layers coalesce in a final design of a patient room. The room depicted is from Essentia Health-St. Mary's Medical Center in Duluth, Minnesota, featuring: a private room with dedicated family space, delineated by floor pattern, that accommodates sleeping and personal lighting; clinical clearances also delineated by floor pattern to provide sufficient space for medical

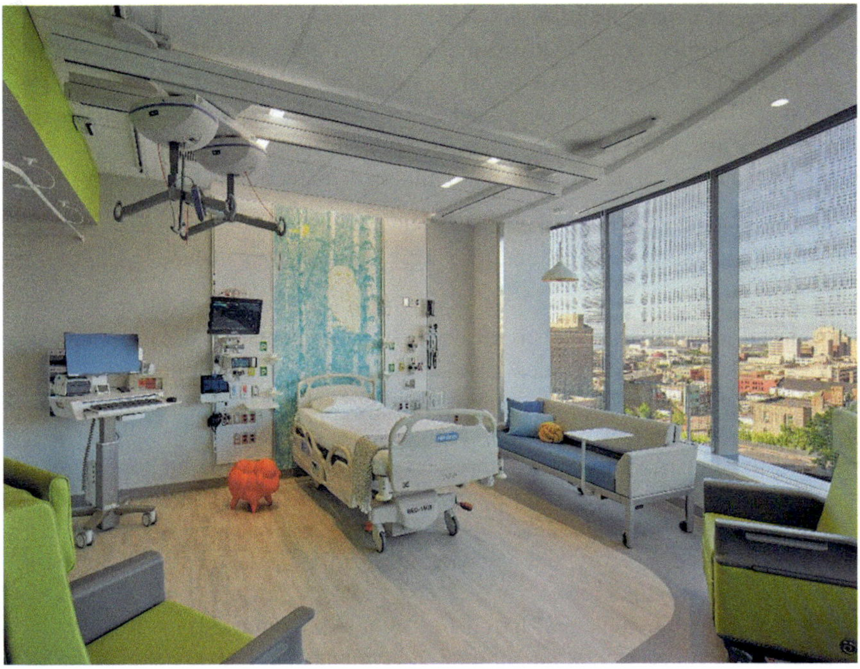

Fig. 6 Patient room design comprised of integrated "design layers" to create a comprehensive family-centered care experience. (Photo Source: Halkin Mason)

equipment and patient care; large window with a view visible by the patient; medical gases located on either side of the bed and no services behind the patient; access to ceiling-mounted patient lift to support staff safety; and a movable staff computer. Additionally included but not shown in Fig. 6 were wide doors for ease of transporting equipment in and out of the room and enabling quick ingress of large emergency response teams; a private bathroom, nested between patient rooms; and decentralized nursing station just outside the room with a window to provide visibility of the patient's head from the station.

Balancing Feasibility, Innovation, and Safety

Healthcare environments are regulated by legal codes, that are in place to set safety standards and vary based on location. The codes are focused on life safety, building safety, and healthcare operations. The disciplines of the interdisciplinary design team have respective codes for which they are responsible. Coordination amongst all is required to produce a feasible design that supports safety while providing opportunities for innovation.

Architectural and Engineering Guidelines For the architecture and engineering design teams, the Guidelines for Design and Construction are a foundational starting point for determining the minimum requirements for a healthcare construction project [50]. Commonly referred to as "The Guidelines," this reference establishes the minimum requirements for metrics such as key square footage for key room types, bed and door clearances and mechanical air ventilation requirements, among many other detailed design and process requirements. The series of Guidelines are currently published by a not-for-profit organization, named the Facility Guidelines Institute (FGI). Many US states adopt the Guidelines as a standard for regulatory approval. The recommendations are currently updated every 4 years and include chapters for pediatric specialties [50].

Safety Risk Assessment A safety risk assessment is a key process in identifying risks that can be mitigated by planning physical conditions and operations that address hazards. In the context of pediatric environments, considerations could include the following [3, 10, 50, 51]:

- **Infection Control**: Protocols to accommodate family members in the room with an infectious patient under Airborne Infectious Isolation or an immunosuppressed patient under Protective Environment precautions. These protocols can influence procedural space or patient rooms through the integration of positive- and negative-pressure air flow, filters, ventilation, and anterooms.
- **Patient Handling**: Adult-sized changing tables, sometimes with a ceiling-mounted patient lift, are being considered in pediatric environments for children older than infants or toddlers for those with complex medical conditions that require a larger changing surface as well as space to navigate a wheelchair transfer. Door and elevator sizes are also key considerations to ensure appropriate space to accommodate necessary equipment and care team during patient transport.
- **Medication Safety**: Options to maintain the visibility of the patient, while having access to an uninterrupted and well-lit space to focus on dosing accuracy. It is important to have easy access to proper disposal of excess medication because smaller patients receive smaller doses, so wasting excess medication is a frequent occurrence.
- **Falls**: Features typically not considered hazardous for adult facilities, such as retaining walls, half-height partitions, and countertops, must be evaluated for climbing potential. The position of the patient's toilet room relative to the patient's bed can support fall prevention by providing a direct path of travel, avoiding turns or navigation around furniture and medical equipment.
- **Behavioral Health**: Safety features intended to prevent self-harm or harm to others should be integrated, such as ligature-resistant elements and weighted furniture that cannot be thrown. Thought should also be given to the location of de-escalation spaces and what type of stimulation needs are required, ranging from a need to be active and expel energy to minimizing stimuli and activity to reintroducing stimuli.

- **Security**: Keeping children out of spaces, preventing them from leaving spaces, and restricting unit access including guardian identification, such as in custody disputes, are important pediatric-specific considerations.

This assessment is developed by an interdisciplinary team including clinicians, facility management, quality improvement, safety specialists, security, human factors specialists, architects, interior designers, and engineers. The earlier in the process the assessment is started, the less impact desired implementations will have on design changes and cost, and the more likely ideal safety solutions will come to fruition [50].

Designing for Flexibility While the future is unpredictable, there are influences that can help designers anticipate how healthcare may change, such as trends in patient volumes and acuity. Patients are becoming more complex, so future healthcare will likely see acute patients continuing to be treated within hospitals, while less acute patients will be treated at home or through community-based healthcare facilities [3]. Healthcare systems will likely become more distributed; systems will have a centralized building with complex technology for intensive care and decentralized satellite facilities in the surrounding area, closer to patients' homes [3].

While focusing design on process, user, and patient experience, it is also important to consider how processes and experience goals may be different in the future. A few strategies for consideration that could ease future adaptation include [11, 42, 44]:

- **Avoid over-customization**: Building custom millwork to support a current process or enclose the latest technology will be expensive and require a redesign and construction project to update the space when the process or technology is upgraded. Furniture designed for a specific need is easier and less expensive to replace than custom construction. Similarly, using universal room design or otherwise designing spaces and units for flexible use facilitates future updates as models of care, patient populations, and technology change.
- **Plan for growth**: Designing soft spaces such as offices and generously sized Waiting Rooms adjacent to service lines with growth targets or open land and shell space for future building floors makes it easier to grow in place and prevents the need to relocate completely and rebuild in entirety.
- **Be strategic in design feature placement**: Including current color palettes in long-term features such as millwork, poured flooring and exterior cladding is likely going to remain in place for the life of the material. Furniture, wallcovering, signage, and paint have a shorter lifespan, so including trends in those features provides a more frequent cycle of renewal, making it easier for an environment to maintain a fresh appearance.
- **Apply lessons learned**: Referencing how existing locations and other healthcare organizations and facilities have evolved over time can provide insight into how much flexibility is desired over time. Inevitably, renovations in response to changes in needs will be required in time. In particular, shifting perspectives in infection control, age-agnostic design, cultural shifts and advancements in technology will lead to building updates within the longer-standing building lifespan.

Assessing Design Effectiveness

Maintaining confidence in design decisions, especially given the absence of prior research on the exact design and the scarcity of relevant studies with numerous confounding variables, poses a significant challenge [3, 10].

The human-centered design process is structured to facilitate direct learning from individuals, fostering a wide range of creative ideas that are subsequently refined to focus on solutions that are desirable, feasible, and viable for the target audience. Consideration given to the technical feasibility and financial viability of proposed solutions helps to strike a balance between innovation and practicality. By prioritizing human needs and aspirations and iteratively assessing designs with the target audience, human-centered design ensures that solutions are aligned with the preferences of the intended beneficiaries, even after changes have been made [17–19].

Engaging in participatory design throughout the project not only validates the design as it is created but also facilitates the identification of risks, discrepancies in expectations, and the acquisition of stakeholder buy-in and consensus [18, 19].

Following the completion of a building, design professionals can conduct a post-occupancy evaluation (POE) to measure the performance of the physical and operational design. A POE tests design hypotheses that were generated as part of the overall design and decision-making process; the results of a POE feed back into the body of evidence-based design for application on future projects [17].

Ideally, POE planning starts at the beginning of a design project to enable comparison of before-and-after performance metrics. However, only characterizing performance and satisfaction after construction is still useful. POE studies are conducted 6, 12, and/or 24 months after the building is occupied and operational. This delay between building activation and the study enables users to undergo a familiarization period, which can counteract overly optimistic opinions that can result simply from interacting with something new (i.e., "the honeymoon effect") and overly pessimistic opinions that can result from unfamiliarity with new operations and design elements [17].

Pediatric Care Informing Adult Care

Several healthcare designs originally developed for pediatrics are being adapted for adult populations due to their effectiveness and potential benefits. Some examples include:

- **Sensory-friendly environments**: Hospital designs that incorporate sensory-friendly elements, such as calming colors, adjustable lighting, and quiet spaces, initially created for pediatric patients with sensory sensitivities or autism spectrum disorders, are now being adapted for adult populations [10]. These environments can benefit adult patients with sensory sensitivities, mental health conditions, or neurodevelopmental disorders, providing a more comfortable and supportive healthcare experience. Additionally, enabling control over the environment is a way to increase patient satisfaction and reduce overall stress [3, 10, 33]. Integration of passive positive distractions is being considered rather than active distractions. For instance, television in waiting rooms has been shown to increase reported stress levels due to the inability to control the stimulus in a public setting [4]. Instead, positive distractions could include murals, virtual ponds, aquariums, or art in waiting areas and procedural spaces, ideally enabling the patient to choose their preferred distraction or adjust settings to a preferred stimulus level [1].

- **Interactive technology and gamification**: Interactive technology and gamified healthcare interventions, often developed to engage, educate, and motivate pediatric patients in their treatment and rehabilitation, are increasingly being used with adult populations [57]. Virtual reality therapy, mobile health apps, and wearable devices designed to promote physical activity, monitor health metrics, and track medication adherence can help adults manage chronic conditions, recover from injuries, and improve their overall health outcomes [57]. Technologies, particularly three-dimensional ones that incorporate sound, lights, and simulated experiences, have been shown to increase engagement and recall during health education sessions, as well as reduce anxiety ahead of procedures [56, 57].
- **Family-centered care models:** Family-centered care models, which prioritize the involvement of family members and caregivers in the healthcare decision-making process and provide support to patients and their families throughout the care journey, were initially developed for pediatric healthcare settings. These models are now being adapted for adult populations, recognizing the important role of family support in promoting patient well-being, adherence to treatment plans, and successful recovery [54].

Bridging the Gap Between Built Environment and Technology

Critical to space sizing is knowledge of what technologies or care models are on the horizon through monitoring of scientific and technological advancements, therapeutic breakthroughs, and healthcare policy [46]. Encouraging internal innovation in healthcare systems has gained traction over the past decade, spurred by evolving trends and competitive markets [46–49]. The passage of the Affordable Care Act prompted a shift towards value-based and accountable care, necessitating novel approaches to care delivery [49], and the COVID-19 pandemic necessitated innovative thinking to adapt quickly to large-scale social, infection control, and organizational changes [39].

Healthcare organizations recognize the benefits, both financial and organizational, of driving innovation from within their organization [46–49]. Many institutions are establishing their own labs to foster innovation and address systemic challenges; as of 2017, over 70% of large (>400 bed) health systems had either built or were planning to build an innovation lab [47, 48]. These centers typically engage a diverse range of stakeholders and apply clinical, operational, and technological expertise to drive the evaluation and implementation of healthcare innovations, aiming to improve clinical, operational, and financial outcomes [47]. For instance, institutions like Boston Children's Hospital, the University of California, San Francisco (UCSF), and the University of Pennsylvania have established programs and centers to support clinicians and scientists in developing innovative ideas [49]. Programs such as the Brigham Care Redesign Incubator and Startup Program (BCRISP) have facilitated the design and testing of solutions by frontline clinical employees, resulting in enhanced patient engagement and improved care transitions [49].

The optimal forms and impact of these innovation centers are still emerging, as health systems adapt to the post-pandemic era and navigate the evolving landscape of medical, technological, and fiscal demands [47, 48]. Existing centers vary in terms of funding sources, internal staffing and skill sets, external partnerships, and facility design [48]. Specific examples of facility design elements are spaces dedicated to rapid prototyping, collaborative workshops, usability testing, cadaver labs, wet labs, lectures, and clinical research administration [47–49].

Conclusion

General trends in pediatric healthcare spaces have shifted over the years from carbon copies of adult settings to bespoke designs catered to the varying ages of pediatrics, their support systems, and their wider communities. As healthcare progresses, the trends of family- and patient-centered care will continue to dominate pediatric healthcare as policymakers, clinicians, and advisory groups advocate for children's and families rights.

Following suit, designers will continue to integrate the voices of pediatric stakeholders in an effort to create "a second home" rather than a purely functional healthcare facility. Facilities should provide spaces and activities that allow hospitalized children to forget momentarily their medical condition and engage in activities typical of childhood, fostering a sense of normalcy and socialization. Furthermore, the focus on creating pleasant and welcoming environments in children's hospitals is recognized as potentially beneficial not only for patients but also for adults and older individuals, contributing to improved user experiences and potentially aiding in recovery processes.

Acknowledgments In appreciation of their support, the authors would like to thank Ashlee Caruolo for the graphic development of figures and case studies as well as Zach DeRitis and Joe Donahue for their overall guidance and proofreading for this chapter.

Competing Interests Author Leslie Kipps and Author Adrienne Erdman were employed by EwingCole at the time of this writing.

References

1. Halim Babbu A, Haque M. A framework for the design of pediatric healthcare environment using the Delphi technique. Ain Shams Engineering Journal. 2023;14(5):101975. https://doi.org/10.1016/j.asej.2022.101975.
2. Gelb BD. American Pediatric Society 2019 Presidential Address: Striving to be a lasting blessing to the community. Pediatric Research. 2019;86(4):428–31. https://doi.org/10.1038/s41390-019-0470-0.
3. Pinhão, C. S. F. (2016). *Children's Hospitals - The role of architecture in children's recovery and development* [Master of Science in Architecture, Técnico Lisboa]. https://doi.org/10.13140/RG.2.2.16572.03200
4. Ulrich, R. S. (1991). Effects of interior design on wellness: Theory and recent scientific research. *Journal of Health Care Interior Design: Proceedings from the ... Symposium on Health Care Interior Design. Symposium on Health Care Interior Design, 3,* 97–109.
5. Dwiputri A, Swasty W. Color Coding and Thematic Environmental Graphic Design in Hermina Children's Hospital. Journal of Visual Art and Design. 2019;11(1):35–44. https://doi.org/10.5614/j.vad.2019.11.1.3.
6. Radbill SX. A History of Children's Hospitals. A.M.A American Journal of Diseases of Children. 1955;90(4):411–6. https://doi.org/10.1001/archpedi.1955.04030010413005.
7. Johnson, B., Abraham, M., Conway, J., Simmons, L., Edgman-Levitan, S., Sodomka, P., Schlucter, J., & Ford, D. (2008). *Partnering with Patients and Families to Design a Patient- and Family-Centered Health Care System: Recommendations and Promising Practices* (p. 107). Institute for Patient- and Family-Centered Care.
8. *The EACH Charter with Annotations: Promoting Children's Rights and Needs in Healthcare* (p. 19). (2022). European Association for Children in Hospital.
9. Verschoren, L., Annemans, M., Van Steenwinkel, I., & Heylighen, A. (2015, July 27). Designing Child-Size Hospital Architecture: Beyond Preferences for Colours and Themes. *Proceedings of the 20th International Conference on Engineering Design (ICED15).* International Conference on Engineering Design (ICED15), Politecnico di Milano, Italy.
10. Litwin S, Clarke L, Copeland J, Tyrrell J, Tait C, Mohabir V, Campbell F. Designing a Child-, Family-, and Healthcare Provider–Centered Procedure Room in a Tertiary Care Children's Hospital. HERD: Health Environments Research & Design Journal. 2023;16(3):195–209. https://doi.org/10.1177/19375867231161097.
11. Buie, A. (2019). Healthcare Facilities for Children: Designing for Distinct Age Groups. In *Architecture and Health: Guiding Principles for Practice* (pp. 25–42). Routledge.
12. Barahona, L. F. (2001). *New trends in health architecture for children and the effects of the built environment on young patients* [Master of Architecture, Florida International University]. https://doi.org/10.25148/etd.FI14050424
13. Nanda U, Chanaud CM, Brawn L, Hart R, Hathorn K. Pediatric Art Preferences: Countering the "One-Size-Fits-All" Approach. HERD: Health Environments Research & Design Journal. 2009;2(4):46–61. https://doi.org/10.1177/193758670900200403.
14. Ali H, Fatemi Y, Cole A, Tahat S, Ali D. Listening to the Voice of the Hospitalized Child: Comparing Children's Experiences to Their Parents. Children. 2022;9(12):1820. https://doi.org/10.3390/children9121820.
15. Ulrich RS. View through a window may influence recovery from surgery. Science. 1984;224(4647):420–1. https://doi.org/10.1126/science.6143402.
16. Zhang X, Liu C, Zhang X. Research on the Experience of Influencing Elements and the Strategy Model of Children's Outpatient Medical Services under the Guidance of Design Thinking. Sustainability. 2023;15(12):9383. https://doi.org/10.3390/su15129383.
17. Augustin, S., & Coleman, C. (2012). *The designer's guide to doing research: Applying knowledge in practice for design excellence*. John Wiley & Sons.

18. Melles M, Albayrak A, Goossens R. Innovating health care: Key characteristics of human-centered design. International Journal for Quality in Health Care. 2021;33(Supplement_1):37–44. https://doi.org/10.1093/intqhc/mzaa127.
19. IDEO (Ed.). (2015). *The field guide to human-centered design: Design kit* (1st. ed). Design Kit.
20. Human Development. (2010). In M. S. Gazzaniga, T. F. Heatherton, & D. F. Halpern, *Psychological science* (3rd ed, pp. 469–521). W. W. Norton & Co.
21. U.S. Department of Health and Human Services, Office of Disease & Prevention and Health Promotion. Retrieved. (n.d.). Social Determinants of Health Literature Summaries—Healthy People 2030. Retrieved 28 March 2024 from https://health.gov/healthypeople/priority-areas/social-determinants-health/literature-summaries
22. Rice, L. (2019). A health map for architecture: The determinants of health and wellbeing in buildings. In M. Jones, L. Rice, & F. Meraz, *Designing for Health & Wellbeing: Home, City, Society* (pp. 155–186). Vernon Press.
23. ArcGIS. (n.d.). Introduction to Social Equity Analysis. Retrieved March 28, 2024, from https://doc.arcgis.com/en/arcgis-solutions/11.2/reference/introduction-to-social-equity-analysis.htm
24. Perkins & Will. (2023). PRECEDE – Public Repository to Engage Community and Enhance Design Equity. https://precede.perkinswill.com/
25. Tyler, S. (2020). *Human Behavior and the Social Environment.* Pressbooks. https://uark.pressbooks.pub/hbse1/chapter/physical-development-in-early-childhood_ch_13/
26. Kopec, D. A. K. (2012). *Environmental psychology for design* (2. ed). Fairchild Books.
27. Vollmer TC, Koppen G. The Parent–Child Patient Unit (PCPU): Evidence-Based Patient Room Design and Parental Distress in Pediatric Cancer Centers. International Journal of Environmental Research and Public Health. 2021;18(19):9993. https://doi.org/10.3390/ijerph18199993.
28. *Guide 1: An Introduction to Evidence-Based Design: Exploring Healthcare and Design* (3rd ed., Vol. 1). (2015). The Center for Health Design.
29. Bishop K. The Role of Art in a Paediatric Healthcare Environment from Children's and Young People's Perspectives. Procedia - Social and Behavioral Sciences. 2012;38:81–8. https://doi.org/10.1016/j.sbspro.2012.03.327.
30. Lambert V, Coad J, Hicks P, Glacken M. Young children's perspectives of ideal physical design features for hospital-built environments. Journal of Child Health Care. 2014;18(1):57–71. https://doi.org/10.1177/1367493512473852.
31. Malkin, J. (1990). *Medical and dental space planning for the 1990s.* Van Nostrand Reinhold.
32. Goldman DP, Vaiana M, Romley JA. The Emerging Importance of Patient Amenities in Hospital Care. The New England Journal of Medicine. 2010;363(23):2185–7. https://doi.org/10.1056/NEJMp1009501.
33. Nourmusavi Nasab S, Karimi Azeri AR, Mirbazel S. Ideal physical features of environmental design in children's hospital: Using children's perspectives. Facilities. 2020;38(5/6):445–66. https://doi.org/10.1108/F-03-2019-0032.
34. Tian Y. A review on factors related to patient comfort experience in hospitals. Journal of Health, Population and Nutrition. 2023;42(1):125. https://doi.org/10.1186/s41043-023-00465-4.
35. Lehrich BM, Goshtasbi K, Brown NJ, Shahrestani S, Lien BV, Ransom SC, Tafreshi AR, Ransom RC, Chan AY, Diaz-Aguilar LD, Sahyouni R, Pham MH, Osorio JA, Oh MY. Predictors of Patient Satisfaction in Spine Surgery: A Systematic Review. World Neurosurgery. 2021;146:e1160–70. https://doi.org/10.1016/j.wneu.2020.11.125.
36. Killam, K., & Kawachi, I. (2022). Social Capital and Community Design. In N. Botchwey, A. L. Dannenberg, & H. Frumkin (Eds.), *Making healthy places: Designing and building for well-being, equity, and sustainability* (Second edition, pp. 129–142). Island Press.
37. Aboelata, M. J., & Bains, J. K. (2022). Community Engagement for Health, Equity, and Sustainability. In N. Botchwey, A. L. Dannenberg, & H. Frumkin (Eds.), *Making healthy places: Designing and building for well-being, equity, and sustainability* (Second edition, pp. 371–387). Island Press.

38. Jiang S. Positive Distractions and Play in the Public Spaces of Pediatric Healthcare Environments: A Literature Review. HERD: Health Environments Research & Design Journal. 2020;13(3):171–97. https://doi.org/10.1177/1937586720901707.
39. Kleinsmann M, Sarri T, Melles M. Learning histories as an ethnographic method for designing teamwork in healthcare. CoDesign. 2020;16(2):152–70. https://doi.org/10.1080/15710882.2018.1538380.
40. Katkin, J. P., Kressly, S. J., Edwards, A. R., Perrin, J. M., Kraft, C. A., Richerson, J. E., Tieder, J. S., Wall, L., Task Force on Pediatric Practice Change, Alexander, J. J., Flanagan, P. J., Hudak, M. L., Quinonez, R. A., Shenkin, B. N., & Smith, T. K. (2017). *Guiding Principles for Team Based Pediatric Care. Pediatrics, 140*(2), e20171489. https://doi.org/10.1542/peds.2017-1489
41. Privitera MB. Do we really know our users? A scientific approach to defining user groups for human factors studies. Proceedings of the International Symposium on Human Factors and Ergonomics in Health Care. 2020;9:139–43. https://doi.org/10.1177/2327857920091038.
42. Hicks C, McGovern T, Prior G, Smith I. Applying lean principles to the design of healthcare facilities. International Journal of Production Economics. 2015;170:677–86. https://doi.org/10.1016/j.ijpe.2015.05.029.
43. Prentice D, Moore J, Fernandes B, Larabie E. Nursing Care Delivery Models and Intraprofessional Collaborative Care: Canadian Nurse Leaders' Perspectives. SAGE Open Nursing. 2022;8:23779608221133648. https://doi.org/10.1177/23779608221133648.
44. Hollander, H. (2020). Defining a Project Method: Ensuring Project Success with Pre-Design Planning. In J. J. Wilhelm & D. Battisto (Eds.), *Architecture and health: Guiding principles for practice* (pp. 169–180). Routledge, Taylor & Francis Group.
45. Battisto, D., & Wilhelm, J. J. (2020). Introduction: Discovering an Architecture for Health. In J. J. Wilhelm & D. Battisto (Eds.), *Architecture and health: Guiding principles for practice.* Routledge, Taylor & Francis Group.
46. Bohl, M. A., Mooney, M. A., Sheehy, J., Morgan, C. D., Donovan, M. J., Little, A., & Nakaji, P. (2018). The Barrow Innovation Center: A Novel Program in Neurosurgery Resident Education and Medical Device Innovation. *Cureus.* https://doi.org/10.7759/cureus.2142
47. Bhatia, A., & Thomas, M. (n.d.). *How to Build an Innovation Center at Your Institution.* Washington University School of Medicine, Healthcare Innovation Lab.
48. Bhattacharyya O, Shapiro J, Schneider EC. Innovation Centers in Health Care Delivery Systems: Structures for Success. Journal of Medical Internet Research. 2022;24(2):e33961. https://doi.org/10.2196/33961.
49. Rotenstein LS, Wickner P, Hauser L, Littlefield M, Abbett S, Desrosiers J, Bates DW, Dudley J, Laskowski KR. An Academic Medical Center-Based Incubator to Promote Clinical Innovation and Financial Value. The Joint Commission Journal on Quality and Patient Safety. 2019;45(4):259–67. https://doi.org/10.1016/j.jcjq.2018.12.004.
50. *Guidelines for Design and Construction of Hospitals.* (2022). https://shop.fgiguidelines.org/products/digital-2022-user-hospital
51. The Center for Health Design. (2014, 2017). Safety Risk Assessment Toolkit | A Process to Mitigate Risk [CHD Tools]. Retrieved29 March 2024 from https://www.healthdesign.org/sra
52. Institute for Patient- and Family-Centered Care. (n.d.). What is PFCC? Retrieved 30 April 2024 from https://www.ipfcc.org/about/pfcc.html
53. Young C, Chen X. Patients as Consumers in the Market for Medicine: The Halo Effect of Hospitality. Social Forces. 2020;99(2):504–31. https://doi.org/10.1093/sf/soaa007.
54. Clay AM, Parsh B. Patient- and Family-Centered Care: It's Not Just for Pediatrics Anymore. AMA Journal of Ethics. 2016;18(1):40–4. https://doi.org/10.1001/journalofethics.2016.18.1.medu3-1601.
55. Freihoefer K, Kaiser L, Vonasek D, Bayramzadeh S. Setting the Stage: A Comparative Analysis of an Onstage/Offstage and a Linear Clinic Modules. HERD: Health Environments Research & Design Journal. 2018;11(2):89–103. https://doi.org/10.1177/1937586717729348.

56. Urlings J, Sezer S, Ter Laan M, Bartels R, Maal T, Boogaarts J, Henssen D. The role and effectiveness of augmented reality in patient education: A systematic review of the literature. Patient Education and Counseling. 2022;105(7):1917–27. https://doi.org/10.1016/j.pec.2022.03.005.
57. Olsen, E. (2021, October 11). Philips' new tools help gamify the MRI experience for kids. *Mobi Health News.* Retrieved 27 February 2024 from https://www.mobihealthnews.com/news/philips-new-tools-help-gamify-mri-experience-kids

Bioethics and the Life of the Child

Brooke M. Ellison and Jamie Leigh Wells

Fundamentals of Medical Ethics

A Brief History

As a field, medical ethics as we know it today has its origins in the 1950s and 1960s. At this time, particularly in the United States, there were increasing calls for greater civil rights protections afforded to groups like women and members of racial minorities. However, the precise origin of the bioethical movement is far less clear or certain. By some accounts, the bioethical movement followed the atrocities perpetrated during the Holocaust [1]. While it is not certain that this catastrophic event provided the genesis of the field of bioethics, it did, undoubtedly, contribute a great deal to how the field has developed the principles on which it stands.

Brooke M. Ellison, MPP, PhD, CPHQ passed away while completing this work. Please refer to the ***Editor's Note*** at the end of this chapter.

B. M. Ellison
Stony Brook University, Stony Brook, NY, USA

Technology and Innovation, United Spinal Association, Kew Gardens, NY, USA

Harvard Kennedy School of Government Carr Center for Human Rights Policy, Cambridge, MA, USA

World Economic Forum, Cologny, Switzerland

Harry Truman National Security Project, Washington, DC, USA

Suffolk County Human Rights Commission, Hauppauge, NY, USA

J. L. Wells (✉)
Ventnor, NJ, USA
e-mail: drjamiewells@gmail.com

J. L. Wells (ed.), *A Clinical Lens on Pediatric Engineering*,
https://doi.org/10.1007/978-3-031-84655-7_4

Ethical Principles

As it is often practiced, bioethics is based on a set of principles: a series of critical concepts and ideas on which bioethical challenges are debated and resolved. From this vantage point, bioethics takes a *principlism* approach. In this section, we will discuss the guiding principles that have sat at the heart of the bioethics movement. These principles apply to care for patient populations spanning the entire life cycle.

Autonomy

The Nuremberg Code of Ethics was presented in 1947 and serves as a framework for physicians to conduct research in an ethical manner. The introduction of the code states, "The judgment by the war crimes tribunal at Nuremberg laid down 10 standards to which physicians must conform when carrying out experiments on human subjects [2]." The atrocities perpetrated against people were perpetrated against their will and without their consent, so prioritizing choice or *autonomy* in decision-making became a matter of the highest prominence.

According to the British Medical Association (2024), autonomy in ethical medical practice refers to, "the right of competent adults to make informed decisions about their own medical care [3]." This quote refers to the fact that adult patients who are in a competent state have the absolute right to choose or deny medical treatment. The decision that one makes on one's own supersedes any decision that others feel to be a better choice. Similarly, it does not matter the reason an individual makes the decision he or she makes so long as it has been made without coercion. In other words, autonomy means that competent adults are free to make decisions that are perceived by others as self-destructive or inappropriate.

The centrality of autonomy largely comes into play in instances of willingness or denial of treatment or in end-of-life decision-making. In these instances, there can often be differences in what a healthcare provider believes to be the best course of action and what course of action a patient, in fact, takes.

Example(s) *A patient with end-stage renal disease is receiving dialysis and awaiting a transplant. He has been on the transplant waiting list for a year. No living donor is available. He decides that the burden is too great and may opt for a more palliative than curative route. In this instance, despite the fact the healthcare professional may believe differently, the choice belongs to the patient. The patient knows what values they hold and what priorities they have for their lives, and those guide the decision that is ultimately made.*

Autonomy for the patient is a paramount principle in bioethics, and the ability of patients to make their own medical decisions is well respected within the medical profession. There are circumstances in which autonomous decision-making at the individual level might be problematic. The first is adults who are deemed not capable or competent to make decisions on an autonomous basis, and this population is beyond the scope of this chapter. Another population, however, that warrants additional consideration is composed of children and young people. Minor status, requirements of parental or legal guardian consent and variability in pediatric age-specific levels of cognitive development are a few factors that contribute to the multi-layered complexity.

Example(s) *An adolescent desires treatment (and confidentiality), but the parent or legal guardian refuses. Traditionally, the default position has favored the latter. However, there are certain situations (e.g., emancipated minor, pregnancy, sexually transmitted infections, drug dependency) where the minor (individual < 18 years old) can consent on their own. Depending on the state, mature minor laws may be in place to facilitate a teenager's autonomy in medical decision-making. These scenarios are much more nuanced than described here—consider the experimental nature of the Covid-19 vaccine during a global pandemic and the debate and legislation that ensued surrounding adolescent autonomy and the parental consent requirement.*

Nonmaleficence

Another pillar on which the notion of bioethics stands is that of nonmaleficence. Nonmaleficence is, perhaps, the oldest and most well-established ethical principle in the bioethical canon. Before entering the medical field, many medical students swear the Hippocratic Oath, a 2500-year-old pledge written by the Greek philosopher, Hippocrates. The Hippocratic Oath requires a physician to "swear upon a number of healing gods that he will uphold a number of ethical standards [4]." Although this document does not contain the exact phrase, it is often credited as the origin of the idea "first, do no harm."

Nonmaleficence is the core principle that demands physicians not harm their patients. In practical terms, according to the National Institutes of Health (NIH), this principle translates into the moral obligation to not kill, not cause pain or suffering, not incapacitate, not offend, or not otherwise inflict harm on a patient [5]. The principle of nonmaleficence, in other words, demands that the healthcare practitioner consider the risks and benefits, potential consequences, and similar burdens of all interventions and treatments before administering them. While the principle of nonmaleficence is appropriate in all facets of medical practice, it is of particular relevance in situations that involve end-of-life decision-making or the need to either administer or withdraw life-sustaining interventions. Nonmaleficence is the bioethical imperative that forces physicians to consider deeply whether the actions they take will bring about or remove the harm or pain that the Hippocratic Oath demands that they do not inflict.

Example(s) *When treating a pediatric patient for cancer or alternate acute or chronic disease, it is essential to focus not only on the benefits of the cure or treatment modalities but also on their unintended harms and actively consider how to mitigate them. If disease surveillance or procedures entail frequent CT scanning,* *then there is a need to reduce harmful effects of cumulative radiation exposure that may predispose to a future cancer(s) or fertility issue. Using protective lead gear and adjusting to pediatric doses of radiation as well as continually reassessing whether this course is in the patient's best interest are some ways to ensure the treatment is not worse than the disease. Stopping a chemotherapy that has proven not to work and is causing adverse effects would be another example.*

The imperative to do no harm, which lies at the heart of nonmaleficence, seems like a straightforward and obvious obligation for healthcare practitioners and a responsibility that is a basic premise of the field. However, there are many instances in which the line between the removal and perpetuation of harm is difficult to detect. These instances include the side effects associated with different medications, the encouragement of interventions without knowing a broader social or personal context, and when to continue or remove life-sustaining treatment.

Beneficence

The corollary of nonmaleficence is the bioethical principle of *beneficence*. While nonmaleficence demands that healthcare providers not inflict harm, beneficence takes this idea a step further and calls for doing good, creating an overall net benefit to patients, and ensuring their general welfare [5]. Beneficence often comes into play when healthcare professionals must protect the well-being and rights of their patients, remove those conditions that might create harm, or provide assistance to individuals with disabilities. In some respects, beneficence can be thought of as the positive or proactive configuration of nonmaleficence, both originating in Greek philosophical thought.

Example(s) *There are many avenues to enhance a child's (or any patient's) well-being. Assist them in tasks they cannot perform themselves. Incorporate child life services into their care, especially when hospitalized. They can provide pet therapy and other forms of emotional support to assuage suffering and alleviate anxiety over treatments and painful procedures to entertainment that is customized to a patient's interests and development level. Avail patients of technology required to stay current in their schoolwork and present in their social sphere or professional obligations when admitted. A physician should meet a child where they are and take the time to explain therapies and why they are being done so they can be in agreement with the plan. Coordinate team meetings and appointments when parents and guardians are available to attend. Sort out and make any transportation, medication procurement, communication or dietary issues easier. Provide materials to inform about the research being done on the condition. Basically, do as much good as you can whenever you can.*

Justice

The fourth of the four primary bioethical principles is *justice*. Generally, and historically speaking, justice refers to fair and equitable treatment of people, but in many healthcare circumstances, the concept of justice requires more comprehensive thinking. When it comes to medical treatment and services, it is important to think about how these resources are distributed across the population—in essence, that is what, why, and in what order. Making these decisions and the outcome of these decisions is known as *distributive justice*. There is a great deal of philosophical debate to determine what an appropriate allocative framework ought to look like and under what set of guiding principles.

Several key frameworks under which the distribution of resources might be considered just include justice as fairness theory, equality of resources theory, common ownership theories, and entitlements theory. Each of these takes a different perspective from which to assess a just way to allocate resources that can be scarce. *Justice as fairness* is a framework that argues the distribution of goods and services should only be just if it maximizes the opportunity for those who are least well-off [6].

Example(s) *All eligible patients should be treated equally when it comes to being prioritized for a new treatment or alternate therapy in short supply. For patient recruitment in clinical trials to be just, researchers should not take advantage of the most vulnerable populations like those desperate for a cure of a fatal disease nor exclude out of biased reasoning those who really could benefit and would make good candidates. There should be fairness at the core of resource allocation and the ability to pay or discrimination can erode the concepts of equitable distribution and accessibility.*

The aforementioned is a basic primer on the fundamental principles of bioethics. There is a litany of other essential standards beyond the scope of this chapter that cement trust in the medical professional and patient relationship. These tenets include but are not limited to confidentiality, veracity and fidelity. Furthermore, specific professional codes of conduct are actively in place for the wide array of medical professionals and other stakeholders involved in patient care (e.g., physicians, nurses, biomedical engineers, clinicians, researchers, Institutional Review Boards (IRBs) charged to uphold regulatory and ethical standards for patient protection when testing human subjects, hospital ethics committees, pharmacists, laboratory scientists).

Ethical Frameworks

Ethics provides us with a lens through which to disentangle complex situations that test our sense of morality. Morality gives us a collectively agreed-upon structure by which to determine actions that are "right" versus those that are "wrong." Ethics is a discipline that provides the tools to explore the situations in which rightness and wrongness are less clear-cut. However, making assessments regarding what ought to be considered "right" as opposed to what ought to be considered "wrong" can be quite difficult and fraught with intellectual challenges. Fortunately, there are theoretical frameworks, built upon centuries of philosophical thought, that can provide structure in our thinking to make these normative decisions. There are many theoretical frameworks from which to investigate problems; we will focus our attention on a few of them.

Deontology

Deontology is a framework of moral philosophy that focuses on which decisions are "morally required, forbidden, or permitted [7]." As Alexander explains, those who favor a deontological perspective are far less concerned about the consequences of actions and, instead, argue that there are some actions that cannot and should not be taken no matter what the outcome might be [7]. According to a deontological perspective, a correct choice is one that conforms to a generally agreed-upon set of moral norms.

Consequentialism

In seeming juxtaposition to deontology, there is another theoretical perspective, known as consequentialism. As its name implies, a consequentialist favors a decision that brings about the best outcome or greatest amount of good. One of the most prominent forms of consequentialism is utilitarianism. Utilitarians found their origin and home in the age of the Enlightenment, when visions of the Good came less from religious sources and more from secular perspectives. Utilitarianism argues, then, that the "right" course of action is the one that creates the greatest pleasure, happiness, or units of welfare for the greatest number of people.

Due to the inherent complexity in ethical decision-making in medical practice, it is not uncommon that these and other frameworks may cause friction when employed. While deontology tends to be patient-centered and utilitarianism emphasizes the greater good, triage choices in emergent scenarios and the allocation of scarce resources might be in conflict. Ethical conundrums are just that, rarely simple [8].

Example(s) *Chukwuneke and Ezenwugo pose the example of the elderly patient in need of a kidney transplant versus vaccinating one hundred children lacking financial resources. They note the clinician has a duty to treat their patient and the patient is entitled to proper medical care. Further, the authors maintain the utilitarian would forgo providing the kidney surgery in favor of immunizing the children "because of the good of the majority, whereas the deontologists argue that the quality-adjusted life years for a 90-year-old after a kidney transplant cannot be compared with lifesaving immunization for young school children [8]." Let us also* *not forget that the key bioethical principle of justice would need to be balanced so that discrimination would not be a factor in medical decision-making.*

Natural Law

Natural law theory (NLT) revolves around morality intrinsic to human nature from which we reason and enact behavior. It is predicated on the notion human beings overall inherently understand "right" from "wrong" and share key principles to flourish; these "basic goods usually include items such as life and health, knowledge, work and play, the appreciation of beauty, friendship, and religion. A key question for NLT is whether the basic goods can be measured or compared in terms of their value, and, relatedly, whether there is any kind of order or hierarchy of value [9]."

Example(s) *Steffen suggests, "Medical futility illustrates how a natural law ethic works in the bioethics arena. Reasonable people of good will agree that persons facing distressful end-of-life medical conditions should receive care, but futility raises the issue about the extent of that care [*10*]." He features the case of Baby K—an infant born with anencephaly who was missing a large portion of her brain, describing her situation as "medically catastrophic, and she had no reasonable prospect of flourishing, that is, of enjoying the goods of life [*10*]." The family and hospital were at odds over the extent of her care with the latter wanting to limit such provisions to those palliative in nature, ultimately taking the former to court. Baby K's mother prevailed with the courts and the hospital was obligated to provide emergency interventions which did prolong Baby K's life.*

Virtue Ethics

Virtue ethics, as a theory, centers around the core character of the individual and less so on their actions. For, if the person is continually developing themselves by nurturing traits like courage, fairness and honesty, then they are virtuous and inclined to make ethical decisions. This is often applied to professional conduct and development.

Bioethics and Pediatrics

The ethical principles and frameworks described above relate to care applied to the general population. When taking into consideration the unique experiences and needs of the pediatric population, we also need to incorporate different factors that can change the nature of the decisions that we make or the parameters within which we make them. Children are not simply little adults but, rather, face a very different set of needs or demands. We will explore these additional considerations in the upcoming Section.

Ethical Decision-Making and the Child

Decision-making is a central part of ethics in the healthcare setting. The choices that we make are a representation of how we uphold principles like autonomy, beneficence, nonmaleficence, and justice. When it comes to decisions we make in the healthcare setting, the principle of autonomy is paramount, in that patients have a right to make their own determinative medical decisions, free of coercion or contempt. Allowing people to make their own decisions with regard to what happens to their bodies. The manner by which the principle of autonomy is enacted is through *informed consent.*

According to the National Institutes of Health (NIH), informed consent is defined as "the process in which a healthcare provider educates the patient about the risks, benefits, and alternatives to a given procedure for intervention [11]." The Joint Commission stipulates that there ought to be 5 specific elements to informed consent, which include:

1. A discussion of the nature of the procedure
2. Articulation of the risks and benefits of the procedure
3. Clarification of reasonable alternatives
4. Description of risks and benefits of alternatives
5. An assessment of the patient's understanding of steps 1 through 4

In the pediatric sphere, these issues become more complex. Brain physiology and cognitive stage evolve along a developmental spectrum while maturity and level of insight are also integrally at play in mooring ethics to clinical decision-making. Patterns of communication and language, analysis and reasoning are dynamic as are levels of dependence, resilience and an ability to abstract thought. For instance, a younger child possesses a more concrete rationale when assessing an expression like "people in glass houses should not throw stones" possibly wondering why a house would be all glass or what color the stone is. An older child with the capacity for abstract reasoning will get to the true meaning of being wary not to criticize others, especially when they are not perfect themselves. The variability and heterogeneity at baseline are vast and unique to this population when compared to any other. The nature of the procedure and its future ramifications—reversible or irreversible—will not be fully understood by the infant, toddler and school-age child like it might for the older, more autonomous adolescent or young adult.

Additionally, there may be even more gradations within these subsets. While one teenager may demonstrate a profound sense of responsibility and maturity, another might be more impulsive and require greater support. Extrapolate further how these distinctions could apply to (or not) entrusting independent medication or therapeutic administration outside the presence of a parent or guardian. Miscalculations in this arena, depending on the particular substance or intervention, can lead to misuse and unintentional or intentional errors that prompt deleterious effects with short- and long-term consequences. Striking the ethical balance of empowering individual autonomy, first doing no harm and promoting the continued welfare of a child albeit with this example, or when delivering pediatric medical care in general to such a diverse age span, making informed decisions, employing novel technology and enrolling in experimental treatments and trials can prove challenging.

Compounded by these foundational physical, emotional, psychological, social, environmental and biological ever-transforming differences is the reality that those under the age of 18 years old are minors restricted by laws, with some exceptions, to make medical decisions without the consent of their parent or legal guardian—with the latter often positioned to override the former, especially when the patient cannot speak for themselves. Hence, at every opportunity, it is a physician or caregiver's duty to ensure the pediatric patient is optimally informed via unique language and at a level they can understand with a focus on their best interest (now and into the future). For a child to be a research participant, for example, the most ethical path is to obtain *assent*. This is further addressed in the upcoming section on "Ethical Standards and Clinical Trials." For as much as the pediatric patient has the capacity to contribute to discussions surrounding their care, it is most beneficial to their care and the therapeutic plan, affirming their autonomy and fostering trusted relationships between them and their caregivers.

Operationalizing Bioethics in Pediatrics

In section "Fundamentals of Medical Ethics," we discussed several of the foundational concepts related to medical ethics. These concepts, however, are abstractions unless they can be placed in a practical, real context. In section "Operationalizing Bioethics in Pediatrics," we will look at the applications of medical ethics and how medical ethics can be borne out and operationalized through parts of the patient-provider encounter.

Power Asymmetry

Before diving more deeply into the ethical implications of the patient-provider encounter, let us begin by addressing the nature of the patient-provider relationship. The relationship between patients and their healthcare providers can be filled with a plethora of dynamics. First and foremost, there is a discrepancy of knowledge, and therefore power, between the healthcare provider and the patient, which can create an asymmetry that is felt on both sides. Historically speaking, the nature of the relationship has been complex. Physicians and other healthcare providers enter a healthcare encounter with a great deal more knowledge about diagnoses, symptomatology, courses of treatment, and medical prognoses. Patients, then, arrive at the encounter in their times of medical need, experiencing some kind of ailment or impairment that can leave them feeling vulnerable, weakened, or incapacitated. Yet, it is in these times when the power balance or knowledge asymmetry comes into keen focus.

The power asymmetry between patient and physician is considerable when this relationship is between two adults. The imbalance is even more pronounced when the relationship is between an adult and a child. There is a power and knowledge disparity that exists between children and adults already, and when this relationship is placed within a medical context, the disparity becomes that much more evident.

Shared Decision-Making (SDM)

It is with this relationship in mind that the value of shared decision-making becomes center stage. The concept of shared decision-making in healthcare is a relatively new one and exists as a veritable response to the bioethical prioritization of *autonomy* among other ethical principles. Autonomy places the enormity of healthcare decisions on the shoulders of patients, allowing them to make healthcare decisions on their own, free of coercion, and guided by the process of informed consent. While the principle of autonomy is essential to maintaining patients' notion of dignity and self-governance, it can also leave some individuals without direction or support in times of their greatest uncertainty and vulnerability. Shared decision-making is designed to alleviate some of this decision-making burden.

Shared decision-making is a relatively new healthcare practice in which healthcare professionals assist patients in their decisions. According to Edwards and Elwyn, shared decision-making can be described as "a 2-way communications process between patients and professionals, where both collaborate in making healthcare decisions [12]." Developed in the 1980s and 1990s, shared decision-making has become a critical component of healthcare quality and delivery.

Just as there is a process related to informed consent, there is a process for shared decision-making, which was articulated by Quill and Suchman in 1993. According to these researchers, shared decision-making involves the following steps: "The clinician shares information about the illness and treatment options. The patient contributes her expertise about her own goals, attitudes towards risk, and the value she places on various outcomes. Then, combining their perspectives, they can negotiate a solution that is uniquely suited for this patient [13]." When "the patient" is a child, the circle of those involved in the shared decision-making process must expand.

In a systematic review of the data and literature, Wyatt and colleagues state, "SDM aims to engage patients and clinicians in a partnership to make medical decisions that are supported by the best available evidence and aligned with patient's values, preferences, and treatment goals. A reasonable extension of this idea to pediatrics should include the involvement of parents [14]." By "parents" Wyatt refers to biological parents, legal guardians, and other caregivers with medical decision-making responsibilities. In their review, the researchers assessed a total of 53 studies utilizing shared decision-making in the pediatric population. In the majority of these studies, the targeted relationship focused on the parents alone, under 10% targeted the pediatric patient alone, and just over 10% targeted the patient and the parent. Given this diversity in approach to how best to study shared decision-making in the pediatric population, Wyatt and colleagues [14] concluded, and rightly so, that SDM in pediatrics remains an underexplored and poorly defined area, specifically as it relates to the roles that children and their parents play in this dynamic.

Ethical Standards and Clinical Trials

One of the foundational steps in the translation of biomedical research to patient use lies in the clinical trial. Clinical research is all types of medical research that involves people, and clinical trials are a critical part of this research. According to the NIH, clinical trials explore new strategies to prevent, detect, or treat disease, including the testing of new drugs, new surgical procedures, medical devices, new ways to use this experience, new ways to change behaviors to improve health, and ways to improve the quality of life for people with chronic illness [15].

Clinical trials provide the evidence to determine whether new medical interventions are safe and effective. Historically speaking, clinical trials have been fairly homogeneous in terms of their recruitment, largely involving white, middle-class adults. This state of affairs is problematic for many populations, among those are children. However, over the past several decades, it has become evident that "efforts to protect children by avoiding clinical trials in this population were actually putting children at risk because the vast majority of drugs pediatric patients received had never been tested in this age group [16]."

As a result of this realization, the American Academy of Pediatrics (AAP) developed a series of guidelines on the ethical conduct of pediatric clinical trials [17]. The guidelines created by the AAP call for higher standards than those related to adults. For instance, as Kuehn explains, the pediatric guidelines recommend that "drugs undergo at least initial testing results before they are tested in children [16]." In addition, the AAP guidelines [17] advise for the following things:

- That all pediatric clinical trials be registered at www.clinicaltrials.gov. The purpose of this guideline is to promote the education of trials and to ensure that the results of the trials are made public.
- That investigators conducting clinical trials involving children disclose any and all potential conflicts of interest to both patients and their parents or guardians.
- That the informed consent for pediatric trials require the consent of a parent or guardian.
- That, for trials involving children over the age of 7, there is assent from the patient.

Despite these guidelines from the AAP, there is still a significant lack of understanding among child patients and their parents regarding the nature of clinical trials. Researchers have called for greater involvement of participants from historically marginalized groups in the decision-making process to maintain that these patients have a voice in decisions that are made on their behalf.

In recent years, the countries of the European Union came together to develop a framework related to ethical, legal, and social issues embedded in clinical trials involving children [18]. As Berde and colleagues discuss, children are particularly vulnerable research participants, often unable to articulate their own views and self-interests [19]. With this in mind, there are special ethical clinical considerations that must be taken into account when designing clinical trials that involve minors. According to Berde and colleagues, when it comes to clinical trials involving children, there must not be more than minimal risk; anything more than minimal risk must be justified by the potential direct benefit to the participating child [19].

What becomes difficult are instances in which there is disagreement among family members regarding the best or most appropriate course of action for a child's prospective involvement in a clinical trial. To this end, the National Academies of Science and Institute of Medicine has provided recommendations for strategies to design clinical trials when there is disagreement between the participating children and their parents [20]. Experts contributing to these recommendations suggest that, in these circumstances, the research investigators and institutional review boards should "aim to create assent processes that consider and respect the child in the family as a unit as well as individually [20]." However, in one scoping review of the literature by Cayouette and colleagues, the steps by which to operationalize this recommendation for research participation in the pre-adolescent set revealed universally inconsistent and often vague processes [21]. They concluded, "We found significant variation in criteria used for assessment of patient capacity, delivery of information used to obtain assent and documentation of the assent process. While we acknowledge that individual children, settings and jurisdictions may require different approaches to obtaining assent, there should be agreement on important principles to be followed with resulting common guidance on assessing capacity, delivering information and documentation of the assent process for publication." While age was a frequent determinant, the level of cognition and the dynamic nature of clinical stability further impacted meaningful assent procedures [21].

Fundamentally, there is a need for increased ethical standards when experimental medical subjects are pediatric patients (e.g., channels of consent). This reality, among other distinct challenges (e.g., small population sizes, compliance, problematic trial design, inappropriate endpoints, need for developmental- and age-specific nuance), has often impeded pediatric clinical trial success [22]. Given many biomedical pathways, discoveries and protocols are designed for the adult and adapted for the child, there can be much lost in the translation. Fortunately, the need for procurement of high-quality pediatric data has taken a spotlight and the pendulum is swinging to employ a pediatric lens first when innovating. For example, there is emphasis now on the adjunctive inclusion of real world evidence (RWE) and efforts in clinical trial redesign.

Applications Specific to Pediatrics: Novel Technologies and More

Bioethical considerations abound in the field of pediatrics. We have discussed a mere few clinical, research and multi-stakeholder scenarios that barely scratch the surface. The broad scope influences the developing child and health outcomes in incalculable ways (e.g., inpatient, outpatient, on an individual level and in the family unit, extent of adult disease and prognosis, experimental and standard treatments), especially when integrating the latest technologies whose ultimate applications are not always initially identifiable and often intended for adults.

For example, medical uses for artificial intelligence and machine learning are only as promising as the quality of data put into these systems. The lack of high-quality pediatric data has made advances in this realm problematic for children, but actionable efforts are underway to right the ship. Missteps have occurred that have harmed children due to poor regulation of product marketing. Look no further than vaping technology and how it was introduced to the market, intentionally advertising to youth, which promulgated an epidemic of teen nicotine addiction and a subsequent U.S. Food and Drug Administration (FDA) crackdown. Ethical practices should be at play to eliminate such dangers in the direct-to-consumer marketing domain. The electronic health record, as initially implemented, did not reflect the unique demands of children or address the needs of the medical professionals who care for them.

Ethical obligations in biomedical terrains ideally align to serve the best interest of the child. However, due to the expanding stakeholder involvement in modern medicine and its rapid progress, situations can vary from clear to murky even with the best of intentions. Impossible choices are rampant. Look no further than the case of 22-month-old conjoined twins, documented in the *New England Journal of Medicine* [23], whereby the parents, medical team and ethics committee convened to evaluate the utility of separation surgery versus no intervention. In this case, doing nothing meant the likely death of both twins while separation surgery meant the precipitous death of one. Ultimately, the latter path was selected and the expected intraoperative death of one twin took place. The report maintains the other is thriving.

Pediatrics often requires navigating a special ethical landscape that is not always unambiguous and is routinely bespoke. From prenatal screening and understanding the current limitations of certain genetic probabilities to end-of-life challenges and the fluid reality of disease distinct to the infant, child or adolescent when compared to the adult, biomedical solutions can have high stakes. One case that became an international scandal involved the unethical practices of a scientist in China who utilized breakthrough CRISPR technology for germline gene editing on healthy human embryos that resulted in live twins and a subsequent third child. This type of genetic alteration is carried on to all future offspring, the safety and risks of which are presently unknown. At the time, given the gravity of the ramifications to humanity, calls were made for a global moratorium on the technology by many leading scientific communities until universal guidelines could be established. Others worried such a drastic response could stifle continued discovery of this promising technology that has hope for eradicating heritable forms of lethal diseases. Worldwide condemnation of the scientist was swift focusing, in part, on issues of informed consent (or lack thereof) and suspected coercion—for the parents and babies. He was convicted and has served his prison sentence. The costs of his actions may not be known for generations.

Often, there is no well-defined road map in this population given the historical propensity of just extrapolating (with minor tweaking) what is done for adults. Given the warp speed of new biomedical discovery and technological advancement, this strategy of playing catch-up is routinely insufficient. Children are at the start of a hopefully long life trajectory. Thankfully, an ethics-centered pediatric lens in research, design and medical care is an ever-evolving current trend and recognized clinical need.

Conclusion

Bioethics is integrally linked to the highest quality care in pediatrics; sacrificing it halts innovation. Readers of this chapter will be able to discuss the foundational principles of medical ethics and the frameworks for ethical decision-making; articulate how infants, toddlers, children, and adolescents are uniquely affected by medical ethics and how this changes throughout development; apply the foundational principles of medical ethics to healthcare encounters involving children and adolescents; and relate medical ethics to the lives of children and adolescents. These tools are necessary to facilitate a mindfulness toward ethics which, in turn, helps us engage in the difficulties that come with traversing healthcare challenges and operationalizing compassionate environments.

Editor's Note

Image 1: Editor, Jamie L. Wells, MD, FAAP (left); First author, Brooke M. Ellison, MPP, PhD, CPHQ (right)

Dr. Ellison and I originally met when I, as a prior winner, delivered the keynote address at the Washington, DC, ceremony honoring the top twenty high school students in the nation who were featured in *USA Today* and awarded their prestigious academic scholarship. To say it was a privilege and life highlight to cross paths with her and, subsequently, share in her orbit for many years beyond then would be a massive understatement. Her star only continued to rise and the reverberating impact of her inspiring legacy in advancing disability rights and science will meaningfully influence generations to come.

Dr. Ellison put considerable time and thought into writing this chapter. She persevered with this project regardless of her challenging circumstances. Sadly, she passed away before the completion of her manuscript. It is an honor to include her final published work here. With reverence, it was my goal to preserve her original voice and maintain her vision as much as possible. She wrote the title, most of the abstract and a lot of the chapter from the start through much of the clinical trials section. Our conversations, emails and her outline guided me to fill gaps and supplement when necessary (e.g., incorporate examples, add paragraphs).

It is most fitting to end this tribute in her own words,

> "*It has always been fascinating to me how the way we are viewed by the world can only vaguely resemble the way we view ourselves. I am many things and am seen in many, sometimes contradictory ways. I am a PhD and a Professor at Stony Brook University, where I teach and research disability and medical ethics. This is where I make a living, but it is also the very same place that saved my life. In September 1990, I was hit by a car while walking home from school. I was crossing a major highway nestled in the Long Island suburbs and, in that accident, suffered innumerable injuries that have affected my life in immeasurable ways. But, I was a fighter. I am a fighter. My disability in no way compromises that. In fact, my disability amplifies that.*
>
> *Details of my life and experiences that my family have undergone have been chronicled and relayed in narrative and cinematic form. I received my undergraduate degree in cognitive neuroscience from Harvard University, and then returned to Harvard to do my Master's in Public Policy at the Harvard Kennedy School of Government. A lover of mobilizing change, I ran for New York State Senate in 2006 and followed that with the creation of a nonprofit organization, the Brooke Ellison Project, designed to provide education about stem cell research. In 2012, I received my doctorate in Sociology from Stony Brook University. I'm now a scholar. I'm now a researcher. I am now a professor.*
>
> *But, I am a fighter. My life has been shaped by challenges and adversity of all different kinds, but it has not been characterized by them. I am not defined by them. For every obstacle, I have sought opportunity. In each instance of sorrow, I have sought strength. When there has been challenge and adversity, I have looked for resilience and hope. That is what I want people to know about me, as that is how I see myself. That is also what I want people to know about themselves, as that is how I view the world.*"

Competing Interests The authors have no conflicts of interest to declare that are relevant to the content of this chapter.

References

1. Schicktanz S, Stoff H. The legacy of the Holocaust in bioethics. Bioethics. 2021;35:497–8. https://doi.org/10.1111/bioe.12909.
2. The Nuremberg Code (1947) *BMJ* 1996; 313:1448. https://doi.org/10.1136/bmj.313.7070.1448
3. British Medical Association. (Updated 2024, June 28). *Communicating with patients and autonomy*. URL: https://www.bma.org.uk/advice-and-support/international-doctors/life-and-work-in-the-uk/toolkit-for-doctors-new-to-the-uk/communicating-with-patients-and-autonomy#:~:text=Patient%20autonomy&text=Autonomy%20is%20the%20right%20of,about%20their%20own%20medical%20care.
4. National Library of Medicine. *Ancient Greek Medicine, "I swear by Apollo Physician...": Greek Medicine from the Gods to Galen*. The content of this website is drawn from a display held in the History of Medicine Division of the National Library of Medicine, March 6 through June 15, 2008. The exhibit was curated by Michael North who was then Head of Rare Books at NLM. URL: https://www.nlm.nih.gov/hmd/topics/greek-medicine/index.html#case1
5. Varkey B. (2020, June 4). Principles of Clinical Ethics and Their Application to Practice. *Med Princ Pract*. 2021;30(1):17-28. https://doi.org/10.1159/000509119
6. Allingham, M. (2004). *Distributive Justice*. Internet Encyclopedia of Philosophy. URL: https://iep.utm.edu/distributive-justice/
7. Alexander, L. and Moore, M. "Deontological Ethics", *The Stanford Encyclopedia of Philosophy (Winter 2021 Edition)*, Edward N. Zalta (ed.), URL: https://plato.stanford.edu/archives/win2021/entries/ethics-deontological/
8. Chukwuneke, Felix N.; Ezenwugo, Anthony C. (2022, Jan–Mar). Deontology vs. Utilitarianism: Understanding the Basis for the Moral Theories in Medicine. *International Journal of Medicine and Health Development* 27(1): p 19-23. https://doi.org/10.4103/ijmh.IJMH_57_20
9. Shea M. Value Comparability in Natural Law Ethics: A Defense. J Value Inquiry. 2024;58:383–402. https://doi.org/10.1007/s10790-022-09895-5.
10. Steffen, L. (2016, April-June) Core values in bioethics: A natural law perspective. *Ethics, Medicine and Public Health*, Volume 2, Issue 2, Pages 170-180. https://doi.org/10.1016/j.jemep.2016.03.009
11. Shah P, Thornton I, Turrin D, *et al.* [Updated 2023, Jun 5]. Informed Consent. In: *StatPearls* [Internet]. Treasure Island (FL): StatPearls Publishing; 2024 Jan-. Available from: https://www.ncbi.nlm.nih.gov/books/NBK430827/
12. Edwards A, Elwyn G, editors. *Shared decision-making in health care: Achieving evidence-based patient choice*. 2nd ed. Oxford: Oxford University Press; 2009.
13. Quill TE, Suchman AL. Uncertainty and control: learning to live with medicine's limitations. Humane medicine. 1993;9(2):109–20. PMID: 11656250
14. Wyatt, K. D., List, B., Brinkman, W. B., Prutsky Lopez, G., Asi, N., Erwin, P., Wang, Z., Domecq Garces, J. P., Montori, V. M., and LeBlanc, A. (2015, Nov-Dec). Shared Decision Making in Pediatrics: A Systematic Review and Meta-analysis. *Academic Pediatrics*, 15(6), 573–583. https://doi.org/10.1016/j.acap.2015.03.011
15. National Institute of Mental Health (Last reviewed April 2024). Clinical Trials – Information for Participants. *U.S. Department of Health and Human Services, National Institutes of Health*. Retrieved October 28, 2024. URL: https://www.nimh.nih.gov/health/trials
16. Kuehn BM. Reports Promote High Standards for Ethical Conduct of Pediatric Clinical Trials. JAMA. 2010;303(18):1800. https://doi.org/10.1001/jama.2010.579.
17. Shaddy, R.E., and Denne, S.C. (2010, April). The Committee on Drugs and Committee on Pediatric Research; Guidelines for the Ethical Conduct of Studies to Evaluate Drugs in Pediatric Populations. *Pediatrics*; 125 (4): 850–860. https://doi.org/10.1542/peds.2010-0082
18. Lepola P, Kindred M, Giannuzzi V, et al. Informed consent and assent guide for paediatric clinical trials in Europe. Archives of Disease in Childhood. 2022;107:582–90. https://doi.org/10.1136/archdischild-2021-322798.

19. Berde, C.B., Walco, G.A., Krane, E.J., Anand, K.J.S., Aranda, J.V., Craig, K.D., Dampier, C.D., Finkel, J.C., Grabois, M., Johnston, C., Lantos, J., Lebel, A., Maxwell, L.G., McGrath, P., Oberlander, T.F., Schanberg, L.E., Stevens, B., Taddio, A., von Baeyer, C.L., Yaster, M., and Zempsky, W.T. (2012, February). Pediatric Analgesic Clinical Trial Designs, Measures, and Extrapolation: Report of an FDA Scientific Workshop. *Pediatrics*; 129 (2): 354–364. https://doi.org/10.1542/peds.2010-3591
20. Institute of Medicine (US) Committee on Clinical Research Involving Children; Field MJ, Behrman RE, editors. *Ethical Conduct of Clinical Research Involving Children.* Washington (DC): National Academies Press (US); 2004. 2, *The Necessity and Challenges of Clinical Research Involving Children.* Available from: https://www.ncbi.nlm.nih.gov/books/NBK25553/
21. Cayouette F, O'Hearn K, Gertsman S, et al. Operationalization of assent for research participation in pre-adolescent children: a scoping review. BMC Med Ethics. 2022;23:106. https://doi.org/10.1186/s12910-022-00844-2.
22. Green, D. (2017, September 8). *Pediatric Trial Design and Endpoint Considerations* [PowerPoint slides]. Presented at U.S. Food and Drug Administration (FDA) workshop: Pediatric Trial Design and Modeling: Moving into the Next Decade. Office of Clinical Pharmacology, Office of Translational Sciences, Center for Drug Evaluation and Research (CDER). URL: https://www.fda.gov/files/about%20fda/published/Pediatric-Trial-Design%2D%2D-Endpoint-Considerations.pdf
23. Cummings BM, Gee MS, Benavidez OJ, Shank ES, Bojovic B, Raskin KA, Goldstein AM. Case Records of the Massachusetts General Hospital: Case 33-2017—22-Month-Old Conjoined Twins. N Engl J Med. 2017;377:1667–77. https://doi.org/10.1056/NEJMcpc1706105.

Part II
Facilitating Progress from Bench to Bedside: Wins, Cautionary Tales, and Lessons in Translation

Advancing Pediatric Healthcare: A Guide to Medical Device Innovation from Conceptualization to Clinical Use

R. Brandon Hunter, Jhalak N. Mehta, Christine Luk, and Chester J. Koh

Introduction

Pediatric medical devices account for only 25% of medical devices approved in the United States [1]. Moreover, many of these approved "pediatric" devices are not applicable to children under 18 years old. They only qualify as pediatric devices due to the fact that the U.S. Food and Drug Administration's (FDA) definition of the pediatric age group is individuals 21 years old and younger, so even devices only suitable for those 18 years of age and older could technically qualify [1]. There are numerous reasons for this shortage of pediatric devices:

- ***Anatomical and physiological constraints***: Pediatric patients exhibit distinct anatomical and physiological features compared to adults. Some examples of these distinct features include an increased surface area to weight ratio compared to adults in infants, leading to reduced heat retention, and narrower pediatric airways, increasing obstruction risks, larger head-to-body ratios leading to a higher risk of cervical ligamentous injury [2]. Smaller physical size of pediatric

R. B. Hunter (✉) · C. J. Koh
Southwest-Midwest Pediatric Device Innovation Consortium, Houston, TX, USA

Baylor College of Medicine, Houston, TX, USA

Texas Children's Hospital, Houston, TX, USA
e-mail: rxhunter@texaschildrens.org

J. N. Mehta
Southwest-Midwest Pediatric Device Innovation Consortium, Houston, TX, USA

Baylor College of Medicine, Houston, TX, USA

C. Luk
Southwest-Midwest Pediatric Device Innovation Consortium, Houston, TX, USA

Proxima Clinical Research, Houston, TX, USA

J. L. Wells (ed.), *A Clinical Lens on Pediatric Engineering*,
https://doi.org/10.1007/978-3-031-84655-7_5

anatomy can lead to technical design constraints. Finally, children grow and develop over time so devices must be adaptable to this change.

- ***Small patient populations and clinical trial enrollment constraints***: Significant chronic illness in children is less common than in adults, leading to smaller and often more geographically dispersed patient populations for any given disease [3]. This population characteristic can significantly hinder the execution of adequate clinical trials to obtain the safety data needed for pediatric labeling. Failure to recruit enough patients is the most common reason for pediatric clinical trial failure, accounting for ~40% of all failed trials [4].
- ***Financial constraints***: In the United States, per capita expenditures on pediatric medical care are less than half of that of adults as children generally have fewer health issues and, consequently, utilize healthcare services less frequently than adults [5]. There are also no standard benchmarks for pediatric reimbursement rates. Medicare provides reimbursement standards for older adults while Medicaid provides reimbursement standards for children. Since Medicare is managed by the federal government, it offers a national reimbursement standard. Medicaid, in contrast, is managed by each individual state, which leads to variability in reimbursement rates [6]. This lack of national standards makes it difficult for innovators to generate accurate predictions about financial returns during development, which can impede the acquisition of early-stage investments.
- ***Ethical constraints***: Ethical concerns with enrolling children in clinical trials may exist because they are not able to give fully informed consent based on their ever-evolving capacity for understanding during the pediatric period. Most clinical studies with more than minimal risk involving children require consent from parents or guardians over the age of 18 years old. In addition, they require assent, or agreement to participation in age-appropriate terms, from children as young as 6–7 years old [7]. As such, the legal and ethical complexity of exposing a child to an intervention when there is true equipoise in the medical literature poses significant barriers.

Overall, these and other factors lead to a situation where many pediatric medical devices are used "off-label," are repurposed, or modified from adult devices.

Various national initiatives have been employed to overcome these barriers to pediatric device innovation. The FDA's Office of Orphan Product Development offers grants to support the development of devices used for treating rare diseases, defined as those that affect 8000 or fewer individuals in the United States per year, which is relatively common in pediatric illness [8]. The Humanitarian Device Exemption (HDE) is a regulatory pathway established by the FDA to facilitate the review of devices designed to treat or diagnose rare diseases [9]. Finally, in 2009, the FDA initiated the Pediatric Device Consortia (PDC) Grants Program after receiving Congressional approval, which led to the creation of nonprofit pediatric device accelerators at children's hospitals to offer non-dilutive regulatory, clinical, and research support for pediatric medical device innovation [10].

Despite these developments, pediatric medical device innovators still encounter distinct challenges. In this chapter, we outline a bench-to-bedside roadmap that starts with an overview of the medical device design process and describes needs identification and concept generation in the "Identify" and "Invent" sections. In the "Implement" section, we describe the regulatory landscape and provide advice on how to navigate clinical trials and submissions. In the "Investment in and Returns on Pediatric Medical Device Research" section, we explore strategies for securing funding to achieve a viable return on investment and discuss the unique features of the pediatric market. In the "Current Trends in Pediatric Innovation" and "Organizations Dedicated to Advancing Pediatric Innovation" sections, we describe emerging trends and organizations contributing to accelerating pediatric innovation. Finally, in "Looking to the Future: Moving Toward a National Ecosystem for Pediatric Medical Devices," we discuss two initiatives that stand out as pivotal to the future success of pediatric device innovation. Throughout this chapter, we will present real-world examples and provide innovators with the knowledge needed to navigate the landscape of pediatric medical device innovation.

Pediatric Medical Device Biodesign Process

The Biodesign process is one of the most popular approaches to medical device design and was created by Drs. Paul Yock and Josh Makower at Stanford in 2000. Yock and Makower also co-founded the Stanford Byers Center for Biodesign and created the Center's core curriculum. The Biodesign process requires multidisciplinary collaboration and was designed based on the belief that "innovation is a process that can be learned, practiced, and perfected" [11]. This process is iterative and consists of three phases: identify, invent, and implement. The next section presents an overview of each step.

Identify: Need Finding, Validation, and Screening

The identify phase is the most critical because it lays the foundation for the device life cycle, so we will describe it in more detail than the other phases. The Biodesign process is guided by the philosophy that "a well-characterized need is the DNA of a great invention" [11]. Thus, the purpose of this stage is not to identify a space in which an existing technology will fit but rather to identify a need that, if met, will have a lasting impact. Without a deep understanding of the clinical needs of customers and end users, innovators could spend significant resources developing a solution to a problem only to find out that the problem is not big enough to drive the purchase of their solution. This phase is especially important in designing pediatric devices because markets and company runways are often smaller.

Needs Finding and Validation

Innovators must first identify the most important unmet healthcare needs. They often start by identifying problems in a general need area based on the interests and skill sets of those in the innovation group. Needs finding involves literature research along with observing and interviewing potential end users and anyone involved in patient care, such as physicians, patients, nurses, and technicians. Customers may be different from end users and could be stakeholders such as insurance companies, hospital system representatives, or other administrators. After identifying many clinical needs (the Biodesign process suggests innovators identify at least 200 clinical needs), innovators should validate them [12].

Need validation has multiple components:

- ***Define the need statement***: Articulate the need in a specific, actionable, and focused statement. This need statement typically includes descriptions of the problem, affected population, and desired outcome.
- ***Gather evidence***: Next, gather evidence to validate the need. This process involves collecting data and insights from various sources, including literature reviews, expert interviews, and clinical observations, to understand the scope and impact of the need. Engage with various stakeholders, such as patients, healthcare providers, potential customers, and regulatory bodies, to understand their perspectives and requirements.
- ***Assess the market and existing solutions***: To understand the current market landscape, analyze existing solutions and their limitations, determine whether there is a service gap that a new device could fill, ascertain the potential market size, and discern the economic feasibility of entering that market.
- ***Refine the Needs Statement***: Based on the gathered data and analysis results, refine the need statement to better capture the essence of the problem and innovation opportunity.

Gathering Evidence: Clinical Observations

One of the most effective ways to identify and validate needs is to conduct clinical observations in various healthcare settings. Such observations generally occur over several weeks and involve shadowing healthcare stakeholders, such as nurses, physicians, home health workers, and caregivers, to identify their clinical challenges and practices. This process helps innovators understand their pain points better [11]. It is important to note that stakeholders will not always verbalize the complications, challenges, and worries they face, so skillful observation is needed to identify underlying sources of problems that may be different than what the clinician or patient may identify when asked.

An example of a medical device created using clinical observation is Pediatric Advanced Weight Prediction in the Emergency Room (PAWPER) tape that was invented by Dr. Mike Wells. Wells conducted clinical observation during his daily work as an emergency room physician. Children's significantly varying sizes make accurate weight-based dosing essential for effective and safe treatment. At the time of his observation, the most common method for estimating children's weight was to use Broselow tape, which estimates weight based on length alone and is frequently inaccurate. Overdosing or underdosing medications could have serious clinical consequences, so Wells developed a device known as PAWPER tape that more accurately estimates both total and ideal body weight. Fast, reliable calculation is vital to avoiding unnecessary delays and errors, especially in situations involving life-saving resuscitation. The most recent iteration of the PAWPER tape, the PAWPER XL-MAC, is performed in three steps which consider the child's length and mid-arm circumference (Fig. 1). This addition of mid-arm circumference causes weight estimations using PAWPER tape to be more accurate than those using Broselow tape [13]. By identifying this clinical need through direct observation,

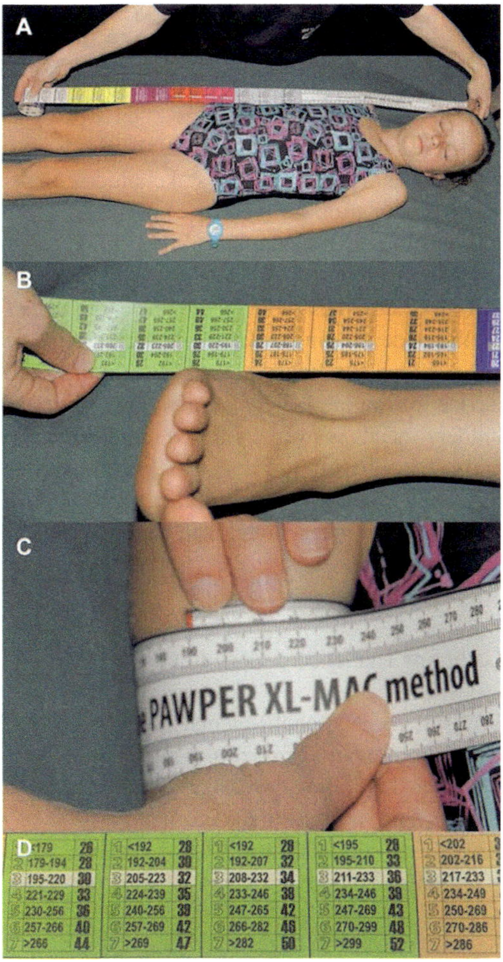

Fig. 1 The PAWPER XL Tape. The first step is to measure the length of the child from his/her head to his/her heel (panels **a** and **b**). The user thus identifies the length segment in which the child falls. The second step is to measure the child's mid-arm circumference using the tape (panel **c**). The final step is to read off the estimated weight in the appropriate length segment based on the measured mid-arm circumference (panel **d**) [14]

Wells was able to innovate effectively, demonstrating the importance of close, insightful observation when identifying a need.

Gathering Evidence: Stakeholder Interviews

Stakeholder interviews are another important part of needs validation. *The Mom Test* by Rob Fitzpatrick is a guide on how to conduct customer (or, in our case, stakeholder) interviews to acquire the most valuable and truthful feedback [15]. The title of the book reflects the notion that if you ask your mom whether your business idea is good, she is likely to say "Yes" to spare your feelings, akin to how stakeholders may provide overly positive feedback about an idea so as not to hurt the innovator's feelings or appear rude. This guide emphasizes the importance of properly framing interview questions to elicit truthful and useful responses rather than kind but misleading affirmations. Fitzpatrick lays down three core rules:

1. Talk about their life instead of your idea
2. Ask about specifics in the past instead of generics or opinions about the future
3. Talk less and listen more [15]

These rules align with those of the Biodesign process and fundamentally encourage innovators to focus on stakeholders' experiences and challenges rather than seeking validation for their ideas.

Consider a hypothetical SleepSound Monitor, a pediatric medical device designed to monitor and analyze children's sleep patterns to help diagnose sleep disorders at home. Table 1 contains examples of biased and preferred questions.

When developing pediatric medical devices, it is essential to apply *The Mom Test* rules. Given the financial, ethical, clinical, and anatomical barriers that pediatric device innovators face, accurately identifying a problem and understanding stakeholders' needs are crucial to avoid additional challenges and mitigate preventable harm. By adhering to Fitzpatrick's rules during stakeholder interviews, innovators can glean valuable insights, making it more likely that their solution addresses a real problem and meets a market need, significantly reducing the risk of wasting resources.

Gathering Evidence: I-Corp Interviews

In 2011, the National Science Foundation (NSF) launched the I-Corps program, which trains teams of scientists and engineers in determining the commercial potential of technologies developed in university laboratories. Since 2012, more than 2500 teams have participated, nearly 1400 of which have launched startups that have raised a cumulative $3.16 billion in funding [16]. The I-Corps program has four goals:

Table 1 Examples of biased and preferred questions for a hypothetical interview between the innovator of a novel SleepSound Monitor and a potential customer

Related rule	Question type	Question	Likely answer	Interviewer conclusion
1	Biased	Do you think a device that monitors your child's sleep would be useful to you?	Yes, that sounds helpful; I always wonder if my child is getting adequate sleep or if something else is going on	I'm onto something! This seems like a real problem for parents
	Preferred	Have you had concerns about your child's sleep patterns? *[Does not lead the interviewee or describe the proposed solution]*	Last year, I thought this a bit because of his snoring, but he was just sick, and the issue resolved on its own. I didn't think it was a cause for concern. Should I be worried?	This parent does not have concerns about their child's sleep patterns and may not consider it a significant problem
2	Biased	Our team is considering developing a monitor that tracks your child's sleep patterns. Would you buy this device for your child in the future?	Yes, I would. It seems like it could offer me some peace of mind	We have a customer lined up!
	Preferred	Have you ever sought out professional advice or purchased solutions for your child's sleep issues in the past? *[Asks about a specific and tangible past behavior]*	No, not that I can recall. My child usually sleeps well	Without any specific purchasing behavior in the past, it is not apparent that this parent is a potential customer

1. Train an entrepreneurial workforce to bring I-Corps knowledge to their industries and academic institutions.
2. Bring cutting-edge technologies to market that are likely to have a positive economic and social impact.
3. Leverage federal research investments by advancing the commercialization of research outcomes.
4. Increase the economic impact of federally funded research [17].

Besides on-site learning and lectures, the I-Corps program helps participants identify stakeholders and potential customers to understand their specific innovation landscape. By the end of the experience, teams are expected to have engaged in at least one hundred face-to-face interviews with potential customers and partners [18].

For example, Pediametrix, an I-Corps graduate, is a pediatric startup that developed SoftSpot, the first and only FDA-cleared mobile application that measures infant skull shape using Artificial Intelligence (AI) and computer vision (Fig. 2). Through their engagement in I-Corps and numerous potential customer interviews,

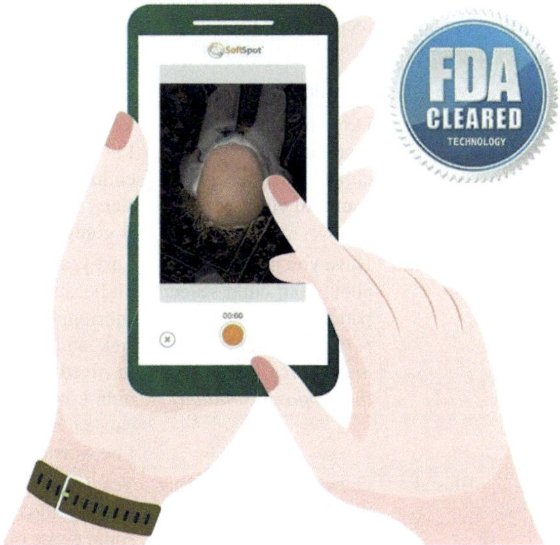

Fig. 2 A visualization of the SoftSpot mobile application. (Image source: https://softspot.ai/)

they identified that both parents and pediatricians had a need to measure infants' head shapes objectively at the point of care and thereby designed a solution that uses smartphones to assess skull deformities such as plagiocephaly. Traditionally, these evaluations were done by visual inspection because objective tools were unavailable. Because these evaluations are provider-dependent, there exists the possibility of provider error or missed diagnoses. In response to this need, Pediametrix created SoftSpot, an application that objectively evaluates pediatric cranial shape to detect deformities and pathology and hopes to improve diagnostic yield. In addition, interviews conducted during their I-Corps participation served as opportunities to meet potential clinical and business advisors and partners [19]. This example underscores the value of customer-focused research in driving meaningful pediatric medical device innovation.

Gathering Evidence: Customer and End User Surveys

Another needs validation method is customer or end-user surveys, which can generate robust quantities of information about a large audience for relatively low financial and temporal costs. However, it is important not to rely solely on surveys to evaluate the behavior and viewpoints of potential customers and end users. Word selection for their question-and-answer design is powerful because it can directly impose bias, making responses susceptible to misinterpretation and assumptions. Respondents may rush through surveys and misunderstand or only answer certain questions, resulting in misleading data.

Moreover, it is difficult for respondents to express pain points in writing or through multiple-choice questions, limiting the specificity of their responses. Surveys can be used to generate quantitative insights, but they cannot provide a comprehensive understanding of a problem. For instance, hypothetically, a survey about challenges with using CPAP (continuous positive airway pressure) machines for the management of sleep apnea might show that 60% of respondents lack a bedside table for their CPAP machine or 90% of respondents rate their CPAP comfort as 3 out of 5. However, understanding the challenges associated with not having a bedside table or understanding why respondents rated their comfort as 3 out of 5 requires more information than surveys can generate. Thus, it is important to complement survey data with interviews to understand a customer or end user's perspective fully.

Market Analysis

Market analyses are conducted to understand the competitive landscape for a device and, ultimately, to create a value proposition for it based on its commercial appeal. To conduct such an analysis, innovators should study existing solutions and their limitations to further identify unmet needs. It is not necessary for this analysis to be in-depth, but it should generate a rough estimate of the potential market. For example, innovators can determine an unmet need's healthcare costs or the number of possible customers facing it.

Unfortunately, the market for pediatric devices is not large because children constitute a relatively small and healthy proportion of the population when compared to older people, so pediatric device innovations do not frequently attract a significant amount of attention from investors. For instance, in the United States each year the number of in-hospital cardiac arrests among pediatric patients is 15,200, compared to 292,000 in adults [20]. This 20× difference in event frequency and market size explains why large companies innovating in this space have favored adult devices. In the event of cardiac arrest, chest compressions are required to pump blood to vital organs, but this can cause significant fatigue over time for the chest compressor. Stryker has attempted to address this need with the LUCAS chest compressor system [21]. The LUCAS device performs uniform chest compressions to a specified depth for adult patients, allowing for controlled and consistent chest compressions over time. No such device exists in pediatrics, and while it can be used off-label in children, there are considerable risks because the depth of compression required is less in small children and infants.

To overcome the challenge of small markets, innovators can pursue certain strategies to increase the potential return of devices, such as targeting niche sub-markets with few competitors or ensuring that a proposed device can be used in both children and adults. These strategies are further explored in the "Investment in and Returns on Pediatric Medical Device Research" section.

Needs Screening and Selection

During needs screening, innovators systematically evaluate which identified needs are worth pursuing further. This process involves several key considerations that vary according to the problems in question. Many of these considerations overlap with those of needs validation. A few to consider are:

- **Clinical impact**: Assess the potential clinical significance of solutions, including how they might improve patient outcomes, reduce complications, increase quality of care, or fill a gap in existing treatments or diagnostic procedures.
- **Market viability**: Evaluate the market potential for solutions by examining various factors, such as target patient population size, current and future market trends, competition, and potential revenue.
- **Feasibility of development**: Consider the technical and logistical feasibility of developing solutions, such as by assessing technological challenges, resource requirements, and development timelines.
- **Stakeholder alignment**: Ensure that solutions align with the interests of key stakeholders, such as patients, healthcare providers, customers, and regulatory bodies.
- **Regulatory and reimbursement pathways**: Understand solutions' regulatory landscape and reimbursement potential.

A statement should then be drafted about the most promising needs that will drive the rest of the Biodesign process. This needs statement should be clear and concise and have three parts: the problem, the affected population, and a measurable outcome if the problem is solved. This statement should not discuss the solution. The need statement should follow this format:

A way to [address the problem] in [a specific population] to [achieve a desired outcome] [11].

It cannot be overemphasized how critical this phase is. Establishing a well-validated, pressing clinical need is a key task for medical device innovators. To construct a needs statement effectively, needs must be translated into measurable variables. These measurable variables are also called needs criteria and fit into two categories. Must-have criteria are essential to addressing the need. Nice-to-have criteria describe elements of a solution that are attractive but not imperative. For example, for a new diagnostic screening test, an example of must-have criteria would be a sensitivity of >97% to ensure a low risk of false negatives. A nice-to-have criterion for the same test could be setup and diagnosis process that takes less than 15 minutes. The needs identification research is used to set acceptable goals [8].

Some goals, such as comfort, are difficult to quantify and vary from patient to patient. In the case of comfort, examine factors that cause discomfort. For example, children's skin tends to be more sensitive than adults' skin and they can experience discomfort due to itchiness. The goal could then be "material used does not cause itchiness or skin irritation." Although this goal is not quantitative, it can be determined whether a solution qualifies or not. The Biodesign process is iterative, so innovators can return to their observations, customer interviews, and clinical research to identify more specific needs even after drafting an initial needs statement.

Invent: Concept Generation and Screening

This phase is used to develop solutions. It consists of two steps: concept generation and concept screening.

Concept generation involves generating as many potential solutions as possible during multiple brainstorming sessions without initially judging or prioritizing any of them. There are multiple brainstorming approaches the team can take. For example, teams can spend the first 5 min of sessions silently writing down as many ideas as possible. Then, each member shares their ideas and others write them down and discuss them to create more ideas. Another method is to generate them rapidly and write down the most creative solutions as possible for 10 min and then later review them. Regardless of the specific method used, these sessions should be energetic, fun, productive, inspiring, and non-judgmental.

Concept screening involves identifying the most feasible possible solutions, which can be screened by the presence of competing intellectual property, technical feasibility, potential regulatory pathways, or business models [11]. These filters are important in deciding on possible solutions, but they should never lead a team to revise their initial criteria developed during needs screening. This process is just as rigorous and important as needs identification and validation, so the same amount of time and effort should be employed. The goal of this process is to identify one or two solutions that can be developed into prototypes. As before, it is critical to continually engage key stakeholders, customers, and end users when screening possible solutions.

Ember Sleep, a startup conceived during the Texas Medical Center Biodesign Fellowship program, is a testament to the efficacy of rigorous solution screening. After exploring dozens of healthcare needs, the team settled on the problem of current approaches to the treatment of sleep apnea. The current standard of care for the non-surgical treatment of sleep apnea is the use of continuous positive airway pressure (CPAP) machines, which are often uncomfortable and have rates of noncompliance consistently >30% [22]. End-user and stakeholder interviews suggested that patients and clinicians desired a solution that improved end-user compliance that was comfortable and noninvasive. This team evaluated various solution concepts modifying the delivery of positive airway pressure, repositioning of the mandible, and other non-surgical procedures. Must-have criteria included a high degree of clinical efficacy, high rates of user comfort and compliance, a noninvasive approach, and low cost for widespread adoption. Ultimately, they selected the solution of noninvasive stimulation of pharyngeal tone, which met all must-have criteria. This decision was driven by the solution's technical feasibility, clear regulatory pathway, and patient and clinician interviews. It exemplifies the importance of thorough solution screening.

Intellectual Property Analysis

Intellectual Property (IP) analysis consists of screening concepts according to whether they are sufficiently different from already-patented solutions. In general, for a new solution to be sufficiently different and to be granted a patent, it must have utility and novelty and not be an obvious variation of an existing product. A patent grants an exclusive right to produce a new product or a new way of producing an existing product [23]. Once patents are obtained, innovators are granted freedom to operate, which is the ability to make, use, or sell a product without infringing upon other patents or having legal liabilities to third parties.

To determine whether innovators would have the freedom to operate for a new solution, they must analyze related patents to determine whether their proposed solution would infringe on existing IP. There are many tools available online that innovators can use to do a reasonable review of the IP landscape. While applying for a patent requires IP expertise and should be done with experienced counsel, at this stage innovators only have to determine whether their proposed solution would infringe on other patents. To perform a basic search, innovators should first brainstorm alternative terms for or ways of referring to their concept device. For example, sphygmomanometer, blood pressure cuff, and blood pressure gauge could all refer to devices intended to estimate arterial blood pressure. Next, they can use a public patent database such as the United States Patent and Trademark Office's (USPTO) Public Search (https://ppubs.uspto.gov/pubwebapp/static/pages/landing.html) or Google patents (https://patents.google.com) to search for patents of similar devices. Innovators should conduct detailed reviews of these similar patents and applications, including their drawings, specifications, and claims, to compare to the device concept in question. The front page of a patent also includes cited references, which can be very helpful for innovators to review and understand the landscape of similar technology.

Regulatory Affairs

A solution's regulatory pathway is a mechanism that allows a medical device to be used clinically and includes the clinical trials that are needed before being marketable. These trials may be burdensome and increase the financial investment required to get a product to market. Solutions with burdensome regulatory pathways but small final market sizes may be screened out because it would be difficult to obtain sufficient early investment without a large windfall in earnings once the product is used clinically. Early regulatory pathway identification and discussions with the FDA are essential to reducing the occurrence of unexpected hurdles. A more detailed explanation of regulatory pathways is given in the "Regulatory Pathways and Clinical Evidence Requirements for Pediatric Medical Devices" section.

Business Model Viability

Ultimately, a concept without a viable business model is doomed to fail. A business model describes how a venture generates revenue or captures market value and considers elements such as the value a product or service offers, the target customer, the size of the market, and a reasonable market share that can be obtained. Healthcare business models can be complicated because end users, customers, and beneficiaries of a solution may all be different. For example, in a simple business-to-consumer model where a business is selling orthotic shoes, the customer, end-user, and beneficiary are the same person—i.e., the person who pays for the goods is the same one who uses it and benefits from it. In healthcare, it can be much more complex. For example, with a new medical device used in the inpatient setting, hospitals (and insurance companies) are the customers, clinicians are the end users, and patients are the beneficiaries.

Evaluating the overall size of the market and the degree of competition is key. Both tasks can be challenging, especially in the early stages of an innovation. Estimating market size is usually best done by evaluating the number of relevant patient encounters per year in the target market, usually the United States. This data may occasionally be difficult to locate outside of paid market research companies but may be available in peer-reviewed publications on PubMed in reviews of the epidemiology and incidence of a given medical condition. Innovators should also evaluate the degree of market saturation, the success of similar products, and the magnitude of competition. Understanding who the major players are in each market is critical. If there are dominant device companies, an innovator must be able to define why a customer would switch from their current supplier to a new one, which often comes with considerable costs in changing practice. Intense competition could indicate that it is better not to proceed with development.

Business models must also account for reimbursement strategy, which is where a third party, such as an insurance company, covers some portion of a patient's costs. Here, it is critical to understand healthcare payment systems. In-depth reimbursement analysis requires extensive research, but an important early step is identifying how existing technologies or procedures are reimbursed. Solutions for which the primary principles of reimbursement (coding, coverage, and payment) are provided are more likely to be used, so the solution should be described using similar language as existing reimbursable solutions. Doing so requires a thorough understanding of where the solution will be used clinically (inpatient, emergency department, outpatient) and the corresponding diagnoses, procedures, and payment codes [11]. For example, for an innovator developing a new pediatric otoscope to diagnose ear infections in the office setting, it would be important to understand Current Procedural Terminology (CPT) codes such as CPT 92504—Otolaryngologic and Binocular Microscopy Procedures, which describes a diagnostic procedure where the provider uses a microscope to examine the ear (or nose) of a patient [24]. Understanding how visits are reimbursed for common conditions where an otoscope is used would also be important, such as ICD-10 codes related to otitis media.

Technical Feasibility

Technical feasibility is assessed through prototyping. A prototype is a representation of a design idea, regardless of the medium. Prototypes are used for verification and validation testing and the feedback from such testing is incorporated into the next wave of device development. Innovators should evaluate the design and functionality of prototypes as soon as possible to eliminate risks. Prototyping is iterative and prototypes are developed in low-, medium-, and high-fidelity stages (Table 2) [25].

Rapid prototyping is an especially critical part of pediatric innovation because pediatric devices tend to have greater physical constraints than adult devices with smaller physical dimensions and should accommodate changing anatomy. A valuable prototyping tool is 3D printing, which is an inexpensive, effective, and fast additive manufacturing method. 3D printers use computer-aided design (CAD) models, which are 3D digital drawings of prototypes used to guide printers in prototype construction layer by layer. There are many types of 3D printers with varying durability, speed, parts, and input materials, so they are used in various industries and create a wide range of objects, from toys to entire organs.

3D printing is being used in pediatric specialties, such as surgery, dentistry, drug delivery, orthotics and prosthetics development, organ and tissue treatment, ventilation mask production, and robot and manipulator interactive interface production [26]. 3D printers can create objects tailored to patients' bodies [26]. For example, a team at the University of Michigan utilized 3D printing to treat pediatric airway disorders. Pediatric tracheobronchomalacia is a condition in which airways collapse excessively during respiration. One common treatment is tracheostomy tube placement with prolonged mechanical ventilation, but tracheal stenting can result in recurring airway obstruction because airway stents frequently fail due to their inability to accommodate children's airway growth. Therefore, the team designed a personalized bioresorbable medical device using an image-based computer-aided

Table 2 A description of low-, medium-, and high-fidelity prototyping (modified from Niinikoski et al. [25])

Prototype fidelity	Description	Sample tools and materials used
Low	Early-stage prototypes with limited details and functionality that vary in terms of how closely they resemble the final product and are used for initial testing and feedback	Low-cost everyday materials such as paper, cardboard, glue, coffee filters, and buttons
Medium	Early-stage prototypes with more details and realistic functionality that more closely resemble the final product and are used for testing and user evaluation	3D printers, CNC machines, laser cutters, electronic equipment, fabric
High	Late-stage prototypes with more interactive and advanced features that closely resemble the final product in terms of functionality and appearance and are used for thorough user testing and user evaluation	Plastic, rubber, metal, printable biomaterials, and electronic components

design and manufactured it using a 3D printer. The device accommodated airway growth and prevented both external compression and internal collapse [27].

Implement: Clinical Testing and FDA Approval

After thorough preliminary research during the Identify and Invent stages, the leading one or two concepts have been vetted and refined. During the implementation phase, a strategy for advancing each concept is developed, encompassing a grasp of the IP landscape, a plausible strategy for reimbursement, and insights into the technical viability of the product. A protocol for testing the device is established, which may include preclinical or large animal studies, clinical trials, and quality control measures. A robust business model is created that factors in sales channels, economic projections, investment approaches (including venture capital, industry sponsorship, and government funding), and a marketing and stakeholder engagement plan, centering on a compelling value proposition. Establishing a competitive edge is fundamental to a successful market entry, and exploring various avenues for bringing the product to market is also crucial. While we will not deeply discuss these strategies in this section, we will discuss clinical and FDA testing and pediatric innovation-specific regulatory requirements [28].

Pediatric Medical Device Definition and Classification

Medical devices are regulated by the FDA Center for Devices and Radiological Health (CDRH). It defines a medical device as an item recognized in official medical references intended for diagnosing, treating, or preventing disease or affecting bodily structure or function without achieving its primary purposes through chemical action or metabolism (Fig. 3) [29]. Of note, software intended to be used for medical purposes that is not integrated into hardware is defined as Software as a Medical Device (SaMD) and is subject to FDA regulatory oversight [30]. The FDA defines pediatric medical devices as those that treat or diagnose diseases and conditions from birth until, but not including, the 22nd birthday [31].

Medical devices are classified as Class I, II, or III depending on risks associated with their use and the regulatory requirements necessary to determine their safety and effectiveness. In general, lower-risk devices require less clinical evidence before clearance. Class I devices represent ~47% of all medical devices and are the lowest risk category. As such, 95% of Class I devices are exempt from regulatory processes. Examples include tongue depressors, gauze, elastic bandages, or enema kits. Class II devices have a moderate risk of harm and include devices like needles, syringes, wheelchairs, or pregnancy test kits. Class III devices are the highest risk and usually sustain or support life, are implanted, or present a potentially unreasonable risk of illness or injury with their use. Examples include pacemakers, breast

Per Section 201(h)(1) of the Food, Drug, and Cosmetic Act, a device is:
An instrument, apparatus, implement, machine, contrivance, implant, in vitro reagent, or other similar or related article, including a component part, or accessory which is:

- *(A) recognized in the official National Formulary, or the United States Pharmacopoeia, or any supplement to them,*
- *(B) intended for use in the diagnosis of disease or other conditions, or in the cure, mitigation, treatment, or prevention of disease, in man or other animals, or*
- *(C) intended to affect the structure or any function of the body of man or other animals, and which does not achieve its primary intended purposes through chemical action within or on the body of man or other animals and which is not dependent upon being metabolized for the achievement of its primary intended purposes. The term "device" does not include software functions excluded pursuant to section 520(o).*

Fig. 3 FDA's definition of a medical device

implants, and many implanted prosthetics. These devices have the highest burden of required clinical evidence [32, 33].

Basic FDA Regulatory Pathways

The FDA has three basic pathways for obtaining medical device marketing eligibility certification: 510(k) premarket notification (PMN), de novo requests, and premarket approval (PMA). The **Premarket Notification (PMN) pathway**, also called the 510(k) pathway, is generally for Class I or II devices where there is a legally marketed device that is substantially equivalent to the proposed device in question. Substantial equivalence of a predicate does not mean that the device must be identical but must show similarity regarding intended use, design, energy used or delivered, materials, performance, biocompatibility, or other applicable characteristics [34]. The testing burden of the 510(k) pathway, including benchtop, animal, and clinical evidence requirements, is significantly reduced if innovators can demonstrate that their device is substantially similar to an approved device and no additional safety risks have been identified.

De novo pathway requests are appropriate for Class I and II devices that cannot follow the 510(k) pathway because there is no identified substantially similar FDA-approved device. The goal of de novo classification is to avoid the stringent clinical evidence requirements for Class III devices to speed up regulatory approval.

The **premarket approval (PMA) pathway** is the most stringent and is required for Class III devices for which there is no identified substantially similar equivalent. PMA requires high levels of clinical evidence, generally consisting of the results of rigorous human, randomized, controlled trials. This requirement may be a major barrier to entry for innovative pediatric medical devices because it can be difficult

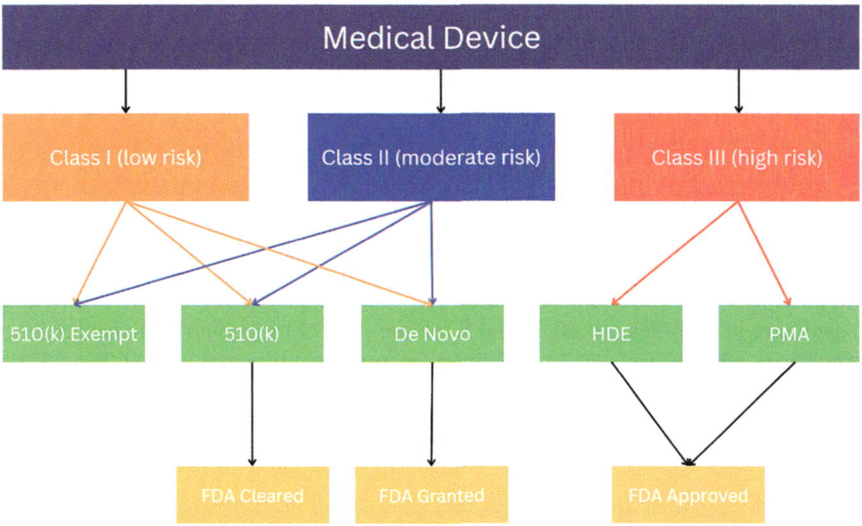

Fig. 4 An overview of medical device classifications and common pathways

to get a suitable number of participants, so the FDA developed the humanitarian use device (HUD) designation for devices that benefit patients with rare diseases or conditions, defined as those with 8000 or fewer children in the United States per year. HUD devices are eligible for the humanitarian device exemption (HDE) pathway, which allows for more lenient clinical trial enrollment practices and requirements for market approval (Fig. 4).

Alternative Pathways and Designations

A HUD is a medical device intended to benefit patients with a rare disease or condition, defined as affecting fewer than 8000 individuals in the United States per year. HUDs are eligible for the **Humanitarian Device Exemption (HDE) pathway**, which expedites the device's approval process if it meets certain safety and probable effectiveness criteria [9]. This pathway is particularly relevant for pediatric medical devices because rare diseases or conditions disproportionately affect children and children make up a smaller proportion of the population than adults, so pediatric medical devices have relatively high chances of qualifying as HUDs. In addition, given smaller markets and potentially lower returns, pediatric medical device innovators must pursue maximum efficiency, making expedited FDA regulatory approval critical.

The **Breakthrough Device Designation (BDD)** is awarded to devices that target life-threatening or irreversibly debilitating conditions and is intended to speed up review for PMA, PMN, or De Novo authorizations. It is reserved for devices that have breakthrough technology that could provide more effective treatment or help healthcare providers make more accurate diagnoses than existing solutions, where rapid availability of the technology is in the best interest of patients. Innovators can apply for BDD from the Center for Devices and Radiological Health and, if awarded, receive priority review and additional attention from the FDA to facilitate regulatory approval [35].

The **Safer Technologies Program (STeP)** expedites the assessment and review of medical devices reasonably expected to significantly improve safety or performance of currently available treatments or diagnostics but that target a disease or condition that is less severe than those targeted with the BDD. Through STeP, innovators can obtain FDA expert feedback during the premarket review phase and expedite regulatory review. BDD and STeP are critical for pediatric medical device innovators who often face limited financial runways compared to adult medical device innovators [36].

Real-World Evidence

Pediatric medical devices used in clinical practice are frequently only officially approved for adult indications and are used off-label in children by necessity. Sufficient evidence has been generated through clinical research to support FDA approval in adults, but insufficient evidence exists in children. In certain specialties, like pediatric interventional cardiology, as many as 63% of devices are used off-label in situations where no suitable approved pediatric devices exist [37]. In the course of using these devices in usual clinical care for children, data is generated about the effects they have on patients, which could be useful in determining their safety and efficacy.

Real-World Data (RWD) describes data relating to patient health status or healthcare that is routinely collected from a variety of sources, such as electronic health records, insurance claims, and patient registries. Real-World Evidence (RWE) refers to the clinical evidence about usage and potential benefits, i.e., safety and efficacy of products currently being used and generating RWD. Historically, the FDA has used RWD to monitor and evaluate the post-market safety of approved drugs and devices. With ever-expanding access to clinical data, and because so many adult devices are used off-label in pediatrics, the FDA has recognized that RWE can be used to establish safety and efficacy outside of the usual FDA regulatory pathways, which usually require formal clinical trials [38]. In addition, several FDA-funded Pediatric Device Consortia (PDCs) [See "Organizations Dedicated to Advancing Pediatric Innovation" for further detail on PDCs] are advancing initiatives to establish frameworks for collecting RWE about pediatric devices. RWE is vital to understanding pediatric devices' potential risks and benefits due to the difficulty in executing large-scale pediatric clinical trials.

Q-Submission Program: The Importance of Early and Frequent Engagement with the FDA

The **Q-Submission (Q-Sub) Program** offers an opportunity for medical device innovators to engage with the FDA review team during the product development process. The most common type of meeting is the Pre-Submission (Pre-Sub) meeting, which permits companies to receive guidance prior to an official premarket submission. These meetings are critical as they offer a chance for innovators to ask questions directly to the FDA review team. These meetings usually occur later in the design process, but as soon as a company is near to a design freeze on a device, that is, that there are no further significant modifications planned, it is best to engage with the Q-Sub program. Pre-Sub meetings, while technically voluntary, can shorten total review times and expedite regulatory pathways, build a working relationship with the FDA, and confirm FDA expectations for the device to ensure that the company has chosen the appropriate pathway and level of evidence required for approval. As such, they are a critical component of a successful and efficient regulatory approval process.

AdvaMed offers an effective list of best practices when preparing for a Pre-Sub meeting with the FDA, summarized here [39]

1. Adhere to FDA guidelines while preparing the Pre-Sub package and include all required information that could aid the FDA in addressing your questions.
2. Formulate your questions for the FDA thoughtfully, aiming for clarity and completeness. Ideally, limit questions to 3–5 per meeting, favoring multiple meetings over an overcrowded single session.
3. Approach the Pre-Sub meeting with the same rigor as any other FDA application. Ensure it is well-organized, clear, and logical.
4. Be aware that this first contact with the FDA is an opportunity to make a positive or negative impression!
5. If a pre-scheduled meeting becomes unnecessary due to clear written feedback from the FDA, consider holding it to save time and resources.
6. Meticulously document the Pre-Sub meetings. Record the FDA's commitments and statements accurately and submit the draft meeting minutes within 15 days post-meeting.
7. Set clear objectives for each Pre-Sub meeting, directly linked to the questions posed in your application.
8. Allocate sufficient time for discussion and questions during each meeting.
9. Use Pre-Sub meetings as an opportunity to build a positive and effective working relationship with the FDA review team.

Table 3 FDA fees for fiscal year 2024 (October 1, 2023 through September 30, 2024)

Application type	Standard fee	Small business fee[a]
510(k)[b]	$21,760	$5440
513(g)	$6528	$3264
PMA, PDP, PMR, BLA	$483,560	$120,890
De Novo classification request	$145,068	$36,267
Panel-track supplement	$386,848	$96,712
180-Day supplement	$72,534	$18,134
Real-time supplement	$33,849	$8462
BLA efficacy supplement	$483,560	$120,890
30-day notice	$7737	$3869
Annual fee for periodic reporting on a class III device (PMAs, PDPs, and PMRs)	$16,925	$4231

[a] Small Business Fee: For businesses certified by the Center for Devices and Radiological Health (CDRH) as a small business
[b] 510(k) Fees: All types of 510(k)s (Traditional, Abbreviated, and Special) are subject to the user fee. However, there is no user fee for 510(k)s submitted to the FDA on behalf of an FDA-accredited third-party reviewer

FDA Fees

FDA fees can be burdensome for early-stage companies with limited funding and financial runway (Tab. 3). To encourage pediatric innovation, the FDA offers fee waivers for devices intended solely for pediatric use. Specifically, submissions for 510(k), De Novo, and PMA are exempt from user fees. If a device's intended use is changed from pediatric to adult, a new submission is required and will incur user fees. No waivers are available for 513(g) submissions, which are requests for information about the regulatory classification of a device (Table 3) [40].

Investment in and Returns on Pediatric Medical Device Research

A 2010 survey of more than 200 companies showed that the mean cost to bring a medical device from concept to FDA clearance was $31M for low-to-moderate-risk 510(k) devices and $94M for higher-risk devices requiring PMA [41]. These high costs make clear that it would be nearly impossible for most medical devices to make it to market using public grant funding alone. To attract private investment, device companies must show that they will allow investors to achieve a certain rate of return on their investments if they can achieve sufficient market capture.

This dynamic presents a considerable problem for pediatric medical device innovators. Healthcare dollars spent in pediatrics account for only 8.4% of healthcare spending in the United States. This fact, along with challenges already discussed such as difficulty with clinical trial enrollment, ethical concerns, and unique engineering constraints, leads to hesitation to invest in pediatric devices [42]. Pediatric innovators often utilize one of three justifications to convince investors that an invention will provide a sufficient return: the pediatric sub-market being targeted is sufficiently large, the device could achieve an overwhelming market share of a small pediatric market, or the device could be used for both children and adults.

Limbix

The larger pediatric sub-markets, such as for well-newborn care, attention-deficit/hyperactivity disorder (ADHD), and mental health treatment, are all reasonably sized in and of themselves. For example, the adolescent mental healthcare market is projected to reach $26 billion by 2027 [43]. As such, even achieving a modest market share could result in significant returns for investors. Limbix is a company that created SparkRx, which delivers cognitive behavioral therapy to teens through a mobile application. Limbix secured significant funding of over $30 million as they were targeting a large pediatric market, identified a real clinical need, and developed an effective solution to meet that need. The company's focus on adolescent depression aligns with the growing recognition of mental health issues among children. In 2023, it was acquired by Big Health, a leader in digital mental health therapeutics. Limbix's experience shows that identifying a real need and an effective solution for a large pediatric sub-market can enable effective fundraising and the building of a successful business.

Tueo Health

Tueo Health developed sensor-based technology that could monitor and manage asthma symptoms overnight, which is a valuable tool for both parents and healthcare providers. Their founder noted, "When we looked at the competitive landscape, we found plenty of solutions but none that were addressing the core problems parents face. Families want to know if their child's asthma is under control and what to do when it's not. There was no good marker of asthma control, and if you don't have that, all other solutions are just Band-Aids." Although the pediatric asthma treatment market was $11.2 billion in 2022, Tueo Health's focus on a niche problem with no major solutions gave them a significant advantage [44]. Apple acquired Tueo in 2019. This acquisition validates the significance of Tueo Health's innovation and underscores the potential for pediatric medical device companies to succeed even within smaller market segments.

NeuroPace

NeuroPace is a company that treats epilepsy in both children and adults. Its Responsive Neurostimulation System treats idiopathic generalized and other types of epilepsy, such as focal epilepsy, and Lennox-Gastaut Syndrome, in those 12 years old and older. Lennox-Gastaut Syndrome is a severe form of epilepsy that is characterized by repeated seizures, cognitive impairment, and abnormal electrical brain activity [45]. The Responsive Neurostimulation System continuously monitors brain activity, recognizes unique seizure patterns, and responds to prevent seizures before they start. Despite the pediatric epilepsy market's relatively small size, projected to be $1.5 billion by 2026 when combined with adults, the total market is nearly >$10 billion, which makes it a more attractive target for investors [46, 47]. As such, NeuroPace has managed to attract significant investment, securing $67 million in funding as of 2020. Their success highlights the potential for success by developing innovations that cater to both child and adult populations.

The investment histories of Limbix, NeuroPace, and Tueo Health reflect three strategies that pediatric device innovators can use to secure private funding, which has become critical for successful device development: target a sufficiently large pediatric sub-market, dominate a niche within a smaller pediatric sub-market, or create devices applicable to both children and adults. Their successes exemplify how well-articulated value propositions coupled with strategic market approaches can entice investment and accelerate the transition from concept to clinic, ultimately advancing pediatric healthcare.

De-risking Pediatric Devices Prior to Private Investment

Academic medical centers play a vital role in developing pediatric medical devices, particularly given their potentially limited returns. In addition to the three strategies discussed in the previous section, another strategy for attracting private investment is making pediatric devices less risky than their adult counterparts. Academic centers are crucial for leveraging public grants and non-dilutive funding sources and allowing for further product development. Since they often have access to significant research, legal, clinical, and commercialization resources, the longer that innovation can stay within the walls of an academic institution, or the more that a small business can partner with such an institution, the better.

Making use of public funding early in the development of an innovation can be very helpful for innovators as they can generate research that builds technical credibility for their idea without giving up equity of their future company. When originating within an academic institution, many early discoveries are made by traditional academic National Institutes of Health (NIH) funding routes through K- or R- programs for academic clinical researchers. One critical mechanism that innovators should be aware of is the R41/42 and R43/44 grants, which represent the Small

Business Technology Transfer (STTR) and Small Business Innovation Research (SBIR) programs, respectively [48]. These grants offer non-dilutive government funding to small businesses to support research and development. These programs provide funding in distinct phases. In Phase I, ~$300,000 is provided over 6 months to 2 years to assess the technical merit, feasibility, and commercial potential of the business's proposed research and development efforts. Phase II provides up to ~$2,000,000 over one to three years based on the business's scientific merit and commercial potential as determined in Phase I. Pediatric device innovators should strongly consider seeking out this funding [49].

Current Trends in Pediatric Medical Device Innovation

We highlight four notable trends in medical device innovation that will have a dramatic impact on pediatrics in the coming years: artificial intelligence, regenerative medicine, mental health, and wearable and home diagnostics.

Artificial Intelligence

Artificial intelligence (AI) generally refers to computers performing tasks that typically would require human intelligence, including learning from experience, reasoning, and problem-solving. AI has been increasingly applied in medicine with a diverse array of use cases over the last two decades, including applications in drug discovery, genomics, clinical decision support, and more. Figure 5 shows the explosion of AI-related publications on PubMed over the last 20 years. Pathology and radiology have seen an explosion of exploration of AI-assisted image recognition

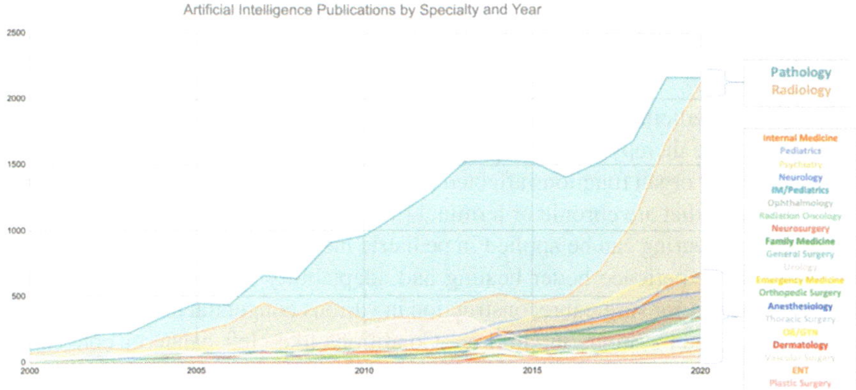

Fig. 5 Overall trend in AI publications from 2000-2020 (reproduced from Perera et al. [50])

and diagnosis. Pediatrics was notably the specialty with the fourth most AI-related publications in 2020 [22].

Computer Vision (CV) is an interdisciplinary field of AI/ML (machine learning) that enables computers to derive information from images and videos, allowing devices to gain meaningful information from their environment. It has driven much of the research in radiology and pathology seen in Fig. 5. There are dozens of companies attempting to convert these research findings into products that can aid in diagnosis of radiological images, pathology slides, or pictures of dermatologic conditions [51–53]. A prominent example of the application of this type of AI technology is Google Lens, a visual search tool that allows you to search based on images taken with a phone's camera. In a recent release, Google demonstrates the application's ability to search skin conditions that are visual matches to what one sees on their own or their child's skin. Although many of these tools are at the stage of demonstrating technical feasibility, they will likely be implemented into clinical workflows in the coming years as performance continues to evolve [54].

AI-driven diagnostics and therapeutics are frequently regulated by the FDA under the umbrella SaMD, which refers to software and applications used for medical purposes that are not part of a medical device's hardware. In 2021, the FDA released an action plan for AI regulation. In it, they note that the traditional PMN, de novo, and PMA regulatory pathways were not intended to evaluate adaptive AI technologies, especially where software is used to enhance or modify the function of existing medical devices. Most AI algorithms function differently over time, usually with better performance as more data is accrued. The FDA has suggested a potential framework where device innovators would submit anticipated modifications and associated methodology being used to implement these changes in a controlled manner. Through transparency and real-world performance monitoring, the FDA could evaluate and monitor a software product from premarket development to post-market performance. This would allow it to embrace the iterative improvement of AI-based software and still maintain patient safety [55].

Regenerative Medicine

Regenerative medicine includes disciplines such as stem cell biology, tissue engineering and gene therapy [56]. The purpose of regenerative medicine is to restore or repair tissue and organ functions affected by disease, injury, or aging to manage and cure conditions that are chronic or terminal [57].

Tissue engineering can be applied in pediatric medicine especially because children typically experience better healing and adaptability than adults. Dunn et al. describe the need for surgical reconstruction in various congenital anomalies affecting different organ systems in children. Conditions like cleft palate or congenital heart disease benefit from tissue regeneration [58]. However, a common challenge among pediatric patients is accommodating a child's changing anatomy and continuous development, and the same challenge applies when designing tissue-engineered organs. To address this challenge, studies are conducted on the use of 3D bioprinting to be used to fabricate tissue constructs such as self-expandable and biodegradable polymer stents with the ability to grow with the patient [59]. Exemplifying this innovative approach, Renata Medical, founded by Dustin Armer and Eason Abbott innovated the first patented growth stent that can be delivered in an extremely low-profile system in infants while being re-expanded over the course of a lifetime to adult vessel sizes. This eliminates the need for multiple open-heart surgeries.

The FDA has released guidelines that outline its approach to facilitating and speeding up the development of regenerative medicine products, which encompass human cells, tissues, and other related products. The documents are "Regulatory Considerations for Human Cell, Tissues, and Cellular and Tissue-Based Products: Minimal Manipulation and Homologous Use" and "Same Surgical Procedure Exception: Questions and Answers Regarding the Scope of the Exception." There are also documents to outline how to bring innovative, safe, and effective products to patients as efficiently as possible. These are "Expedited Programs for Regenerative Medicine Therapies for Serious Conditions" and "Evaluation of Devices Used with Regenerative Medicine Advanced Therapies." These guidelines underscore FDA's dedication to facilitating the introduction of novel and innovative treatment options for patients [60].

While regenerative medicine is an exciting field, there is still a need for further research due to its complexity. However, the future of regenerative science proves to be optimistic and can lead to groundbreaking treatments for children in need.

Mental Health

The National Research Corporation (NRC) Health's patient feedback data finds that mental health hospitalizations of children have increased by 163% from 2020 to 2021. In early 2021, emergency department visits in the United States for suspected suicide attempts were 51% higher for adolescent girls and 4% higher for adolescent boys compared to the same time in early 2019 [61]. Pediatric mental health crisis is a top concern, especially after the COVID-19 pandemic as pediatric emergency departments (EDs) are overwhelmed by such urgent visits [62].

The Centers for Disease Control and Prevention (CDC), which is the national public health agency of the United States, is committed to addressing this issue and optimizing access to mental health care resources. Examples include improving strategies to connect families to mental health care resources, working to understand the gaps in mental health care for pediatrics, learning about funding opportunities for increased mental health care, and understanding social determinants of health so all families can get access to healthcare [63]. Social determinants of health refer to environmental conditions that can have a major impact on people's health, well-being, and quality of life, encompassing factors such as racism, violence, education, housing, air quality, and water access [64]. Utilizing the research and resources provided by the CDC, tech startups are serving to fill this unmet need.

For example, a company mentioned previously, Limbix, provides mental health treatment to young people. Its solution, SparkRx is a prescription-only digital therapeutic intended to provide neurobehavioral intervention to patients 13–21 years of age to treat depression and is intended to be an adjunct to clinician-supervised outpatient treatment. Limbix is introducing accessible interventions for adolescents, leveraging smartphones and at-home completion to mitigate the high costs linked to in-person psychotherapies and delays in treatment access.

Organizations and startups focusing on pediatric mental health problems are playing a crucial role in addressing this critical aspect of children's well-being. Their efforts contribute significantly to reducing the stigma around pediatric mental health, and further progress in this area will allow early intervention, support, and resources for children facing mental health challenges.

Wearable Devices and At-Home Diagnostics

Another emerging trend in pediatric medical devices is the proliferation of wearable sensors and home diagnostics. This evolution reflects a general movement in healthcare, accelerated by the COVID-19 pandemic, toward diagnosing and treating patients outside of doctors' offices and hospitals via telehealth. Dozens of companies now market wearable devices to parents to be used to monitor their children at home. One positive example of applying these wearables is in pediatric electrophysiology. A pediatric cardiology group at Washington University has described the effective utilization of the Apple Watch and KardiaMobile applications to monitor patients for arrhythmias at home. While these devices were used effectively to assess patient heart rate and rhythm, they note that there are significant improvements in reimbursement, workflow, and data integration that are needed for commercial devices to be consistently integrated to the benefit of patients [65].

However, this trend also presents challenges, particularly in the realm of regulation and safety. Over the last decade, a new class of infant physiologic monitors marketed directly to parents for home use has emerged, including sensors built into socks, onesies, and diaper clips, that monitor pulse, respiratory rate, and pulse oximetry. They have been the center of significant controversy among pediatricians. The official American Academy of Pediatrics (AAP) statement has suggested against the use of home physiologic monitoring to prevent sudden infant death syndrome (SIDS), as home monitors have not improved outcomes and may increase misdiagnosis of healthy children and increase parental anxiety. Moreover, there is wide variation in published data, claims made, and accuracy of these devices [66].

The Owlet, for example, is one of the most popular home monitoring brands. There was significant controversy with the FDA as the Owlet smart sock was being marketed to parents without prior FDA approval. The FDA expressed concerns that these devices did not meet the criteria as low-risk, general wellness products and required FDA regulation. In a letter to Owlet Baby Care in October of 2021, the FDA advised ceasing activities related to their smart socks [67]. After collecting appropriate clinical data, Owlet Baby Care did receive FDA clearance in June of 2023 for its BabySat device, a home monitoring device that measures an infant's heart rate and oxygen saturation level and requires a prescription from a physician. In November of 2023, Owlet also received FDA clearance for the Dream Sock, which is the first and only home pulse oximeter and heart rate monitor that is available without prescription [68]. The use of home physiologic monitoring is a critical area of interest in pediatric devices and will shape the delivery of pediatric care in the years to come.

Organizations Dedicated to Advancing Pediatric Innovation

We have already reviewed the many challenges of bringing innovative pediatric devices to market. Therefore, this section highlights organizations that advance pediatric healthcare innovation by providing much-needed assistance to pediatric device innovators.

Pediatric Device Consortia (PDC)

Since 2009, the FDA has funded several PDCs that offer vital support to pediatric device innovators. These nonprofit consortia provide regulatory, business planning, and device development services, including intellectual property advising, prototyping, engineering, laboratory testing, and grant writing to pediatric medical device innovators. They also provide direct device funding to innovators in a non-dilutive manner with no strings attached. The current PDCs for the 2023–2028 cycle are the Southwest-Midwest National Pediatric Device Innovation Consortium, the UCSF-Stanford Pediatric Device Consortium, the Consortium for Technology and Innovation in Pediatrics: CTIP 3.0, the Alliance for Pediatric Device Innovation, and the Midwest PDC [69].

CobiCure

Launched by Advancium Health Network, CobiCure is a nonprofit organization focused on curing rare, life-threatening pediatric diseases. It provides a range of support and services for the full device development lifecycle from ideation through commercialization. CobiCure is associated with Cure Innovation Labs, a 27,000-square-foot nonprofit incubator with cutting-edge technology, lab space, and programs to assist seed and early-stage growth life science companies [70]. As an initial investment, CobiCure provided $2 million to PolyVascular, an early-stage pediatric device company [71]. Polyvascular has created a transcatheter pulmonary valve for children as young as 2 years old. Their aim is to transform the care of children with congenital heart disease by developing an entirely new generation of valves made of medical-grade polymer devoid of any biological tissue. These valves can be placed via a small incision and insertion of a catheter, preventing the need for open heart surgery. It is made of a biocompatible material that reduces the risk of endocarditis that accompanies animal tissue valves. Overall, their system has the goal of reducing pediatric morbidity and extending the time between open heart surgeries [72].

KidsX

KidsX is an influential global pediatric healthcare ecosystem driven by a consortium of leading children's hospitals with the core mission of accelerating pediatric digital health innovation. KidsX fosters collaboration among healthcare organizations, digital health innovators, payors, investors, and businesses to address key challenges in pediatric care delivery. Many of their portfolio companies fall under the category of FDA-regulated products or qualify as SaMD. The KidsX accelerators and innovation programs help companies achieve product and business model validation in the pediatric market. A key feature is giving companies access to leading children's hospitals to validate product/market fit. Recent programs have focused on mental and behavioral health, staffing shortages, care coordination, health equity, and the patient experience [73].

International Society for Pediatric Innovation

The International Society for Pediatric Innovation (iSPI) is an organization dedicated to fostering innovation in pediatric healthcare with a focus on digital health, data informatics, and the application of AI. It serves as an open exchange for innovation leaders from global pediatric hospitals. iSPI facilitates collaboration among healthcare professionals through its various workgroups, each focusing on specific areas of need in pediatric healthcare. These include the iSPI Nurse Innovator Workgroup, which invites nurses interested in innovation to collaborate within the iSPI network, the XR Workgroup, focused on mixed media and XR (extended reality) technologies; and the Wearables Workgroup, aimed at advancing the use of wearable and noninvasive biosensors for children, particularly those with congenital diseases. In addition to these workgroups, iSPI hosts iSPI Spotlights, a series of virtual events discussing relevant topics in pediatric innovation [74].

Pediatric-Focused Venture Capital

Several pediatric-focused Venture Capital groups such as Springhood Ventures, Next Step Ventures, and Health Innovation Capital provide funding and support to startups and established companies developing novel pediatric medical devices, therapies, and digital health solutions. They are generally driven by the so-called "double bottom line" which refers to maximizing both financial returns and healthcare outcomes for children. By investing in pediatric-focused innovations, these firms help bridge the gap in a market often overshadowed by adult healthcare, facilitating the development and commercialization of pediatric innovations.

Johnson & Johnson Innovation

Johnson & Johnson Innovation is a global life science innovation network that provides startups with access to lab space and other resources, including expertise, communities, industry connections, and entrepreneurial programs [75]. Although there are multiple locations throughout the US, the Washington, DC location is dedicated to pediatric innovation, as reflected in its partnership with Children's National to open the Children's National Research & Innovation Campus. Johnson & Johnson Innovation gives innovators access to in-person and virtual residencies and the opportunity to participate in the QuickFire Challenge to receive awards, expertise, and residency.

Looking to the Future: Moving Toward a National Ecosystem for Pediatric Medical Devices

Despite the challenges previously described, the future of pediatric medical device innovation appears promising. The Foundation for the National Institutes of Health (FNIH) is leading a Pediatric Medical Devices Public-Private Partnership (PMD-PPP) with the goal of accelerating pediatric medical device innovation and addressing many of these challenges by coordinating national multi-stakeholder efforts. As background, a System of Hospitals for Innovation in Pediatrics—Medical Devices (SHIP-MD) framework was established in 2021, advocating for a national consortium of hospitals that early-stage devices could leverage to execute streamlined and efficient clinical trials, gain insights into reimbursement and de-risk their products [76].

The FNIH is building on this framework by bringing together resources from U.S. government agencies and private sector organizations, including industry and non-profits to create a comprehensive ecosystem that can allow pediatric device innovators to de-risk their innovations and accelerate clinical development. The

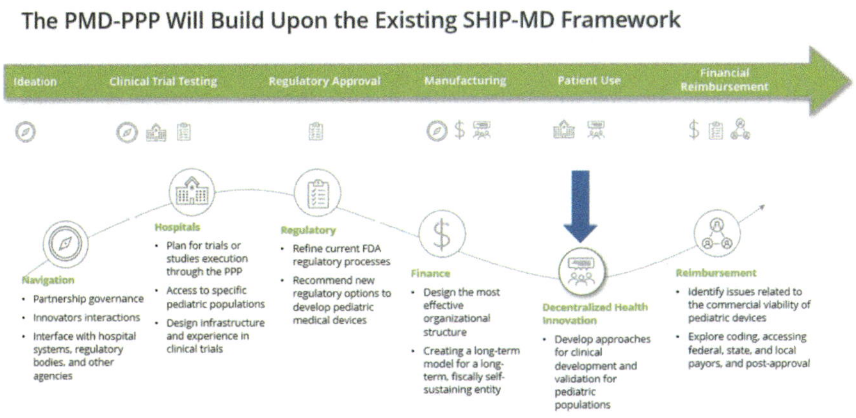

Fig. 6 Overview of the PMD-PPP model, which builds on the SHIP-MD framework [76]

design phase of the PMD-PPP project will last from 2023 to 2025 and will establish a governance structure based on the foundational elements developed during the SHIP-MD 2021 pre-consortium workshop. It will involve integrated workstreams guiding innovators through ideation, clinical trial testing, regulatory approval, manufacturing, patient use, and financial reimbursement (Fig. 6).

The overarching goal of the PMD-PPP is to launch a full-scale partnership by 2025, creating a national ecosystem that optimizes the development, evaluation, and approval of pediatric medical devices and addresses specific challenges directly. Through a network of partner pediatric hospitals, innovators will gain access to expert feedback during ideation and establish clinical partners who will be critical for completing clinical trials. With direct ties to public entities like the FDA, innovators will have early and direct access to ensure they are applying appropriate regulatory pathways. Finally, by partnering with private entities such as reimbursement experts and insurance companies, innovators will be able to make improved projections regarding their commercial potential. This initiative is expected to significantly improve access to medical device options designed and approved for the pediatric population and foster new collaborations in the emerging national ecosystem.

Conclusion

Pediatric device innovators face multifaceted challenges, including technical hurdles due to the small size and evolving anatomy and physiology of pediatric patients, financial constraints stemming from limited investment and small market sizes, clinical challenges in executing effective trials, and complex regulatory landscapes. There has been an ongoing effort to address these challenges by FDA including work by the Office of Orphan Product Development (OOPD) funding of the PDCs. There are also regulatory measures available such as the HUD, HDE, and BDD. In addition, public-private partnerships such as the PMD-PPP and Real-World Evidence further accelerate this progress.

Along with these promising initiatives, it is also essential to reduce disparities in pediatric research to drive the next generation of pediatric innovation leaders. For example, Speer et al. discuss the concentration of funding for pediatric research in a few select institutions, leading to a lack of diversity in research, which affects the pipeline of physician-scientists specializing in pediatrics. Limited support for early-career pediatric physician scientists, along with their career choices, has resulted in an aging workforce in pediatric research. Additionally, despite comparable enrollment in medical schools, women are underrepresented in top positions. Factors contributing to these disparities include implicit gender bias and institutional policies. However, there are continued efforts by organizations such as the National Pediatric Physician-Scientist Collaborative Workgroup, which provides mentorship and networking for physician-scientists, Coalition for Pediatric Medical Research, which trains a diverse next-generation of pediatric researchers, and the NIH Loan Repayment Program, which brings highly qualified health professionals into research careers [77].

Pediatric innovation has come a long way, but there is still a considerable journey to traverse. Pediatric patients deserve equal care, yet specific challenges in developing their devices can hinder innovation in this area. By emphasizing the current initiatives available to address these challenges and discussing ways to accelerate progress, pediatric innovation can prove to be very transformative. Continuous collaboration from government entities, industry leaders, stakeholders, pediatric physicians, and students is crucial to drive this field forward.

We hope that the topics reviewed in this chapter will help innovators in overcoming these challenges. Innovators should follow the Biodesign process in establishing true needs and generating effective concept solutions. By understanding general regulatory requirements and engaging early and often with the FDA, pediatric innovators can increase their chances of quick regulatory approval, which is a critical milestone in the lifecycle of any device company. Innovators should make use of the ever-expanding number of organizations that hope to help catalyze pediatric device innovation. Finally, as we look ahead, we are very excited and hopeful about the future of an integrated ecosystem for pediatric medical device innovation through the FNIH's PMD-PPP initiative. Working together, we see a bright future for pediatric device innovation.

Competing Interests The authors are members of the Southwest-Midwest Pediatric Device Innovation Consortium (SWPDC), a consortium dedicated to advancing pediatric device innovation funded by the FDA P50FD006428 grant.

The authors have no other conflicts of interest to declare that are relevant to the content of this chapter.

References

1. Gottlieb S. Public Meeting on Pediatric Medical Device Development - 08/14/2018. Published April 19, 2019. Accessed December 27, 2023. https://www.fda.gov/news-events/speeches-fda-officials/public-meeting-pediatric-medical-device-development-08142018
2. How Children Are Different - Anatomical and Physiological Differences. Published online 2022.
3. Goodman DM, Hall M, Levin A, et al. Adults with chronic health conditions originating in childhood: inpatient experience in children's hospitals. Pediatrics. 2011;128(1):5–13. https://doi.org/10.1542/peds.2010-2037.
4. Brewster R, Wong M, Magnani CJ, et al. Early Discontinuation, Results Reporting, and Publication of Pediatric Clinical Trials. Pediatrics. 2022;149(4):e2021052557. https://doi.org/10.1542/peds.2021-052557.
5. Children's Health Services 2020 Report. HCCI. Accessed December 27, 2023. https://healthcostinstitute.org/hcci-originals-dropdown/all-hcci-reports/children-s-health-services-2020-report
6. Hoffman ED, Klees BS, Curtis CA. Overview of the Medicare and Medicaid Programs. Health Care Financ Rev. 2000;22(1):175–93.
7. Manti S, Licari A. How to obtain informed consent for research. Breathe (Sheff). 2018;14(2):145–52. https://doi.org/10.1183/20734735.001918.
8. Commissioner O of the. Office of Orphan Products Development. FDA. Published December 13, 2022. Accessed December 27, 2023. https://www.fda.gov/about-fda/office-clinical-policy-and-programs/office-orphan-products-development
9. Health C for D and R. Humanitarian Device Exemption. FDA. Published August 15, 2023. Accessed December 27, 2023. https://www.fda.gov/medical-devices/premarket-submissions-selecting-and-preparing-correct-submission/humanitarian-device-exemption
10. Ulrich LC, Joseph FD, Lewis DY, Koenig RL. FDA's pediatric device consortia: national program fosters pediatric medical device development. Pediatrics. 2013;131(5):981–5. https://doi.org/10.1542/peds.2012-1534.
11. Biodesign Guide – Biodesign. Accessed December 27, 2023. https://biodesignguide.stanford.edu/
12. I.3 A Needs-Based Approach to Health Technology Innovation: The Stanford Biodesign Process Concept Development for Unmet Clinical Needs – ISCTR | International Society for Cardiovascular Translational Research. Accessed January 24, 2024. https://isctr.org/translational-pathways-devices/part-i/a-needs-based-approach-to-health-technology-innovation-the-stanford-biodesign-process-concept-development-for-unmet-clinical-needs/
13. Wu MT, Wells M. Pediatric weight estimation: validation of the PAWPER XL tape and the PAWPER XL tape mid-arm circumference method in a South African hospital. Clin Exp Emerg Med. 2020;7(4):290–301. https://doi.org/10.15441/ceem.19.082.
14. Wells M. A validation of the PAWPER XL-MAC tape for total body weight estimation in preschool children from low- and middle-income countries. PLoS One. 2019;14(1):e0210332. https://doi.org/10.1371/journal.pone.0210332.
15. Rob Fitzpatrick. *The Mom Test: How to Talk to Customers & Learn If Your Business Is a Good Idea When Everyone Is Lying to You.* 1st ed. CreateSpace Independent Publishing Platform; 2013.
16. Impact Data - NSF's Innovation Corps (I-Corps™) | NSF - National Science Foundation. Accessed January 24, 2024. https://new.nsf.gov/funding/initiatives/i-corps/impact-data
17. National Science Foundation Innovation Corps (I-Corps™) Biennial Report. Published online 2021. https://nsf-gov-resources.nsf.gov/2022-06/NSFI-Corps2021BiennialReport.pdf
18. Frequently Asked Questions (FAQs) for I-Corps Teams (nsf17083) | NSF - National Science Foundation. Accessed December 27, 2023. https://www.nsf.gov/pubs/2017/nsf17083/nsf17083.jsp#q15
19. SoftSpot Mobile App | Plagiocephaly and Cranial Measurements App | PediaMetrix. Accessed December 27, 2023. https://www.pediametrix.com/

20. Holmberg MJ, Ross CE, Fitzmaurice G, et al. Annual Incidence of Adult and Pediatric In-Hospital Cardiac Arrest in the United States. Circ Cardiovasc Qual Outcomes. 2019;12(7):e005580.
21. LUCAS - Chest Compression System. Accessed January 8, 2024. https://www.lucas-cpr.com/
22. Trends in CPAP adherence over twenty years of data collection: a flattened curve - PMC. Accessed January 8, 2024. https://www.ncbi.nlm.nih.gov/pmc/articles/PMC4992257/
23. Patents. Accessed December 27, 2023. https://www.wipo.int/patents/en/index.html
24. CPT® Code 92504 - Otolaryngologic and Binocular Microscopy Procedures - Codify by AAPC. Accessed January 8, 2024. https://www.aapc.com/codes/cpt-codes/92504
25. Niinikoski J. The effect of prototype fidelity on usability evaluation in product development. Published online August 28, 2017. Accessed November 2, 2023. https://aaltodoc.aalto.fi:443/handle/123456789/28013
26. Pharmaceutics | Free Full-Text | Medical Device Development for Children and Young People -Reviewing the Challenges and Opportunities. Accessed November 2, 2023. https://www.mdpi.com/1999-4923/13/12/2178
27. Morrison RJ, Hollister SJ, Niedner MF, et al. Mitigation of Tracheobronchomalacia with 3D-Printed Personalized Medical Devices in Pediatric Patients. Sci Transl Med. 2015;7(285):285ra64. https://doi.org/10.1126/scitranslmed.3010825.
28. Schwartz JG, Kumar UN, Azagury DE, Brinton TJ, Yock PG. Needs-Based Innovation in Cardiovascular Medicine. JACC: Basic to Translational Science. 2016;1(6):541–7. https://doi.org/10.1016/j.jacbts.2016.06.011.
29. Health C for D and R. How to Determine if Your Product is a Medical Device. *FDA*. Published online August 14, 2023. Accessed December 27, 2023. https://www.fda.gov/medical-devices/classify-your-medical-device/how-determine-if-your-product-medical-device
30. Health C for D and R. Software as a Medical Device (SaMD). FDA. Published September 9, 2020. Accessed December 27, 2023. https://www.fda.gov/medical-devices/digital-health-center-excellence/software-medical-device-samd
31. Health C for D and R. Pediatric Medical Devices. *FDA*. Published online August 15, 2023. Accessed December 27, 2023. https://www.fda.gov/medical-devices/products-and-medical-procedures/pediatric-medical-devices
32. Health C for D and R. Learn if a Medical Device Has Been Cleared by FDA for Marketing. *FDA*. Published online November 3, 2018. Accessed January 8, 2024. https://www.fda.gov/medical-devices/consumers-medical-devices/learn-if-medical-device-has-been-cleared-fda-marketing
33. Van Norman GA. Drugs, Devices, and the FDA: Part 1: An Overview of Approval Processes for Drugs. JACC Basic Transl Sci. 2016;1(3):170–9. https://doi.org/10.1016/j.jacbts.2016.03.002.
34. Health C for D and R. How to Find and Effectively Use Predicate Devices. *FDA*. Published online August 15, 2023. Accessed January 8, 2024. https://www.fda.gov/medical-devices/premarket-notification-510k/how-find-and-effectively-use-predicate-devices
35. Health C for D and R. Breakthrough Devices Program. *FDA*. Published online September 13, 2023. Accessed January 8, 2024. https://www.fda.gov/medical-devices/how-study-and-market-your-device/breakthrough-devices-program
36. Health C for D and R. Safer Technologies Program (SteP) for Medical Devices. *FDA*. Published online August 15, 2023. Accessed January 8, 2024. https://www.fda.gov/medical-devices/how-study-and-market-your-device/safer-technologies-program-step-medical-devices
37. Sutherell JS, Hirsch R, Beekman RH. Pediatric interventional cardiology in the United States is dependent on the off-label use of medical devices. Congenit Heart Dis. 2010;5(1):2–7. https://doi.org/10.1111/j.1747-0803.2009.00364.x.
38. Commissioner O of the. Real-World Evidence. FDA. Published April 25, 2023. Accessed December 27, 2023. https://www.fda.gov/science-research/science-and-research-special-topics/real-world-evidence
39. AdvaMed Code of Ethics. AdvaMed. Published June 1, 2023. Accessed January 8, 2024. https://www.advamed.org/member-center/resource-library/advamed-code-of-ethics/

40. Health C for D and R. Medical Device User Fees. *FDA*. Published online August 18, 2023. Accessed December 27, 2023. https://www.fda.gov/medical-devices/premarket-submissions-selecting-and-preparing-correct-submission/medical-device-user-fees
41. FDA impact on US medical technology innovation. MedTech Europe. Accessed January 8, 2024. https://www.medtecheurope.org/resource-library/fda-impact-on-us-medical-technology-innovation/
42. Bui AL, Dieleman JL, Hamavid H, et al. Spending on Children's Personal Health Care in the United States, 1996–2013. JAMA Pediatr. 2017;171(2):181–9. https://doi.org/10.1001/jamapediatrics.2016.4086.
43. Exclusive Report From Telosity: New Telosity report reveals 15X growth rate over the last four years in the U.S. mental health market. Published July 25, 2022. Accessed December 27, 2023. https://www.businesswire.com/news/home/20220725005205/en/Exclusive-Report-From-Telosity-New-Telosity-report-reveals-15X-growth-rate-over-the-last-four-years-in-the--U.S.-mental-health-market
44. Pediatric Asthma Treatment Market. Accessed December 27, 2023. https://www.transparency-marketresearch.com/pediatric-asthma-treatment-market.html
45. Lennox-Gastaut syndrome: MedlinePlus Genetics. Accessed December 27, 2023. https://med-lineplus.gov/genetics/condition/lennox-gastaut-syndrome/
46. Global Anti-epileptic Drugs for Pediatrics Market: Industry Analysis and Forecast (2020-2026) By Drug Type, Distribution Class and Region. MAXIMIZE MARKET RESEARCH. Accessed December 27, 2023. https://www.maximizemarketresearch.com/market-report/global-anti-epileptic-drugs-for-pediatrics-market/100398/
47. Epilepsy Therapeutics Market worth $ 15.1 Bn by 2030 at a CAGR of 4.5%. BioSpace. Accessed January 8, 2024. https://www.biospace.com/article/epilepsy-therapeutics-market-worth-15-1-bn-by-2030-at-a-cagr-of-4-5-percent-/
48. Sun RC, Kamat I, Byju AG, et al. Advancing pediatric medical device development via non-dilutive NIH SBIR/STTR grant funding. Journal of Pediatric Surgery. 2021;56(11):2118–23. https://doi.org/10.1016/j.jpedsurg.2021.01.025.
49. Understanding SBIR and STTR | Seed. Accessed December 27, 2023. https://seed.nih.gov/small-business-funding/small-business-program-basics/understanding-sbir-sttr
50. Trends in medical artificial intelligence publications from 2000-2020: Where does radiology stand? Accessed January 8, 2024. https://www.jclinmedimages.org/articles/OJCMI-v2-1052.html
51. Artificial intelligence in radiology - PMC. Accessed January 8, 2024. https://www.ncbi.nlm.nih.gov/pmc/articles/PMC6268174/
52. Försch S, Klauschen F, Hufnagl P, Roth W. Artificial Intelligence in Pathology. Dtsch Arztebl Int. 2021;118(12):199–204. https://doi.org/10.3238/arztebl.m2021.0011.
53. Liopyris K, Gregoriou S, Dias J, Stratigos AJ. Artificial Intelligence in Dermatology: Challenges and Perspectives. Dermatol Ther (Heidelb). 2022;12(12):2637–51. https://doi.org/10.1007/s13555-022-00833-8.
54. 8 ways Google Lens can help make your life easier. Google. Published June 14, 2023. Accessed January 8, 2024. https://blog.google/products/google-lens/google-lens-features/
55. Health C for D and R. Artificial Intelligence and Machine Learning in Software as a Medical Device. *FDA*. Published online October 20, 2023. Accessed January 8, 2024. https://www.fda.gov/medical-devices/software-medical-device-samd/artificial-intelligence-and-machine-learning-software-medical-device
56. Regenerative medicine: current research and perspective in pediatric surgery - PMC. Accessed December 27, 2023. https://www.ncbi.nlm.nih.gov/pmc/articles/PMC10073065/
57. Office USGA. Regenerative Medicine: Therapeutic Applications, Challenges, and Policy Options | U.S. GAO. Published July 18, 2023. Accessed December 27, 2023. https://www.gao.gov/products/gao-23-105430
58. Dunn JCY. Tissue engineering and regenerative science in pediatrics. Pediatr Res. 2008;63(5):459–60. https://doi.org/10.1203/pdr.0b013e3181739fbe.

59. Vijayavenkataraman S, Fuh JYH, Lu WF. 3D Printing and 3D Bioprinting in Pediatrics. Bioengineering (Basel). 2017;4(3):63. https://doi.org/10.3390/bioengineering4030063.
60. Research C for BE and. Framework for the Regulation of Regenerative Medicine Products. *FDA*. Published online October 20, 2020. Accessed December 27, 2023. https://www.fda.gov/vaccines-blood-biologics/cellular-gene-therapy-products/framework-regulation-regenerative-medicine-products
61. Health NRC. New pediatric report unveils NRC Health's 2022 health predictions. NRC Health. Published March 16, 2022. Accessed December 27, 2023. https://nrchealth.com/new-pediatric-report-unveils-nrc-healths-2022-health-predictions/
62. Addressing the Pediatric Mental Health Crisis: Moving from a Reactive to a Proactive System of Care - PMC. Accessed December 27, 2023. https://www.ncbi.nlm.nih.gov/pmc/articles/PMC10181869/
63. CDC. Improving Access to Children's Mental Health Care | CDC. Centers for Disease Control and Prevention. Published November 28, 2022. Accessed December 27, 2023. https://www.cdc.gov/childrensmentalhealth/access.html
64. Social Determinants of Health - Healthy People 2030 | health.gov. Accessed December 27, 2023. https://health.gov/healthypeople/priority-areas/social-determinants-health
65. Roelle L, Dalal AS, Miller N, Orr WB, Van Hare G, Avari Silva JN. The impact of direct-to-consumer wearables in pediatric electrophysiology telehealth clinics: A real-world case series. Cardiovasc Digit Health J. 2020;1(3):169–71. https://doi.org/10.1016/j.cvdhj.2020.09.005.
66. Bonafide CP, Jamison DT, Foglia EE. The emerging market of smartphone-integrated infant physiologic monitors. JAMA. 2017;317(4):353–4. https://doi.org/10.1001/jama.2016.19137.
67. Health C for D and R. Owlet Baby Care, Inc. - 616354 - 10/05/2021. FDA. Published August 10, 2022. Accessed February 19, 2024. https://www.fda.gov/inspections-compliance-enforcement-and-criminal-investigations/warning-letters/owlet-baby-care-inc-616354-10052021
68. FDA Response. Owlet US. Accessed January 8, 2024. https://owletcare.com/pages/fda-response
69. Commissioner O of the. Pediatric Device Consortia Grants Program. FDA. Published September 26, 2023. Accessed January 8, 2024. https://www.fda.gov/industry/medical-products-rare-diseases-and-conditions/pediatric-device-consortia-grants-program
70. Network AH. CobiCure launches to change the course of pediatric healthcare. Accessed December 27, 2023. https://www.prnewswire.com/news-releases/cobicure-launches-to-change-the-course-of-pediatric-healthcare-301573211.html
71. Whooley S. PolyVascular, CobiCure partner on non-surgical pediatric heart valve. MassDevice. Published August 29, 2023. Accessed December 27, 2023. https://www.massdevice.com/polyvascular-cobicure-partner-pediatric-heart-valve/
72. Solutions. PolyVascular Inc. Accessed December 27, 2023. https://polyvascular.com/solutions/
73. Accelerating Pediatric Innovation. KidsX. Accessed December 27, 2023. https://www.kidsx.health/
74. iSPI – International Society For Pediatric Innovation. Accessed December 27, 2023. https://www.ispi4kids.org/
75. Life Science Incubators And Accelerators @ JLABS | JNJ Innovation. Accessed December 27, 2023. https://jnjinnovation.com/jlabs
76. Pediatric Medical Devices Design Phase. FNIH. Accessed January 8, 2024. https://fnih.org/our-programs/pediatric-medical-devices-design-phase/
77. The state and future of pediatric research—an introductory overview | Pediatric Research. Accessed December 27, 2023. https://www.nature.com/articles/s41390-022-02439-4

The Rapidly Evolving Regenerative Medicine Ecosystem: Implications for Infants and Children

Thomas F. Miller, Kanwaljit Singh, Nikki Montgomery, Carole Kenner, Wakako Minamoto Eklund, and Jonathan M. Davis

Introduction

The emerging field of regenerative medicines has provided new hope for breakthrough therapies for countless diseases with limited therapeutic interventions. The life science field is in the early stages of this megatrend that has the potential to transform human health. This encompasses dramatic improvements in a new class of novel therapeutics including cell therapies, gene therapies (GTx), *ex-vivo* and *in-vivo* gene editing (GE) platforms, and other small molecules. The advancement

T. F. Miller
Bayer Healthcare, LLC, Whippany, NJ, USA
e-mail: thomas.miller@bayer.com

K. Singh
Pediatric Programs, The Critical Path Institute, Tucson, AZ, USA
e-mail: ksingh@c-path.org

N. Montgomery
Family Voices, Concord, MA, USA
e-mail: nmontgomery@familyvoices.org

C. Kenner
School of Nursing and Health Sciences, The College of New Jersey, Ewing, NJ, USA

Council of International Neonatal Nurses, Inc. (COINN), Yardley, PA, USA

W. M. Eklund
Pediatrix Medical Group of Tennessee, Nashville, TN, USA

Vanderbilt University School of Nursing, Nashville, TN, USA

J. M. Davis (✉)
Department of Pediatrics, Tufts Medical Center, Boston, MA, USA

Tufts Clinical and Translational Science Institute, Tufts Medical Center, Boston, MA, USA
e-mail: Jonathan.Davis1@tuftsmedicine.org

J. L. Wells (ed.), *A Clinical Lens on Pediatric Engineering*,
https://doi.org/10.1007/978-3-031-84655-7_6

of the field of regenerative medicine has been made possible by simultaneous improvements in diagnostics such as rapid Whole Genome Sequencing (rWGS) [1]. These platforms have diagnostic yields of 40–50% (depending on the study), can produce rapid (<6 h) and high-quality patient-level data, and are becoming less expensive [2]. These two trends are converging in real time, promoting a renaissance of disease-modifying and potentially curative new medicines.

Advancements in the field of gene therapy are primarily due to the development of novel methodologies using viral vectors [3]. There is a long history of humans being exposed to viruses including influenza, polio, and more recently coronavirus. One specific type of virus is adenovirus which comprises a family of known viruses that typically cause the common cold- or flu-like illnesses. Adenovirus was first discovered in 1953 and adeno-associated virus (AAV) was subsequently discovered in 1965 [4]. Today, the utility of AAVs has been harnessed to become an efficient drug development tool and the preferred vector for the delivery of gene therapy (Fig. 1). AAV contains a small, single strand of DNA which effectively infects humans and other vertebrates without causing significant disease. AAV is highly infectious but naturally replication-defective in the absence of a helper virus. Moreover, AAV has a genome that can be readily manipulated, providing for flexible construct development [5].

In parallel with the advancement of the GTx field and *ex-vivo* GE development, *in-vivo* gene editing therapeutic candidates have recently moved into human clinical

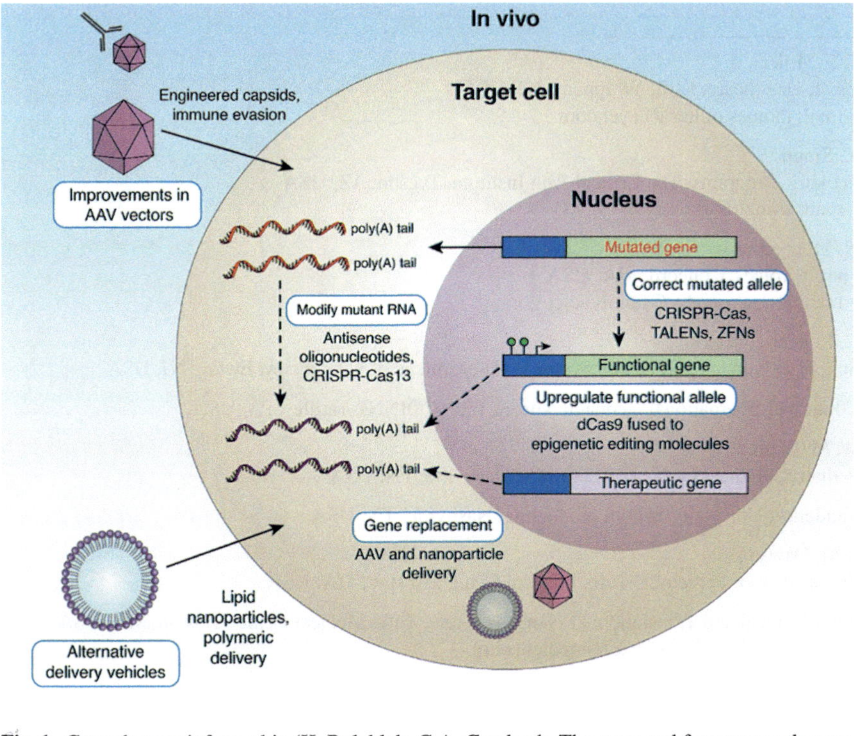

Fig. 1 Gene therapy infographic (K. Bulaklak, C.A. Gersbach. The once and future gene therapy, *Nature Comm.* 11, 5820 (2020))

trials [6]. This exciting new field emerged following the groundbreaking discovery of Clustered Regularly Interspaced Short Palindromic Repeats (CRISPR) which historically has served as a key aspect of innate immunity in certain bacterial species [7]. Today, CRISPR is combined with a CRISPR-associated protein (Cas) to provide drug developers unprecedented precision and the ability to turn on or off target genes *in-vivo* [8] (Fig. 2). When examining available listings on public clinical trial reporting databases, there are approximately 1560 sponsored, interventional gene therapy investigations under way as well as more than 40 gene editing trials. Approximately 90 of these projects allow for the enrollment of children as young as the beginning of infancy [9]. There is a continued focus on improving the safety profiles of early-generation AAV-based delivery technologies as well as the advancement of new delivery systems. There is also an urgent need for continued advancements in inexpensive, accurate, and diagnostically precise rWGS platforms for this field to continue to grow significantly in the coming years.

As of late 2023, 34 regenerative medicines have been approved for human use by the US Food and Drug Administration (FDA). Examples of early entrants in the field include voretigene neparvovec-rzyl (LUXTURNA®) for the treatment of certain inherited retinal disorders (mutations in the *RPE65* genes which can lead to blindness) and onasemnogene abeparvovec-xioi (ZOLGENSMA®) for the treatment of pediatric patients <2 years of age with spinal muscular atrophy (a lethal neuromuscular disorder) with bi-allelic mutations in the survival motor neuron 1 gene. A very important study is currently being conducted using gene therapy in a small number of children with otoferlin gene *(OTOF)*-mediated hearing loss. Preliminary results suggest that directed delivery of the normal *OTOF* gene to the inner ear can significantly improve hearing [10]. Since these patients have very few treatment options, these novel treatments have the potential to revolutionize care in a number of devastating rare diseases. More recently, exagamglogene autotemcel [exa-cel], the first CRISPR/Cas9 genome-edited cell therapy (CASGEVY™), has been approved for the treatment of sickle cell disease in patients >12 years of age

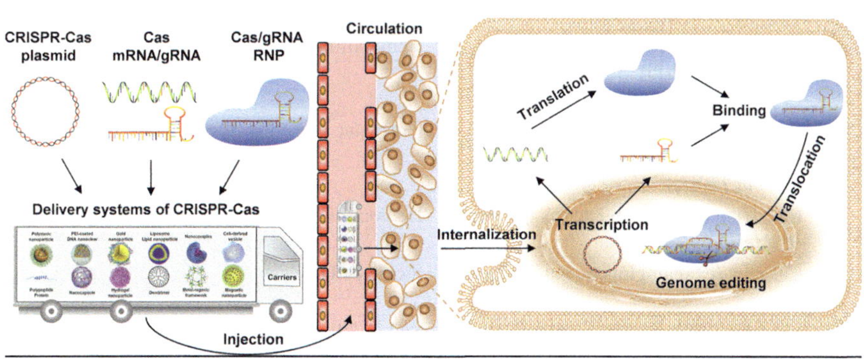

Fig. 2 Gene editing with a CRISPR-Cas delivery system infographic (C.F. Xu, G. J. Chen, Y.L. Luo, et al. Rational designs of *in vivo* CRISPR-Cas delivery systems, *Advanced Drug Delivery Reviews*, 168, 3-29 (2021))

(Table 1). Sickle cell disease occurs primarily in underserved populations and is characterized by abnormally shaped red blood cells, which can lead to significant morbidity and mortality from recurrent vaso-occlusive crises. These recent approvals and studies will undoubtedly and fundamentally change the standards of medical care in many areas of medicine. A curative therapy for sickle cell disease also has the potential to primarily address an issue of health equity and access in an underserved population primarily affected by this genetic disorder. The emergence of the regenerative medicine field also brings unprecedented, new uncertainties for many in the pediatric care ecosystem including prescribers, drug developers, regulators,

Table 1 Regenerative therapies approved by FDA in the last 5 years

Year	Product/developer	Approved indication
2019	Zolgensma (onasemnogene abeparvovedxiol) *Novartis Pharmaceuticals*	Spinal muscular atrophy (SMA) Type 1
2020	Tecartus (brexucabtageneciloleucel) *Kite Pharma, Inc.*	Relapsed or refractory mantle cell lymphoma
2021	Breyanzi (lisocabtagene maraleucel) *Bristol-Myers Squibb*	Relapsed or refractory large B cell lymphoma
	Abecma (Idecabtagene vicleucel) *Bristol-Myers Squibb*	Relapsed or refractory multiple myeloma
	Yescarta (axicabtagene ciloleucel) *Kite Pharma, Inc.*	Follicular lymphoma
	Tecartus (brexucabtageneciloleucel) *Kite Pharma, Inc.*	Relapsed or refractory adult B-cell lymphoblastic leukemia
2022	Carvykti (ciltacabtagene autoleucel) *Janssen Pharmaceuticals*	Relapsed or refractory multiple myeloma
	Kymriah (tisagenlecleucel) *Novartis Pharmaceuticals*	Refractory follicular lymphoma
	Zynteglo (betibeglogene autotemcel) *bluebird bio, Inc.*	Beta-thalassemia
	Skysona (elivaldogene autotemcel) *bluebird bio, Inc.*	Cerebral adrenoleukodystrophy (CALD)
	Hemgenix (etranacogene dezaparvovecdrib) *CSL Bering*	Hemophilia B
2023	Elevidys (delandistrogene moxeparvovec-rokl) *Sarepta Therapeutics*	Duchenne muscular dystrophy
	Roctavian (valoctocogene roxaparvovec) *BioMarin*	Hemophilia A
	Casgevy (exagamglogene autotemcel) *Vertex*	Sickle cell disease
	Lyfgenia (lovotibeglogene autotemcel) *bluebird bio, Inc.*	Sickle cell disease
	Casgevy (exagamglogene autotemcel) [exa-cell] *Vertex*	Sickle cell disease

payers, and most importantly families with young children with significant disease burden.

The International Neonatal Consortium (INC) is a global public-private partnership formed to establish a predictable regulatory path for evaluating the safety and effectiveness of therapies for neonates [11]. INC is a multi-stakeholder consortium including physicians, neonatal intensive care unit (NICU) nurses, parent advocates, regulators, funders (National Institutes of Health, March of Dimes), and life-science industry executives. The motivation for the creation of this consortium stemmed from the recognition that neonatal drug development faces complex and multifaceted challenges. Recognizing this critical need, the US Food and Drug Administration (FDA) collaborated with the Critical Path Institute (C-Path) to create the INC with the goal of advancing medical product development tools and methodological approaches for endorsement by global health authorities. This chapter includes reflections and expert perspectives from INC's multidisciplinary and multistakeholder ecosystem regarding the unique considerations for the development of GTx and GE therapeutics intended for use in human infants.

Nonclinical considerations prior to the evaluation of genetic medicines in young children. Well in advance of enrolling the first patient into a clinical trial, many nonclinical criteria must be fulfilled which presents a number of challenges in young patients. The submission of an Investigational New Drug (IND) application is a prerequisite for initiation of clinical trials in human subjects. While this requirement holds true for investigational regenerative medicines, IND enabling nonclinical packages for a viral vector-based gene therapy are quite different from that of a small molecule or a biologic therapy. Chemistry, manufacturing, and controls (CMC) is an integral aspect of the development of gene therapy, with many challenges due to inherent properties such as shorter shelf life, variability in the use of ancillary materials, manufacturing complexity, logistics, and cost. Key CMC considerations and challenges for regenerative medicines include: (1) the components used in product manufacturing (e.g., vector, promoter, capsid, etc.), (2) the actual manufacturing process, and (3) comparability assessments (e.g., potency assays, analytical procedure validation) for drug product release and stability assurance during labeled shelf-life. Additionally, increased product characterizations and controls are expected as clinical development proceeds and additional product knowledge matures [12].

Multiple nonclinical studies are necessary to determine the safety and efficacy of cell and gene therapy products, including AAV biodistribution, gene vector and transgene biodistribution, gene expression, protein expression, viral shedding, and vector integration [13]. A key and unique component for successful IND filing is the conduct of comprehensive toxicology studies such as cell culture and animal models. Nonclinical toxicology studies with traditional modalities (small molecules, biologics, etc.) typically involve repeat doses that are administered over a protracted period of time. However, studies with gene therapies may involve a single exposure and animals are followed for a period of time that is negotiated with Health Authorities prior to sacrifice. Doses tested in traditional modality nonclinical toxicology studies are often substantially higher than the predicted efficacious exposure. However with gene therapies, toxicology study exposures are typically at or slightly above the anticipated clinically efficacious dose since the viral vectors may be associated with the development of significant inflammation and toxicity [14]. As many therapeutics have shown different safety profiles in pediatric compared to adult patients, age-appropriate toxicology models (one or more different animal species) must be considered and implemented as appropriate [15].

Unique risk/benefit considerations in the pediatric population. AAVs have been established as an early-generation "platform of choice" for *in vivo* gene therapies [16]. AAVs confer a high-efficiency transduction to a wide range of target tissues. They have been shown to integrate reliably within the host cell genome, with gene expression being maintained for a longer period of time. AAV vectors are unable to replicate autonomously without a helper virus and cause relatively less immunogenicity *in-vivo*. Despite these advantages, current-generation AAV vectors do have limitations and convey medical risks related to immunogenicity and genotoxicity [17].

A significant concern about AAV vector safety is the potential for immunogenicity. AAV vectors provoke immune responses, which may lead to the elimination of transduced cells and possible serious adverse effects. Hepatotoxicity, the potential for AAV vectors to cause liver damage due to exaggerated immune responses, is another area of particular concern. AAV vector-related hepatotoxicity may arise from direct and indirect immune responses. Although the direct response involves the activation of innate and adaptive immune cells against the AAV capsid proteins, indirect responses are triggered by the therapeutic transgene's expression [18].

Direct immune responses occur when AAV vectors are identified as foreign substances by the host immune system following administration. This recognition activates innate immune cells such as dendritic cells and macrophages which then produce inflammatory cytokines and chemokines. These inflammatory mediators may contribute to liver damage and facilitate the activation of adaptive immune cells, specifically CD8+ T cells. These cells can recognize and target AAV capsid-derived peptides that are present on transduced hepatocytes' surfaces, leading to their elimination and subsequent liver injury. Indirect immune responses involve the therapeutic transgene's expression which contributes to hepatotoxicity. In some instances, the transgene product may be recognized as a foreign antigen by the host immune system, activating CD4+ and CD8+ T cells. This immune response may result in the elimination of transduced hepatocytes, cause liver damage and reduce the effectiveness of the gene therapy. This risk is especially concerning in neonates (who have very immature and dysregulated immune systems) since the earlier that a patient can be identified and treated, theoretically the better the response should be. Adult tissues tend to replicate slowly, leading to genome persistence and likely stable expression for years or even decades. However, pediatric patients require consideration of liver and other organ development associated with rapid growth over years, with the liver not reaching a stable size until 10–15 years of age. Other factors to consider include liver health (fibrosis, cirrhosis) and pathologic conditions, especially in infants with metabolic diseases that involve the liver. Several strategies are being explored to minimize AAV vector-associated toxicity.

Immune modulation: In pediatric gene therapy clinical trials, immune modulation using drugs such as corticosteroids in advance of test article administration is crucial in attempting to prophylactically reduce the risk of toxicity by suppressing the host immune response. However, the use of these drugs, particularly in an off-label context (not approved for the specific indication by Health Authorities) for young children, demands careful consideration. The primary challenge is the maintenance of the delicate balance between reducing immune-mediated responses against AAV vectors and avoiding excessive immunosuppression, which could lead to increased susceptibility to other infections. This risk is especially heightened in infants due to their developing immune systems and the lack of comprehensive data regarding the long-term effects of immunosuppressants in this vulnerable population [19]. The application of immune modulators in pediatric patients requires a tailored approach, considering each child's specific health condition and the potential impact on the development of their immune system. As all previous immune tolerization protocols used in pediatric gene therapy trials have been deployed without prior Health Authority marketing authorization, a systematic review of the safety and tolerability of various approaches is warranted.

Transient immune cell depletion: In pediatric gene therapy clinical trials, transient immune cell depletion, particularly CD8+ T cells, has been employed to protect transduced hepatocytes from immune-mediated damage. This approach can enhance the efficacy of the therapy while reducing liver damage. However, this can present unique risk tradeoffs in young children since their immune systems are in a critical phase of development and any intervention that alters the immune response can have far-reaching implications. Temporarily depleting immune cells in young patients can potentially leave them more vulnerable to infections and other immune-related challenges as infants may not respond as robustly as an adult would to similar challenges [20]. Furthermore, the impact of transient immune cell depletion on the long-term development of the pediatric immune system is not fully understood. This uncertainty necessitates a highly cautious approach where the potential benefits of improving gene transduction are carefully weighed against the risks of compromising the child's immune defense, particularly in the context of their overall health and any pre-existing conditions.

Improving manufacturing processes: AAV vector manufacturing processes play a crucial role in ensuring their safety and efficacy. Purifying AAV vectors from producer cell lines can lead to contamination with residual host cell proteins and other impurities, triggering immune responses or compromising the vector's stability. Researchers are developing novel purification techniques such as ion exchange chromatography and density gradient centrifugation, to improve AAV vector purity and yield. Furthermore, advancements in producer cell lines and scalable bioreactor systems have significantly enhanced AAV vector production, contributing to their safety and efficacy. A notable innovation in this area entails the incorporation of lipid nanoparticles (LNPs) in gene therapy delivery. LNPs serve as cutting-edge vehicles for delivering genetic material, offering an alternative to traditional AAV vectors and providing enhanced stability, reduced immunogenicity, and improved targeting capabilities. This highlights that LNP technologies currently in development hold the promise to become an attractive construct development option in gene therapy trials, especially for pediatric applications [21].

New capsid engineering: Novel AAV capsid technologies (through rational design or directed evolution) can produce new AAV variants with reduced immunogenicity. This approach may reduce liver damage caused by direct immune responses by minimizing the AAV capsid's recognition by the host immune system.

Promoter optimization: The choice of promoter elements that drive transgene expression can affect the immune response against the transgene product. By selecting tissue-specific promoters, researchers may minimize the risk of inducing an immune response against the therapeutic transgene, thereby reducing hepatotoxicity risk.

Genotoxicity of AAV Vectors: The potential for genotoxicity, or damage to the host genome, is another concern related to AAV vector safety. AAV vectors can sometimes integrate into the host genome, leading to unintended consequences such as the activation of oncogenes or the disruption of tumor suppressor genes which could pose a future risk for cancer. To address this issue, researchers are developing AAV vectors with minimal or no integration capacity. For example, self-complementary AAV (scAAV) vectors have been shown to exhibit lower integration rates than their single-stranded counterparts. Additionally, the use of AAV serotypes with a preference for episomal maintenance (e.g., AAV8) can further reduce integration events. Another strategy to reduce genotoxicity is the incorporation of insulator elements into the AAV vector design, which can help shield the transgene from the regulatory effects of neighboring genes, thereby reducing the risk of unwanted gene activation or silencing [22].

Age-related immune responses: The immune system of pediatric patients, especially neonates and infants, is still developing and can exhibit significantly different responses compared to adults. As a result, any observed immunogenicity attributed to AAV vectors may vary (considerably) between pediatric and adult populations. This variation may depend on factors related to immune system maturation, pre-existing immunity to AAV (in older populations), immune tolerance (in younger populations), and differences in disease progression. A more favorable immune response is definitely associated with better outcomes in younger populations. Tailoring AAV vector administration to account for these age-related differences is essential for ensuring safety and therapeutic efficacy. Furthermore, the evolving nature of a child's immune system means that responses to gene therapy may change over time, necessitating a flexible treatment plan and close long-term monitoring. This dynamic aspect of pediatric immunology not only has the potential to influence immediate treatment outcomes but also impacts the long-term safety and effectiveness of gene therapy, making it imperative to continually reassess and adjust treatment strategies as the child matures.

Tissue growth and development: Rapid tissue growth and cellular proliferation are expected during childhood and can influence AAV vector persistence and transgene expression. For example, the liver grows significantly during childhood, potentially affecting the long-term expression of AAV-delivered transgenes. When designing AAV-based therapies for pediatric patients, it is crucial to consider strategies that ensure stable, durable transgene expression as tissues grow and develop.

In summary, the following ecosystem-wide initiatives are underway with the promise of improving safety, efficacy, and effect durability in the rapidly advancing and evolving field of pediatric gene therapy:

1. Newly engineered AAV vectors with **reduced immunogenicity** and improved tissue-specific targeting.
2. Investigation of **alternative delivery methods** that minimize immune responses and improve transgene expression stability with more precise tissue targeting.
3. Conducting **age-appropriate preclinical studies** to better understand the safety and efficacy of AAV-based gene therapy in pediatric populations in a fundamental way. This is needed to support exposing neonates and infants to a treatment at a time that it can be most efficacious (before significant tissue and/or organ injury can occur) and be maintained for prolonged periods of time.
4. Implementation of **long-term follow-up studies** to monitor and address potential adverse effects and unknown long-term risks associated with gene therapy as well as to assess effect durability as the child grows and develops.

Study design considerations: Designing clinical studies for regenerative medicine in very young patients requires careful consideration of several unique components. These factors have the potential to impact not only the observed safety and efficacy profile for a given gene therapy (and derivatively Health Authority labeling), but also the operational throughput to completion of a clinical development program. Some key study design aspects to consider include:

Trial design and related statistical plans: Clinical studies involving pediatric patients often face challenges in recruiting a sufficient number of participants due to a limited number of patients and the lack of adequate natural history studies for most children affected by certain monogenic, congenital, rare diseases. Researchers should carefully consider the sample size and use appropriate statistical methods to account for potential confounders, missing data, and multiplicity. Since rare pediatric disease clinical trials are notoriously difficult to fully enroll, study teams should carefully consider the feasibility of study conduct for the selected trial design well in advance of trial initiation. Study teams are also encouraged to consider the use of alternative study designs, including those that supplement the prospectively enrolled patient population with real-world data (RWD).

Study teams should explore what role RWD might have at the design stage of a clinical trial. The application of RWD in clinical trials (in support of regulatory decision-making) includes potential study population enrichment, identification of discrete clinical endotypes of disease, and the potential to create either a partial or fully synthetic control group. Employing such a strategy will require that the investigator and team statistician consider alternatives to traditional frequentist-based statistical plans. Bayesian statistical frameworks are used to combine information about the control group from historically published clinical trials by using the probability distribution derived in the meta-analytical prediction as prior information for the control group. A Bayesian analysis offers the option to synthesize evidence for the control group from external, published data from the literature and other potential sources. While a multitude of potential data sources may exist, not all datasets will assure the same degree of data quality control and rigor. Care should be taken to ensure that any "borrowed data" is fit-for-purpose, contemporary, and appropriately quality controlled. Thus, data generation can be focused on filling information gaps on the treatment effect for the investigational agent and putting less weight on the prospectively enrolled control group. Employing such an approach may lead to more expeditious patient enrollment and completion of a clinical trial [23]. This is especially important in pediatric clinical trials where there are high failure rates, often due to difficulties enrolling adequate numbers of patients.

Ethical considerations: Pediatric patients are considered a vulnerable population and their inclusion in clinical studies requires special attention to ethical concerns. Informed consent from parents or legal guardians, along with assent from the child (when appropriate in older children), is necessary. As is always the case with pediatric clinical trials, researchers should consider the balance between potential benefits and risks, aiming to minimize potential harm to the pediatric participants. In addition to traditional expectations on these matters, study teams will be tasked with providing comprehensive information regarding the impact and irreversible change to the child's genome (especially germlines). Additionally, the long-term psychological, social, and emotional impact of gene therapy on pediatric patients and their families should be considered when developing treatment plans.

Age-appropriate dose optimization: In gene therapy clinical trials, traditional pharmacokinetics (PK) and pharmacodynamics (PD) models often do not apply compared to more traditional small or large molecule development programs. This is due to gene therapy trials operating on fundamentally different principles than conventional medications. Instead, factors like hepatocyte turnover, variations in serum volume (and volume growth as the child matures), and the dynamic nature of organ development play a more significant role in determining how these therapies will impact children. For instance, the turnover rate of hepatocytes and the increasing serum volume due to growth can greatly influence the distribution and effect durability of AAV vectors in pediatric patients. These changes are particularly pronounced in infants where rapidly maturing organ systems may lead to the need for repeated administration of the AAV vectors (due to tissue growth or decreases in transgene expression over time).

Selection of endpoints: Selecting appropriate endpoints is crucial in regenerative medicine studies involving pediatric patients. Researchers should choose endpoints that are clinically meaningful, feasible to measure in pediatric populations, and relevant to the specific age group. Following the introduction of the 21st Century Cures Act, the FDA introduced the regenerative medicine advanced therapy (RMAT) designation which has a goal of accelerating the developmental processes leading to marketing authorization. The designation could be considered if the investigational regenerative therapy is intended to treat, modify, reverse, or cure a serious or life-threatening disease or condition and preliminary clinical evidence indicates that the drug or biologic has the potential to address an unmet medical need for the intended condition [24]. The use of surrogate endpoints, biomarkers, or patient-reported outcomes should be carefully considered and validated with a fit-for-purpose approach for any given pediatric population.

Long-term follow-up: Pediatric patients participating in clinical trials typically undergo long-term follow-up (LTFU) to assess the safety and efficacy of any particular investigational medication and to monitor potential late effects such as growth, development, organ dysfunction, or other off-target consequences. The FDA has also issued a guidance document providing various perspectives relating to the design and duration of LTFU, which can range from 5 up to 15 years following the completion of the primary pivotal study [25]. Researchers should design regenerative medicine studies with adequate and carefully planned LTFU (accounting for patients lost to follow up over time) to surveille for effect durability, delayed safety observations, and to capture long-term outcomes relating to developmental milestones (including speech, language, behavior, social interactions, school performance). Additionally, study teams should consider that very young trial participants may become adolescents during this follow-up period, with the potential for changing medical homes.

Patient selection and stratification: Pediatric patients often present with a wide range of disease severity, comorbidities, and epigenetics. Proper patient selection and stratification should include sufficient racial, ethnic, gender, and geographic diversity to ensure that study results are truly generalizable to wider populations. Researchers should also consider the impact of concomitant treatments and previous interventions on study outcomes.

Age-appropriate study procedures: Pediatric patients (and their caretakers) may have difficulty complying with certain study procedures and requirements such as prolonged hospital stays, excessive blood draws, frequent imaging studies, or a perceived invasive nature of gene therapy delivery. Researchers should develop age-appropriate study procedures to minimize discomfort, distress, and inconvenience for pediatric participants and their families. Additionally, the concept of age-appropriateness will change as the child ages (e.g., conflicts with schooling or other activities).

Collaboration and data sharing: Efficient design and execution of clinical studies in the pediatric regenerative medicine field requires collaboration among researchers, clinicians, regulatory agencies, and patient advocacy groups. By sharing data and knowledge, stakeholders can collectively develop more efficient study designs, accelerate the development of regenerative medicine interventions, and improve patient outcomes. A practical example of the utility of this approach is the collaborative conduct of a platform trial in which the same control population data could be shared for use across multiple pediatric gene therapy trials. Such an approach could be particularly useful when studying rare diseases during infancy, where limited patient availability might otherwise drive a prolonged enrollment period.

A family perspective on clinical trials in regenerative medicine: Parents and other caretakers of children with rare genetic conditions are often desperate for a treatment or cure that would alleviate the effects of the condition on the quality and length of their child's life. The challenges of engaging families in the process of diagnosing and treating involve trust—not only in the science supporting the disorder but also in the processes and practices of clinical research. While this is generally true for all drug development initiatives, these complexities are amplified with genetic medicines.

While newborn screening (testing for various metabolic disorders at the time of birth) has become more commonplace in more affluent countries, families are often ambivalent about whether to pursue other types of advanced genomic testing when it is offered. While the idea of genomic sequencing in a newborn to determine current and future medical risks can appear to be a wonderful opportunity for some caretakers, others might be terrified at the prospect of learning about a devastating condition currently affecting their child as well as the potential for other diseases that might develop later in childhood or as an adult. Those who do choose to have additional newborn screening performed (outside of what is mandated by their respective state) might be informed of a diagnosis they have never heard of and suddenly need to decide whether and how to seek treatment if it is even available. In contrast, the promise of identifying and treating the devastating effects of a debilitating and life-threatening condition is quite compelling to some families.

It is important that families understand the purpose of genetic screening and the implications of learning more about a rare condition. Families should also understand what happens to the genetic samples of their children, specifically who owns these samples and how they might be used in the future. This is a very important topic for caretakers, especially for historically marginalized groups with a history of exploitation by the research community. Families will also want to understand how these genetic screening tests and any potential treatments are paid for and whether their health insurance policies will cover the costs.

The promise of genetic medicines for disease-modifying and/or curative potential can seem out of reach for many families—accept this treatment and your child will never experience the negative impacts of the genetic condition. However, there are both medical and social implications that go beyond these therapies. Informing parents about the potential risks of a drug or biologic agent is easy in comparison to providing detailed information about gene therapy. For example, with emerging gene editing technologies, permanently changing a child's genetic code may have risks that are yet undetermined. Any informed consent families provide should include this information and the uncertainty in plain language. Those developing gene therapies and conducting clinical trials should adequately train staff at each site to address family concerns with empathy and candor.

Families are likely to have questions about side effects and risks as well as how long a prevention and/or treatment strategy might remain effective. They will also need to understand that gene therapy treatments are not reversible and what that might mean for young infants and children who receive these treatments. Having a designated point person to address these questions, provide information, and allow families to make their decisions in their own time is key to providing both support and information to help them make these important decisions.

Parents of children with rare genetic conditions make many life-and-death decisions on their child's behalf—about surgeries, medications, and clinical procedures.

None of this is without risk, which can be significant. Families must balance the risk of gene therapies by considering the quality and length of life, needing as much information as possible to make informed decisions. Even the most sophisticated and educated families will likely need many opportunities to ask questions, have discussions, and consider next steps. They hope that the benefits of gene therapies outweigh the risks, but uncertainty can and should be expected. By addressing frequently asked questions with candid answers, researchers and developers can help to inform families about what is known and not known about the therapies being offered. Plain language principles can support health literacy and ensure that families are informed about the risks and benefits in the language they understand.

Patients and their caregivers can be excellent partners in this evolving area of care and treatment. Their voices can help build trust, improve acceptance, and anticipate concerns other families may have. Special considerations of the social and political issues families experience (e.g., costs) can also inform ways to engage families and understand the factors that may influence their acceptance of these novel therapeutics. Moreover, patient advocates (advocacy organizations, parents, and other caregivers) for children with genetic disorders need to be considered as partners in this exciting, yet rapidly evolving new field. Assurance of incorporating the voice of the patient advocate will inevitably add value and improve the quality of these challenging development programs, as well as downstream acceptance of these novel therapeutics as they become more readily available. Ensuring equitable access to these treatments is essential so all families can benefit from these evolving technologies.

The nurse's role in regenerative medicine clinical trials: As described above, parents and caregivers of infants with rare genetic conditions, especially those where effective therapies have not been previously available, are asked to make many decisions related to the care and treatment for their child. Parents are often faced with making crucial decisions on short notice, considering enrollment in trials that they had never anticipated prior to the child's birth. Enrollment in clinical trials that involve novel technologies for screening and treatment is often fraught with questions. Parents and caregivers often turn to nurses when they are frightened or confused about medical terminology—especially lengthy and legally crafted informed consents full of unfamiliar language. Nurses are also consulted when the parents seek further understanding about not only the process or nature of the clinical trials but to better understand the disease or condition itself. Nurses at the bedside often partner with parents and other caregivers to explain in simple language what the short- and long-term objectives and potential outcomes are for gene therapy trials, or they can refer the parents to the research team for relevant follow-up. The bond between parents, caregivers, and nurses is strong and built on trust.

It is important to highlight the need for well-developed processes to adequately inform every nursing staff member who might be exposed to the type of trials that are emerging, or the trials for which their patients may be approached. When nurses are well-informed about the research plan, process, significance, potential outcomes, and parental options, they can help support any enrollment decisions in a professional manner. Nurses can be a supportive asset to the research efforts when they are provided with the necessary information early in the study implementation phase so that they understand the critical nature of the research and its significance to their families. These new therapies or clinical trials involving emerging therapies are not typically addressed in current nursing educational curricula. Additional attention is needed to ensure that bedside nurses who spend the most time with families as well as staff who conduct the research are included early in the research planning process; this concept cannot be overstated.

Conclusions and Future Directions

The emergence of the field of regenerative medicine has brought about unprecedented promise for the future of human health. Exciting advancements in both diagnostics and therapeutics have the extraordinary potential to change the lives of countless children and their families. Although these new therapies have transformative potential, they also bring about several new challenges for drug/biologic development teams. While this chapter attempts to frame the most salient challenges associated with genetic medicine clinical trials in our youngest patients, many additional complexities and likely future advancements have not been fully addressed in this initial digest.

- Exciting advancements in genome diagnostics and genetic therapeutics have unprecedented potential to change the lives of many children and their families.
- Although these therapies have transformative potential, new challenges arise—ethical and financial considerations, expeditious trial conduct, assurance of informed consent, uncertainty with long-term outcome assessment, and equitable access.
- It is critical that all stakeholders in the ecosystem work collaboratively.

Competing Interests Thomas Miller is an employee of Bayer Healthcare, LLC.

References

1. Ouyang X, Zhang Y, Zhang L, et al. Clinical utility of rapid exome sequencing combined with mitochondrial DNA sequencing in critically ill pediatric patients with suspected genetic disorders. Front. Genet. 2021;12:725259.
2. Pagnamenta AT, Camps C, Giacopuzzi E, et al. Structural and non-coding variants increase the diagnostic yield of clinical whole genome sequencing for rare diseases. Genome Med. 2023;15:94.
3. Li X, Le Y, Zhang Z, Nian X, Liu B, Yang X. Viral vector-based gene therapy. Int. J. Mol. Sci. 2023;24(9):7736.
4. Wang D, Tai PWL, Gao G. Adeno-associated virus vector as a platform for gene therapy delivery. Nat. Rev. Drug Discov. 2019;18(5):358–78.

5. Naso MF, Tomkowicz B, Perry WL, Strohl WR. Adeno-Associated Virus (AAV) as a vector for gene therapy. BioDrugs. 2017;31(4):317–34.
6. Lubroth P, Colasante G, Lignani G. In vivo genome editing therapeutic approaches for neurological disorders: Where are we in the translational pipeline? Front Neurosci. 2021;15:632522.
7. Ishino Y, Krupovic M, Forterre P. History of CRISPR-Cas from Encounter with a mysterious repeated sequence to genome editing technology. J Bacteriol. 2018;200(7):e00580–17.
8. Xu Y, Li Z. CRISPR-Cas systems: Overview, innovations and applications in human disease research and gene therapy. Comput Struct Biotechnol J. 2020;18:2401–15.
9. National Library of Medicine. Clinical Trials: gene therapy investigation studies. Available at: https://clinicaltrials.gov/
10. Qi J, Fangzhi T, Zhang L, et al. AAV-mediated gene therapy restores hearing in patients with DFNB9 deafness. Adv. Sci (Weinh). 2024;8:e2306788.
11. Critical Path Institute. International Neonatal Consortium. Available at: https://c-path.org/programs/inc/. (2024)
12. Shoemaker D. The gene therapy product development process. Appl. Clin.Trials. 2019;28(7):10–1.
13. Bolt MW, Whiteley LO, Lynch JL, et al. Nonclinical studies that support viral vector-delivered gene therapies: An EFPIA gene therapy working group erspective. Mol. Ther. Methods Clin. Dev. 2020;19:89–98.
14. Food and Drug Administration. Cellular, Tissue, and Gene Therapies Advisory Committee Meeting #70 [online]. Available at: https://www.fda.gov/media/151599/download. (2021)
15. Food and Drug Administration. Guidance for Industry Nonclinical Safety Evaluation of Pediatric Drug Products. Available at: https://www.fda.gov/media/119658/download. (2006)
16. Pasqualina C, Ronzitti G, Mingozzi F. Emerging issues in AAV-mediated in vivo gene therapy. Mol. Ther. Methods Clin. Dev. 2018;8:87–104.
17. Monahan PE, Négrier C, Tarantino M, Valentino LA, Mingozzi F. Emerging immunogenicity and genotoxicity considerations of adeno-associated virus vector gene therapy for hemophilia. J. Clin. Med. 2021;10(11):2471.
18. Ronzitti G, Alexandre Gross D, Mingozzi F. Human immune responses to adeno-associated virus (AAV) vectors. Front. Immunol. 2020;11:670.
19. Prasad S, Dimmock DP, Greenberg B, et al. Immune responses and immunosuppressive strategies for adeno-associated virus-based gene therapy for treatment of central nervous system disorders: Current knowledge and approaches. Human Gene Therapy. 2022;33(23-24):1228–45.
20. Simon AK, Hollander GA, McMichael A. Evolution of the immune system in humans from infancy to old age. Proc R Soc B. 2015;282(1821):20143085.
21. Cullis PR, Hope MJ. Lipid nanoparticle systems for enabling gene therapies. Mol. Ther. 2017;25(7):1467–75.
22. Browning DL, Trobridge GD. Insulators to improve the safety of retroviral vectors for HIV gene therapy. Biomedicines. 2016;4(1):4.
23. Miller TF, Kaiser A, Schmelter T, Lippert J, Davis JM. Using real-world data in pediatric clinical trials: Lessons learned and future applications in studies of persistent pulmonary hypertension of the newborn. Sem. Fetal Neonatal Med. 2022;27(4):101331.
24. Food and Drug Administration. Regenerative Medicine Advanced Therapy Designation. Available at: https://www.fda.gov/vaccines-blood-biologics/cellular-gene-therapy-products/regenerative-medicine-advanced-therapy-designation. (2023)
25. Food and Drug Administration. Long Term Follow-Up After Administration of Human Gene Therapy Products Guidance for Industry. Available at: https://www.fda.gov/media/113768/download. (2020)

Part III
Pediatric Engineering in Clinical Context: Digital to Device Innovation—Past, Present, and Future

Emergence of Liquid Biopsies in Pediatric Eye Cancers

Shreya Sirivolu, Liya Xu, and Jesse L. Berry

Introduction

Retinoblastoma (RB) is the most common intraocular malignancy in the pediatric population, representing 3% of all childhood cancers [1]. It is diagnosed in nearly 8,000 new cases each year worldwide [2]. The most common first sign of RB is leukocoria, or white reflex in the pupil, as shown in Fig. 1, followed by strabismus, which is a misalignment of the eyes [1]. This cancer can be fatal if left untreated. RB in low-income countries is associated with unfavorable patient survival (approximately 30%); however, survival in high-income countries is over 95% [2–5]. Poorer outcomes in low-income countries are typically caused by late diagnosis, restricted access to RB-specialized care, and socio-economic issues that result in poor compliance or treatment abandonment [2, 6–8]. While the mainstay of treatment is systemic intravenous chemotherapy, treatment of RB is multimodal, with focal treatment options also including trans-pupillary thermotherapy, cryotherapy, laser

S. Sirivolu · L. Xu
The Vision Center at Children's Hospital Los Angeles, Los Angeles, CA, USA

USC Roski Eye Institute, Keck School of Medicine of the University of Southern California, Los Angeles, CA, USA

J. L. Berry (✉)
The Vision Center at Children's Hospital Los Angeles, Los Angeles, CA, USA

USC Roski Eye Institute, Keck School of Medicine of the University of Southern California, Los Angeles, CA, USA

Norris Comprehensive Cancer Center, Keck School of Medicine, University of Southern California, Los Angeles, CA, USA

The Saban Research Institute, Children's Hospital Los Angeles, Los Angeles, CA, USA
e-mail: jesse.berry@med.usc.edu

J. L. Wells (ed.), *A Clinical Lens on Pediatric Engineering*,
https://doi.org/10.1007/978-3-031-84655-7_7

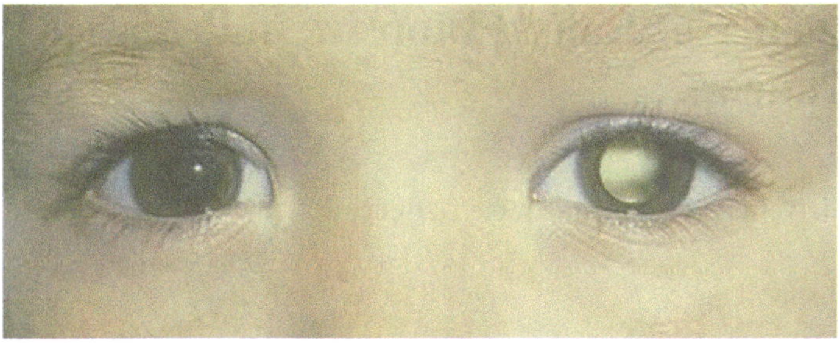

Fig. 1 Child with retinoblastoma presenting with leukocoria [9]. (From Aerts *et al.* Retinoblastoma) (This work is licensed under a Creative Commons Attribution 4.0 International License)

photocoagulation, radiation, surgery, and periocular and intravitreal chemotherapy [1].

This cancer develops from early retinal progenitor cells and is caused by the functional loss of the retinoblastoma protein due to inactivating pathogenic variants in both *RB1* alleles. *RB1* is a tumor suppressor gene that has an autosomal dominant pattern of inheritance. There are heritable and nonheritable forms of RB, with approximately 40% of tumors being heritable [10]. Heritable RB tends to present at an earlier age (ranging from newborn to 12 months) compared to nonheritable RB, which has an average diagnosis age of 24 months [10]. In addition, heritable RB can be unilateral (affecting one eye) or bilateral (affecting both eyes), while nonheritable RB results in only unilateral tumors [11]. Heritability depends on whether at least one of the *RB1* inactivating pathogenic variants is a germline variant occurring in germ cells. Heritable RB results when there is an initial *RB1* germline pathogenic variant and then a second *RB1* pathogenic variant occurring somatically in one or more retinal cells [2]. Most nonheritable RB is caused by biallelic somatic *RB1* pathogenic variants. However, a small fraction of nonheritable RB can also result from *MYCN* amplification, with normal *RB1* [11]. In both heritable and nonheritable forms of RB, somatic copy number alterations, defined as sections of chromosomal gains and losses, have been shown to contribute to subsequent tumorigenesis and progression, and may be prognostic of tumor behavior [12–16].

Although genomics of RB are well defined through extensive research, its application to clinical settings is limited since tumor biopsy for RB patients is contraindicated due to the risk of extraocular spread [17]. As a result, the only way to get a tumor tissue for genomic profiling is from eyes that have been surgically removed. Therefore, precision medicine approaches using tumor-specific genomic information for the development of targeted therapies and earlier detection or elucidation of prognostic factors, for example, have been very limited for this disease – it no longer matters much what alterations are in the tumor when the eye has already been removed. Instead, clinical features of the tumor alone currently facilitate diagnosis, staging, and treatment decisions. The emergence of liquid biopsy solutions for this disease, specifically using blood and aqueous humor, has provided a way to obtain tumor-genomic information safely, paving the way for precision medicine applications in this cancer.

Liquid Biopsies

Liquid biopsies have emerged as a transformative method of capturing tumor information with benefits that go beyond the limitations of tissue biopsies. As tissue biopsies in the pediatric population often require general anesthesia, liquid biopsies, in many cases, can provide an alternative that is non- or minimally invasive [18]. They involve the testing of bodily fluids, including blood, urine, cerebrospinal fluid, and saliva [19]. Unlike a one-time tissue biopsy, repeated liquid biopsies enable the longitudinal sampling of tumor-derived information across different stages of patient management. As a result, they have the potential to provide valuable insights into molecular diagnosis and prognosis, measure treatment response, detect minimal residual disease, track clonal evolution, and provide a method for ongoing post-treatment surveillance for relapse [18, 20–24]. As many tumors are genetically heterogeneous, a single biopsy of one location of the tumor may not provide a comprehensive snapshot of the genetics of different subclones [25, 26]. Another advantage of liquid biopsies is that the tumor information sampled may be able to reflect the mutational landscape of the entire tumor, including novel subclones and potential metastases [18].

In comparison to adult cancers, pediatric cancers are more likely to have chromosomal structural variations such as copy number alterations, defined as gains or losses in sections of the chromosome, as well as translocations and fusions [24, 27]. These genomic aberrations are often detected in liquid biopsies using tumor-derived cell-free DNA (cfDNA), which is fragmented DNA released through apoptosis, necrosis, or active tumor secretion [19]. Both polymerase chain reaction (PCR) and next-generation sequencing (NGS) methods are frequently used for analyzing cfDNA. PCR is a technique used to amplify specific sections of DNA by rapidly making many copies of that segment. Unlike PCR, NGS does not require prior knowledge of mutations and can therefore identify *de novo* mutations, albeit at an increased cost. Using NGS, copy number alterations can be detected with shallow-depth sequencing; however, increased depth is typically needed to identify single nucleotide variants. Additional tumor derivatives that can be detected in liquid biopsies include circulating miRNA, proteins, intact tumor cells, and extracellular vesicles [28].

Liquid biopsies have been shown to have clinical utility in several pediatric cancers, particularly in central nervous system tumors, which constitute the most common solid tumors in children [29]. Certain brain tumors, such as diffuse midline gliomas, cannot be surgically resected and stereotactic or open biopsy are still invasive procedures that carry serious risks [29]. For these reasons, liquid biopsies have provided a solution for minimally invasive molecular profiling of the tumor. In brain tumors, circulating DNA in the blood, CSF, and urine have been shown to monitor disease progression accurately and, on occasion, even identify progression prior to detection using radiography [30]. Liquid biopsies have also shown the potential to be used for risk stratification in pediatric sarcomas, which are a relatively uncommon type of cancer that originates in the bone or soft tissues. For instance, in osteosarcoma the detection of tumor-derived cfDNA itself predicts poorer outcomes, and certain copy number alterations, such as a gain on chromosome 8q can predict poorer 3-year event-free survival [31]. Response to treatment has been shown to track dynamic changes in tumor-derived cfDNA, with concentrations correlating with tumor burden in pediatric cancers such as Wilms tumor (a rare kidney cancer) [32] and hepatoblastoma (the most common type of pediatric liver cancer) [33]. Liquid biopsies have started to become routine companion diagnostic tests in some adult cancers, and while not yet routine in pediatrics, clinically accredited assays are being evaluated [34]. Reasons for this lag are complex and multifactorial. Pediatric cancers are rarer than adult cancers, clinical trials are limited in the pediatric space, and in general for liquid biopsy, there are a complex set of validation parameters that need to be met before clinical use.

The prospect of a liquid biopsy in RB is particularly desirable due to the contraindication of tissue biopsy. Aqueous humor (AH), a transparent fluid in the anterior chamber of the eye, has been shown to be a high-yield source of tumor-derived cfDNA [35]. It is sampled using a simple clear corneal paracentesis, involving extracting approximately 100 μL of fluid from the anterior chamber of the eye using a 32G needle, as illustrated in Fig. 2. Studies have shown that clinically relevant tumor genomic information, such as prognostic somatic copy number alterations and diagnostic single nucleotide variants, can be detected in the AH [36–39]. The presence of tumor-derived cfDNA has also been shown in the blood through the

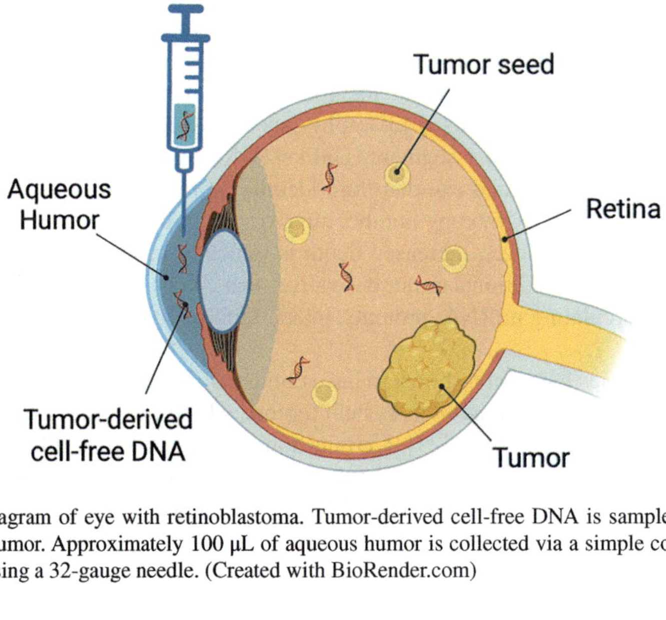

Fig. 2 Diagram of eye with retinoblastoma. Tumor-derived cell-free DNA is sampled from the aqueous humor. Approximately 100 μL of aqueous humor is collected via a simple corneal paracentesis using a 32-gauge needle. (Created with BioRender.com)

detection of somatic *RB1* variants [40, 41]. The emergence of both aqueous humor and blood has brought great excitement into the long-awaited application of precision medicine into various points of RB patient care. While both biopsies are generally considered safe, aqueous humor extraction has a small procedural risk that includes bleeding, infection, cataracts, and iris trauma and should only be done by a trained ophthalmic surgeon. Blood liquid biopsy requires a large volume of blood, often >10 mL, which may be uncomfortable. There are rare but some risks associated with venipuncture, including pain, bruising, and a very small risk of infection. This chapter will discuss the clinical applications of these liquid biopsies for diagnosis, prognosis, treatment, and treatment monitoring.

Diagnosis

While retinoblastoma is a clinical diagnosis that is made through an ophthalmic exam, obtaining a genetic diagnosis of RB is important for directing patient care and counseling at-risk family members. Patients with germline (vs somatic) *RB1* variants are at increased risk of developing additional cancers, including melanomas and sarcomas [42]. Even after successful treatment of RB, these patients still need longitudinal tumor surveillance. For this reason, identifying the mechanism of disease can direct the level of care a patient would need. While most patients with unilateral disease carry somatic *RB1* pathogenic variants, a small percentage (10–15%) carry a germline *RB1* variant [43]. Serum testing for analysis of germline alterations is possible; however, some variants and low-level mosaicism may be missed and is dependent on the sensitivity of testing, which varies by laboratory [43]. A false negative test can be impactful not only to the patient but also to future offspring. For instance, diagnostic screenings for RB and additional at-risk cancers may mistakenly not be indicated, thereby lowering early detection rates and worsening clinical outcomes of these patients. A more robust evaluation for a germline *RB1* variant starts with first identifying the pathogenic *RB1* variant(s) in the tumor and then evaluating if these variants are absent in the blood [44, 45]. This type of analysis requires tumor DNA, which could previously only be obtained in eyes that have been enucleated (surgically removed).

Both blood and AH liquid biopsies provide an avenue to obtain tumor-derived cfDNA without the need for tissue biopsies, providing the much-needed ability to identify somatic *RB1* variants in patients who do not undergo enucleation. CfDNA from both plasma and AH have been used to identify somatic *RB1* variants successfully, with results comparable to current tests on tumor [37, 40, 41, 44, 45]. Liquid biopsies have thereby allowed for genetic diagnosis of patients with RB, regardless of whether their eyes were enucleated or salvaged.

Blood liquid biopsies have also had significant implications on prenatal diagnosis of RB. Familial retinoblastoma is a heritable form of RB passed down in the autosomal dominant pattern, with up to 90% penetrance [46]. Early diagnosis of children at risk of inheriting an *RB1* pathogenic variant can improve clinical outcomes. The majority of genetic testing currently is performed on newborns and postnatal cord blood, with the aim of getting a diagnostic result within two weeks [47]. This has the disadvantages of being psychologically stressful for parents during a sensitive time postpartum. On the other hand, prenatal screening enables parents to have more time for planning decisions and to prepare mentally. A positive diagnosis of RB at the prenatal stage enables additional intervention opportunities even earlier. Such measures can include early induction of a fetus and treatment of the tumor to prevent tumor growth during gestation [48]. A study comparing outcomes of prenatal versus postnatal screening of patients with familial retinoblastoma demonstrated that prenatal molecular diagnosis with early-term delivery increased the likelihood of infants born with no detectable tumors, improved vision outcomes, and the need for less invasive therapy [49].

Prenatal testing can involve both invasive and noninvasive approaches. Invasive methods such as chorionic villus sampling and amniocentesis have been reported not to be favored by parents, with only 12% of patients opting for this service in a study in the Netherlands [50]. While considered a safe procedure, some patients fear the slight (0.5%) chance of miscarriage. Therefore, the development of a method of noninvasive prenatal diagnosis (NIPD) is highly desirable and can encourage patients to opt for prenatal testing. NIPD analyzes cell-free fetal DNA within the mother's blood. The first clinical uses of NIPD were for the detection of fetal sex determination [51] and aneuploidies [52, 53], notably Down syndrome [54].

Gerrish et al. developed an NIPD service for RB that is available as early as 8 weeks gestation, as illustrated in Fig. 3 [47]. In their study, NIPD showed 100% concordance with all postnatal testing performed. In this test, a blood sample is taken from the mother and cfDNA is analyzed using a combination of massively parallel sequencing (MPS) techniques depending on the model of inheritance. Amplicon-based MPS is used to detect paternal or suspected *de novo RB1* variants. For a fetus at risk of maternal inheritance, capture-based targeted sequencing and relative haplotype dosage analysis is performed. NIPD results are diagnostic, with no further testing required, and have started to be implemented clinically in England using a test developed by Birmingham Women's and Children's NHS Foundation Trust [55]. To the authors' knowledge, this is currently not yet being done in the United States or elsewhere.

Prognosis

Approaches to save an eye with RB include systemic and local chemotherapy, and focal therapies such as plaque brachytherapy, cryotherapy, and transpupillary thermotherapy [57]. In cases where treatment fails to preserve the eye, enucleation, or surgical removal of the eye is performed. The prediction of which eyes can be salvaged has commonly been made by using classification systems based on clinical features of the tumor, such as International Intraocular Retinoblastoma Classification (IIRC) [58] and the American Joint Committee on Cancer (AJCC) [59]. These

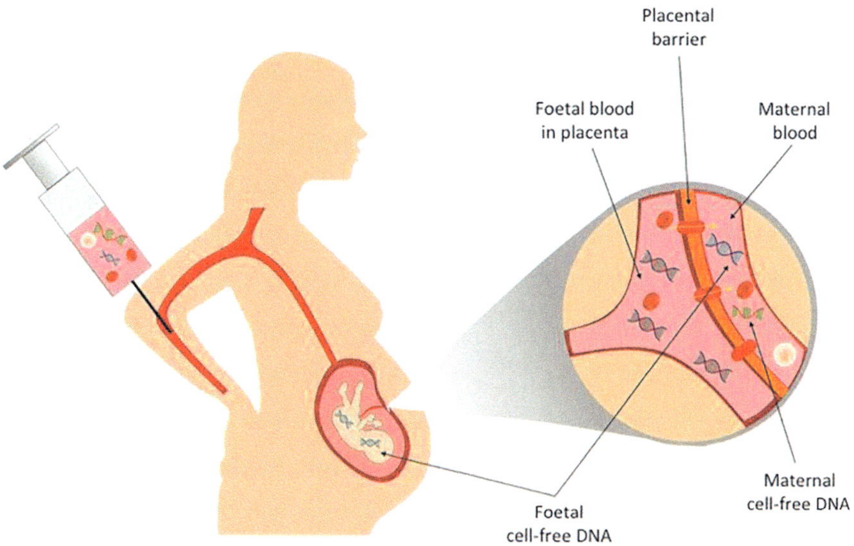

Fig. 3 "Noninvasive prenatal diagnosis (NIPD) [56]. The blood of a pregnant woman contains both maternal and fetal-derived cell-free DNA. Genetic analysis of the cell-free fetal DNA by NIPD can determine whether a fetus has inherited a pathogenic variant. Variants can be detected directly, or an indirect analysis, such as the relative haplotype dosage (RHDO) analysis, can be used." (From Gerrish et al. The Impact of Cell-Free DNA Analysis on the Management of Retinoblastoma) (This work is licensed under a Creative Commons Attribution 4.0 International License)

classification systems use clinical features; however, neither incorporate tumor genomics.

Somatic copy number alterations (SCNA), defined as regions of chromosomal gains or losses, are common in cancer and have been shown to contribute to tumorigenesis by affecting the dosage of multiple genes at the same time [60]. These chromosomal changes are frequent secondary genomic events during tumorigenesis in RB and contribute to disease progression [14, 61, 62]. Studies of enucleated eyes have shown that a gain on chromosome 6p is associated with less differentiated tumors, severe anaplasia, and higher rates of optic nerve invasion [63, 64]. Another SCNA of significance is those involving the focal amplification of the *MYCN* gene on chromosome 2p, which has been shown to be a biomarker for aggressive RB, as well as aggressive disease and poor outcomes for pediatric neuroblastoma (a cancer that originated in early nerve cells called neuroblasts, most commonly in and around the adrenal glands) [13, 65].

Previously SCNA analysis for RB was limited to tumor biopsy after enucleation, thereby providing no prognostic benefit since salvaged eyes could not be analyzed. The emergence of liquid biopsies enables tumor genomic analysis from salvaged eyes. In a retrospective analysis, Berry et al. demonstrated that the detection of SCNAs in the aqueous humor can predict the likelihood of successful salvage of the eye. A gain on chromosome 6p is associated with a 10-fold increased odds of treatment failure and enucleation (surgical removal of the eye), as shown in Fig. 4 [36]. Additionally, a molecular signature of both a 6p gain and *MYCN* amplification is prognostic for a 16.5-fold increased odds of enucleation [38].

In addition to SCNAs, single nucleotide polymorphisms have also been identified as biomarkers of retinoblastoma, including risk of disease, aggressiveness, invasive growth, and age at diagnosis [66–69]. A single nucleotide polymorphism

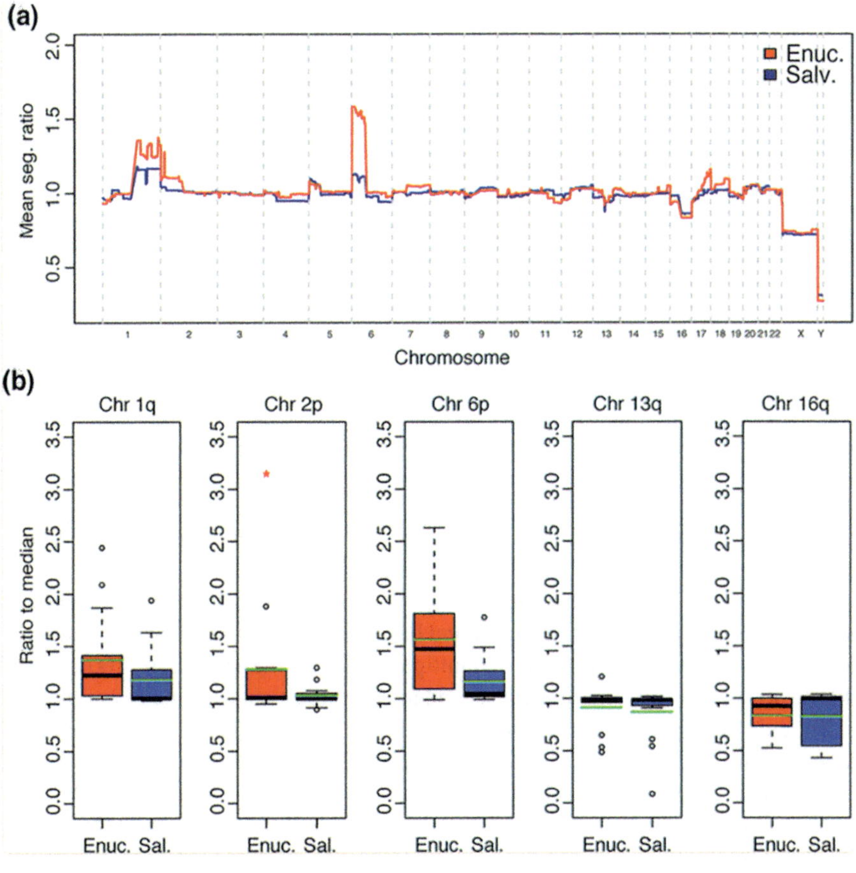

Fig. 4 "(**a**) Composite somatic copy number alteration (CNA) profile from cell-free DNA in the Aqueous Humor (AH) samples from enucleated eyes (Enuc, red) and salvaged eyes (Salv, blue) [36]. (**b**). Box plot demonstrating the range of amplitude changes for the enucleated (Enuc) vs. salvaged (Salv) eyes; the black bar represents the median while the green bar represents the mean (of the ratio to median). The sample with focal *MYCN* gain is shown as a red asterisk in the Chr 2p plot. The median of the ratio to the median amplitude of Chr 6p gain is significantly greater in enucleated eyes ($p = 0.001$), which may be both from the increased copy number of the amplified region and an increase in the total fraction of tumor-derived DNA in the AH of enucleated eyes." (From Berry et al. Genomic cfDNA Analysis of Aqueous Humor in Retinoblastoma Predicts Eye Salvage: The Surrogate Tumor Biopsy for Retinoblastoma)

(SNP) is a genomic variant at a single base position in the DNA in which one base is substituted for another. Gall et al. developed a highly sensitive detection method with a gene panel that can detect SNPs of interest in the aqueous humor, such as variants on the genes *MDM2* and *MDM4* [45]. These variants have been shown to be biomarkers of tumor aggressiveness and increased risk of disease in predisposed patients. It has been reported that MDM2 directly binds p53, a tumor suppressor protein, thereby blocking its activity and promoting its degradation [70]. The exact role of MDM4 is less well understood, but it has been shown to contribute to tumor formation by suppressing p53 and is reported to be amplified in approximately 65% of retinoblastomas [71]. In fact, inhibition of MDM2/MDM4-p53 interaction has become a promising therapeutic approach for RB through the development of small molecule inhibitors, such as nutlin-3a and ALRN-6924 [70, 72].

Using the aqueous humor to detect relevant SCNAs and SNPs can provide a more comprehensive molecular understanding of the tumor and offer prognostic biomarkers with clinical utility.

Evaluating Tumor Response

A benefit of minimally invasive liquid biopsies is that they can be collected longitudinally throughout treatment. This uncovers the possibilities of studying the real-time dynamics of tumor genetic material in response to treatment. One of the measures that has been shown to have clinical potential in evaluating tumor response is calculating the tumor fraction, as shown in Polski et al. [73] In this study, tumor fraction is defined as the proportion of sampled cfDNA that is tumor-derived. In studying 20 eyes that underwent over three AH extractions during intravitreal chemotherapy or at secondary enucleation (surgical removal of the eye after treatment), increases in tumor fraction were associated with disease progression and relative decreases were associated with disease regression. The study reported that a 15% or greater increase in tumor fraction relative to baseline during treatment is associated with over a 90-fold increased likelihood of clinical progression. In blood cfDNA in RB patients, *RB1* variant allele frequency (VAF; a proxy for tumor fraction) was demonstrated to decrease after treatment, suggesting that relative VAF changes can be a biomarker of tumor response. Similar trends of increase in tumor fraction and tumor response have been shown using blood-based liquid biopsies in other cancers, such as prostate and lung [74, 75].

While AH liquid biopsy sampling is currently limited to during intravitreal chemotherapy or immediately after enucleation, this study shows the clinical value of routine sampling of AH after completion of therapy for long-term monitoring for residual or recurrent disease. With multicenter validation, tumor fraction has the potential to become a clinically useful biomarker for RB therapeutic response and surveillance of recurrent disease.

Future Directions

The most established liquid biopsy analyte for RB is cfDNA, and its applications are currently being adopted into clinical practice. For instance, LBSeqKids (liquid biopsy sequencing for kids) is a clinically validated assay to detect chromosomal alterations in cfDNA for various pediatric cancers using blood, cerebrospinal fluid, and AH [76]. This test is being offered at the Center for Personalized Medicine at Children's Hospital Los Angeles and uses combined low-pass whole genome sequencing and targeted sequencing of cfDNA to detect SCNAs and sequence variants.

Additional applications of cfDNA in RB that show potential for clinical utility include epigenetic profiling, which involves the characterization of biochemical modifications of DNA. As it has been reported that 13% of nonhereditary RB tumors have *RB1* promoter hypermethylation and silencing as the initiating cause of disease, epigenetic profiling using liquid biopsies may provide insight into tumorigenesis [77–79]. Additionally, epigenetic deregulation of tumor-promoting pathways has also been associated with further tumorigenesis and disease progression. While previously only detected from tumor tissue harvested from enucleated eyes, the hypermethylation of the *RB1* promoter was reported to be identified in AH cfDNA of a patient [80]. In this study evaluating DNA methylation signatures in AH cfDNA samples of RB patients, paired signatures from AH and primary tumor demonstrated a high degree of concordance. Results further showed that the methylation signatures, which involved 320 differentially methylated probes, could differentiate between eyes that were salvaged with therapy and eyes that were enucleated, with enucleated eyes being correlated with a previously defined subtype of aggressive RB that is associated with dedifferentiation and stemness features [81]. This study suggests that AH is a reliable tumor surrogate from epigenetic profiling with the potential to provide additional prognostic biomarkers.

Further lines of research are under development that go beyond the use of cfDNA analytes in liquid biopsies for RB. One significant analyte is extracellular vesicles. Extracellular vesicles (EVs) are membrane-enclosed particles released by nearly all cells in the body and play a key role in both physiological and pathological processes by mediating cell-cell communication [82, 83]. Cargo carried by EVs includes proteins, nucleic acids, and specialized lipids [84]. By mediating communication between tumor cells and their environment, tumor-derived EVs are involved in tumorigenesis, disease progression, and metastasis [85–87]. For example, in medulloblastoma, the most common malignant pediatric brain tumor, EVs have been shown to contribute to metastasis, with one of its surface proteins degrading the extracellular matrix and creating routes of medulloblastoma cells to invade the surrounding environment [88]. As a result, this EV surface protein was identified as a potential biomarker for metastasis as well as a potential therapeutic target. In RB, cell surface proteins called tetraspanins have been used for phenotypic expression profiling of EVs. Tetraspanins have been shown to have a role in EV biogenesis, cell targeting, and cargo selection [89]. Tetraspanin-positive EVs could be detected in the AH of pediatric patients with various eye diseases, including RB. Through tetraspanin-based profiling of EVs in the study of RB patients, a potential tumor-derived EV subpopulation was identified in the AH. This subpopulation was enriched prior to treatment in eyes with a more significant tumor burden, as defined by more advanced disease classification. Due to the role EVs can have in the tumorigenesis and progression of disease, further molecular profiling of this population may reveal new prognostic biomarkers.

Conclusion

This chapter discusses the rise of liquid biopsies in a much-needed pediatric ocular cancer, retinoblastoma. As tumor biopsy is contraindicated for retinoblastoma, precision medicine applications involving direct tumor profiling are limited in this cancer. In this chapter, research in both aqueous humor and blood-liquid biopsies is summarized, demonstrating the potential of these biofluids in providing tumor-specific information that can inform diagnosis, treatment monitoring, and prognostication.

Although both aqueous humor and blood contain tumor-derived information, each biofluid has specific advantages. The ability to examine fetal-derived cell-free DNA in the maternal blood enables at-risk patients to be diagnosed prenatally, improving clinical planning and outcomes [47]. Further, blood biopsies provide a less invasive sample procedure and enable a larger sample volume than the aqueous humor, for which approximately 0.1 mL is extracted via paracentesis. Although limited to a smaller sample volume, aqueous humor has been shown to be a higher-yield source of cell-free DNA with a higher tumor fraction than blood when evaluating equal volumes of both biofluids during therapy [90]. Aqueous humor also provides the benefit of being eye-specific, which is particularly important for patients with tumors in both eyes as each tumor can have distinct genomic profiles [91]. However, the organ-specific nature of aqueous humor also limits its role in systemic surveillance, whereas blood biopsies can have a role in identifying patients at risk for metastatic disease, surveillance, and evaluating systemic response to treatment. In conclusion, the unique benefits of blood and aqueous humor warrant future prospective, multi-institution studies into both liquid biopsies.

Competing Interests This study was funded by Research to Prevent Blindness/Castle Biosciences Medical Student Eye Research Fellowship in Ocular Cancer (Recipient: Shreya Sirivolu, Mentor: Jesse Berry), Children's Hospital Los Angeles Saban Research Institute (Research Career Development Award), The Knights Templar Eye Foundation, National Cancer Institute of the National Institute of Health Award Number K08CA232344, The Wright Foundation, Children's Oncology Group/St. Baldrick's Foundation, Danhakl Family Foundation, A. Linn Murphree, MD, Chair in Ocular Oncology, The Berle & Lucy Adams Chair in Cancer Research, The Larry and Celia Moh Foundation, an unrestricted departmental grant from Research to Prevent Blindness.

Drs. Jesse L. Berry and Liya Xu have filed a patent application entitled: Aqueous humor cell-free DNA for diagnostic and prognostic evaluation of Ophthalmic Disease.

References

1. Rao R, Honavar SG. Retinoblastoma. *Indian J Pediatr.* 2017;84(12):937–44. https://doi.org/10.1007/s12098-017-2395-0.
2. Dimaras H, Corson TW, Cobrinik D, et al. Retinoblastoma. *Nat Rev Dis Primers.* 2015;1:15021. https://doi.org/10.1038/nrdp.2015.21.
3. Nyawira G, Kahaki K, Kariuki-Wanyoike M. Survival among retinoblastoma patients at the Kenyatta National Hospital, Kenya. Published online August 2013. Accessed February 12, 2024. http://erepository.uonbi.ac.ke/handle/11295/88504
4. Dean M, Bendfeldt G, Lou H, et al. Increased incidence and disparity of diagnosis of retinoblastoma patients in Guatemala. *Cancer Lett.* 2014;351(1):59–63. https://doi.org/10.1016/j.canlet.2014.04.023.
5. Canadian Retinoblastoma Society. National Retinoblastoma Strategy Canadian Guidelines for Care: Stratégie thérapeutique du rétinoblastome guide clinique canadien. *Can J Ophthalmol.* 2009;44(Suppl 2):S1–88. https://doi.org/10.3129/i09-194.
6. Moreno F, Sinaki B, Fandiño A, Dussel V, Orellana L, Chantada G. A population-based study of retinoblastoma incidence and survival in Argentine children. *Pediatr Blood Cancer.* 2014;61(9):1610–5. https://doi.org/10.1002/pbc.25048.
7. Kumar A, Moulik NR, Mishra RK, Kumar D. Causes, outcome and prevention of abandonment in retinoblastoma in India. *Pediatr Blood Cancer.* 2013;60(5):771–5. https://doi.org/10.1002/pbc.24454.
8. Gichigo EN, Kariuki-Wanyoike MM, Kimani K, Nentwich MM. [Retinoblastoma in Kenya: survival and prognostic factors]. *Ophthalmologe.* 2015;112(3):255-260. https://doi.org/10.1007/s00347-014-3123-z
9. Aerts I, Lumbroso-Le Rouic L, Gauthier-Villars M, Brisse H, Doz F, Desjardins L. Retinoblastoma. *Orphanet J Rare Dis.* 2006;1:31. https://doi.org/10.1186/1750-1172-1-31.
10. Mehta M, Sethi S, Pushker N, et al. Retinoblastoma. *Singapore Med J.* 2012;53(2):128–35; quiz 136.
11. Thériault BL, Dimaras H, Gallie BL, Corson TW. The genomic landscape of retinoblastoma: a review. *Clin Exp Ophthalmol.* 2014;42(1):33–52. https://doi.org/10.1111/ceo.12132.
12. Kooi IE, Mol BM, Massink MPG, et al. A Meta-Analysis of Retinoblastoma Copy Numbers Refines the List of Possible Driver Genes Involved in Tumor Progression. *PLoS One.* 2016;11(4) https://doi.org/10.1371/journal.pone.0153323.
13. Rushlow DE, Mol BM, Kennett JY, et al. Characterisation of retinoblastomas without RB1 mutations: genomic, gene expression, and clinical studies. *Lancet Oncol.* 2013;14(4):327–34. https://doi.org/10.1016/S1470-2045(13)70045-7.
14. Corson TW, Gallie BL. One hit, two hits, three hits, more? Genomic changes in the development of retinoblastoma. *Genes Chromosomes Cancer.* 2007;46(7):617–34. https://doi.org/10.1002/gcc.20457.
15. Bowles E, Corson TW, Bayani J, et al. Profiling genomic copy number changes in retinoblastoma beyond loss of RB1. *Genes Chromosomes Cancer.* 2007;46(2):118–29. https://doi.org/10.1002/gcc.20383.
16. Grasemann C, Gratias S, Stephan H, et al. Gains and overexpression identify DEK and E2F3 as targets of chromosome 6p gains in retinoblastoma. *Oncogene.* 2005;24:6441–9. https://doi.org/10.1038/sj.onc.1208792.
17. Karcioglu ZA, Gordon RA, Karcioglu GL. Tumor seeding in ocular fine needle aspiration biopsy. *Ophthalmology.* 1985;92(12):1763–7. https://doi.org/10.1016/s0161-6420(85)34105-2.
18. Andersson D, Fagman H, Dalin MG, Ståhlberg A. Circulating cell-free tumor DNA analysis in pediatric cancers. *Molecular Aspects of Medicine.* 2020;72:100819. https://doi.org/10.1016/j.mam.2019.09.003.
19. Connal S, Cameron JM, Sala A, et al. Liquid biopsies: the future of cancer early detection. *J Transl Med.* 2023;21:118. https://doi.org/10.1186/s12967-023-03960-8.

20. Abbou SD, Shulman DS, DuBois SG, Crompton BD. Assessment of circulating tumor DNA in pediatric solid tumors: The promise of liquid biopsies. *Pediatr Blood Cancer.* 2019;66(5):e27595. https://doi.org/10.1002/pbc.27595.
21. Wan JCM, Massie C, Garcia-Corbacho J, et al. Liquid biopsies come of age: towards implementation of circulating tumour DNA. *Nat Rev Cancer.* 2017;17(4):223–38. https://doi.org/10.1038/nrc.2017.7.
22. Merker JD, Oxnard GR, Compton C, et al. Circulating Tumor DNA Analysis in Patients With Cancer: American Society of Clinical Oncology and College of American Pathologists Joint Review. *Arch Pathol Lab Med.* 2018;142(10):1242–53. https://doi.org/10.5858/arpa.2018-0901-SA.
23. Corcoran RB, Chabner BA. Application of Cell-free DNA Analysis to Cancer Treatment. *N Engl J Med.* 2018;379(18):1754–65. https://doi.org/10.1056/NEJMra1706174.
24. Sundby RT, Pan A, Shern JF. Liquid biopsies in pediatric oncology: opportunities and obstacles. *Curr Opin Pediatr.* 2022;34(1):39–47. https://doi.org/10.1097/MOP.0000000000001088.
25. Gerlinger M, Rowan AJ, Horswell S, et al. Intratumor heterogeneity and branched evolution revealed by multiregion sequencing. *N Engl J Med.* 2012;366(10):883–92. https://doi.org/10.1056/NEJMoa1113205.
26. Bedard PL, Hansen AR, Ratain MJ, Siu LL. Tumour heterogeneity in the clinic. *Nature.* 2013;501(7467):355–64. https://doi.org/10.1038/nature12627.
27. Sweet-Cordero EA, Biegel JA. The genomic landscape of pediatric cancers: Implications for diagnosis and treatment. *Science.* 2019;363(6432):1170–5. https://doi.org/10.1126/science.aaw3535.
28. Armakolas A, Kotsari M, Koskinas J. Liquid Biopsies, Novel Approaches and Future Directions. *Cancers (Basel).* 2023;15(5):1579. https://doi.org/10.3390/cancers15051579.
29. Bonner ER, Bornhorst M, Packer RJ, Nazarian J. Liquid biopsy for pediatric central nervous system tumors. *npj Precision Onc.* 2018;2(1):1–9. https://doi.org/10.1038/s41698-018-0072-z.
30. Greuter L, Frank N, Guzman R, Soleman J. The Clinical Applications of Liquid Biopsies in Pediatric Brain Tumors: A Systematic Literature Review. *Cancers (Basel).* 2022;14(11):2683. https://doi.org/10.3390/cancers14112683.
31. Shulman DS, Klega K, Imamovic-Tuco A, et al. Detection of circulating tumour DNA is associated with inferior outcomes in Ewing sarcoma and osteosarcoma: a report from the Children's Oncology Group. *Br J Cancer.* 2018;119(5):615–21. https://doi.org/10.1038/s41416-018-0212-9.
32. Miguez ACK, de Figueiredo Barros BD, de Souza JES, et al. Assessment of somatic mutations in urine and plasma of Wilms tumor patients. *Cancer Med.* 2020;9(16):5948–59. https://doi.org/10.1002/cam4.3236.
33. Kahana-Edwin S, McCowage G, Cain L, et al. Exploration of CTNNB1 ctDNA as a putative biomarker for hepatoblastoma. *Pediatr Blood Cancer.* 2020;67(11):e28594. https://doi.org/10.1002/pbc.28594.
34. McCabe MG, Geoerger B, Chesler L, et al. Precision Medicine for Childhood Cancer: Current Limitations and Future Perspectives. *JCO Precis Oncol.* 2024;(8):e2300117. https://doi.org/10.1200/PO.23.00117
35. Berry JL, Xu L, Murphree AL, et al. Potential of Aqueous Humor as a Surrogate Tumor Biopsy for Retinoblastoma. *JAMA Ophthalmol.* 2017;135(11):1221–30. https://doi.org/10.1001/jamaophthalmol.2017.4097.
36. Berry JL, Xu L, Kooi I, et al. Genomic cfDNA Analysis of Aqueous Humor in Retinoblastoma Predicts Eye Salvage: The Surrogate Tumor Biopsy for Retinoblastoma. *Mol Cancer Res.* 2018;16(11):1701–12. https://doi.org/10.1158/1541-7786.MCR-18-0369.
37. Xu L, Shen L, Polski A, et al. Simultaneous identification of clinically relevant RB1 mutations and copy number alterations in aqueous humor of retinoblastoma eyes. *Ophthalmic Genet.* 2020;41(6):526–32. https://doi.org/10.1080/13816810.2020.1799417.

38. Xu L, Kim ME, Polski A, et al. Establishing the Clinical Utility of ctDNA Analysis for Diagnosis, Prognosis, and Treatment Monitoring of Retinoblastoma: The Aqueous Humor Liquid Biopsy. *Cancers (Basel)*. 2021;13(6). https://doi.org/10.3390/cancers13061282
39. Xu L, Polski A, Prabakar RK, et al. Chromosome 6p Amplification in Aqueous Humor Cell-Free DNA Is a Prognostic Biomarker for Retinoblastoma Ocular Survival. *Mol Cancer Res*. 2020;18(8):1166–75. https://doi.org/10.1158/1541-7786.MCR-19-1262.
40. Kothari P, Marass F, Yang JL, et al. Cell-free DNA profiling in retinoblastoma patients with advanced intraocular disease: An MSKCC experience. *Cancer Med*. 2020;9(17):6093–101. https://doi.org/10.1002/cam4.3144.
41. Abramson DH, Mandelker D, Francis JH, et al. Retrospective Evaluation of Somatic Alterations in Cell-Free DNA from Blood in Retinoblastoma. *Ophthalmology Science*. 2021;1(1):100015. https://doi.org/10.1016/j.xops.2021.100015.
42. Ketteler P, Hülsenbeck I, Frank M, Schmidt B, Jöckel KH, Lohmann DR. The impact of RB1 genotype on incidence of second tumours in heritable retinoblastoma. *European Journal of Cancer*. 2020;133:47–55. https://doi.org/10.1016/j.ejca.2020.04.005.
43. Skalet AH, Gombos DS, Gallie BL, et al. Screening Children at Risk for Retinoblastoma: Consensus Report from the American Association of Ophthalmic Oncologists and Pathologists. *Ophthalmology*. 2018;125(3):453–8. https://doi.org/10.1016/j.ophtha.2017.09.001.
44. Gerrish A, Stone E, Clokie S, et al. Non-invasive diagnosis of retinoblastoma using cell-free DNA from aqueous humour. *British Journal of Ophthalmology*. 2019;103(5):721–4. https://doi.org/10.1136/bjophthalmol-2018-313005.
45. Le Gall J, Dehainault C, Benoist C, et al. Highly Sensitive Detection Method of Retinoblastoma Genetic Predisposition and Biomarkers. *The Journal of Molecular Diagnostics*. 2021;23(12):1714–21. https://doi.org/10.1016/j.jmoldx.2021.08.014.
46. Rodriguez A, Kelley C, Patel A, Ramasubramanian A. Prenatal Diagnosis of Retinoblastomas: A Scoping Review. *Int J Gen Med*. 2023;16:1101–10. https://doi.org/10.2147/IJGM.S380634.
47. Gerrish A, Bowns B, Mashayamombe-Wolfgarten C, et al. Non-Invasive Prenatal Diagnosis of Retinoblastoma Inheritance by Combined Targeted Sequencing Strategies. *J Clin Med*. 2020;9(11). https://doi.org/10.3390/jcm9113517
48. Richter S, Vandezande K, Chen N, et al. Sensitive and efficient detection of RB1 gene mutations enhances care for families with retinoblastoma. *Am J Hum Genet*. 2003;72(2):253–69. https://doi.org/10.1086/345651.
49. Soliman SE, Dimaras H, Khetan V, et al. Prenatal versus Postnatal Screening for Familial Retinoblastoma. *Ophthalmology*. 2016;123(12):2610–7. https://doi.org/10.1016/j.ophtha.2016.08.027.
50. Dommering CJ, Henneman L, van der Hout AH, et al. Uptake of prenatal diagnostic testing for retinoblastoma compared to other hereditary cancer syndromes in the Netherlands. *Fam Cancer*. 2017;16(2):271–7. https://doi.org/10.1007/s10689-016-9943-z.
51. Hill M, Finning K, Martin P, et al. Non-invasive prenatal determination of fetal sex: translating research into clinical practice. *Clin Genet*. 2011;80(1):68–75. https://doi.org/10.1111/j.1399-0004.2010.01533.x.
52. Fan HC, Blumenfeld YJ, Chitkara U, Hudgins L, Quake SR. Noninvasive diagnosis of fetal aneuploidy by shotgun sequencing DNA from maternal blood. *Proc Natl Acad Sci U S A*. 2008;105(42):16266–71. https://doi.org/10.1073/pnas.0808319105.
53. Lo YMD, Lun FMF, Chan KCA, et al. Digital PCR for the molecular detection of fetal chromosomal aneuploidy. *Proc Natl Acad Sci U S A*. 2007;104(32):13116–21. https://doi.org/10.1073/pnas.0705765104.
54. van der Meij KRM, Sistermans EA, Macville MVE, et al. TRIDENT-2: National Implementation of Genome-wide Non-invasive Prenatal Testing as a First-Tier Screening Test in the Netherlands. *Am J Hum Genet*. 2019;105(6):1091–101. https://doi.org/10.1016/j.ajhg.2019.10.005.

55. NHS rolls out new test to detect eye cancer in unborn babies. NS Medical Devices. Published May 4, 2022. Accessed February 28, 2024. https://www.nsmedicaldevices.com/news/nhs-rolls-out-new-test-to-detect-eye-cancer-in-unborn-babies/
56. Gerrish A, Jenkinson H, Cole T. The Impact of Cell-Free DNA Analysis on the Management of Retinoblastoma. *Cancers (Basel)*. 2021;13(7). https://doi.org/10.3390/cancers13071570
57. Mendoza PR, Grossniklaus HE. Therapeutic Options for Retinoblastoma. *Cancer Control*. 2016;23(2):99–109. https://doi.org/10.1177/107327481602300203.
58. Linn MA. Intraocular retinoblastoma: the case for a new group classification. *Ophthalmol Clin North Am*. 2005;18(1):41–53, viii. https://doi.org/10.1016/j.ohc.2004.11.003.
59. Tomar AS, Finger PT, Gallie B, et al. A Multicenter, International Collaborative Study for American Joint Committee on Cancer Staging of Retinoblastoma: Part I: Metastasis-Associated Mortality. *Ophthalmology*. 2020;127(12):1719–32. https://doi.org/10.1016/j.ophtha.2020.05.050.
60. Harbers L, Agostini F, Nicos M, Poddighe D, Bienko M, Crosetto N. Somatic Copy Number Alterations in Human Cancers: An Analysis of Publicly Available Data From The Cancer Genome Atlas. *Front Oncol*. 2021;11:700568. https://doi.org/10.3389/fonc.2021.700568.
61. Kooi IE, Mol BM, Massink MPG, et al. Somatic genomic alterations in retinoblastoma beyond RB1 are rare and limited to copy number changes. *Sci Rep*. 2016;6:25264. https://doi.org/10.1038/srep25264.
62. Kooi IE, Mol BM, Massink MPG, et al. A Meta-Analysis of Retinoblastoma Copy Numbers Refines the List of Possible Driver Genes Involved in Tumor Progression. *PLoS One*. 2016;11(4):e0153323. https://doi.org/10.1371/journal.pone.0153323.
63. Cano J, Oliveros O, Yunis E. Phenotype variants, malignancy, and additional copies of 6p in retinoblastoma. *Cancer Genet Cytogenet*. 1994;76(2):112–5. https://doi.org/10.1016/0165-4608(94)90459-6.
64. Stålhammar G, Yeung A, Mendoza P, Dubovy SR, Harbour JW, Grossniklaus HE. Gain of Chromosome 6p Correlates with Severe Anaplasia, Cellular Hyperchromasia, and Extraocular Spread of Retinoblastoma *Ophthalmology Science*. 2022;2(1). https://doi.org/10.1016/j.xops.2021.100089
65. Brodeur GM. Neuroblastoma: biological insights into a clinical enigma. *Nat Rev Cancer*. 2003;3(3):203–16. https://doi.org/10.1038/nrc1014.
66. Carvalho INSR, de Oliveira Reis AH, Cabello PH, Vargas FR. Polymorphisms of CDKN1A gene and risk of retinoblastoma. *Carcinogenesis*. 2013;34(12):2774–7. https://doi.org/10.1093/carcin/bgt308.
67. Yu F, Jiang Z, Song A. Association of rs11801299 and rs1380576 polymorphisms at MDM4 with risk, clinicopathological features and prognosis in patients with retinoblastoma. *Cancer Epidemiol*. 2019;58:153–9. https://doi.org/10.1016/j.canep.2018.12.010.
68. Chen R, Liu S, Ye H, et al. Association of p53 rs1042522, MDM2 rs2279744, and p21 rs1801270 polymorphisms with retinoblastoma risk and invasion in a Chinese population. *Sci Rep*. 2015;5:13300. https://doi.org/10.1038/srep13300.
69. Castéra L, Sabbagh A, Dehainault C, et al. MDM2 as a modifier gene in retinoblastoma. *J Natl Cancer Inst*. 2010;102(23):1805–8. https://doi.org/10.1093/jnci/djq416.
70. Pritchard EM, Dyer MA, Guy RK. Progress in Small Molecule Therapeutics for the Treatment of Retinoblastoma. *Mini Rev Med Chem*. 2016;16(6):430–54. https://doi.org/10.2174/1389557515666150722100610.
71. Laurie NA, Donovan SL, Shih CS, et al. Inactivation of the p53 pathway in retinoblastoma. *Nature*. 2006;444(7115):61–6. https://doi.org/10.1038/nature05194.
72. Wilson MW, Stewart E, Dyer MA, Annis A, Brennan RC. Intraocular MDMX Inhibition by ALRN-6924: A Targeted Strategy to Reactivate p53 in Retinoblastoma. *Investigative Ophthalmology & Visual Science*. 2023;64(8):1272.
73. Polski A, Xu L, Prabakar RK, et al. Cell-Free DNA Tumor Fraction in the Aqueous Humor Is Associated With Therapeutic Response in Retinoblastoma Patients. *Transl Vis Sci Technol*. 2020;9(10):30. https://doi.org/10.1167/tvst.9.10.30.

74. Choudhury AD, Werner L, Francini E, et al. Tumor fraction in cell-free DNA as a biomarker in prostate cancer. *JCI Insight*. 2018;3(21):122109. https://doi.org/10.1172/jci.insight.122109.
75. Chen X, Chang CW, Spoerke JM, et al. Low-pass Whole-genome Sequencing of Circulating Cell-free DNA Demonstrates Dynamic Changes in Genomic Copy Number in a Squamous Lung Cancer Clinical Cohort. *Clin Cancer Res*. 2019;25(7):2254–63. https://doi.org/10.1158/1078-0432.CCR-18-1593.
76. Christodoulou E, Yellapantula V, O'Halloran K, et al. Combined low-pass whole genome and targeted sequencing in liquid biopsies for pediatric solid tumors. *npj Precis Onc*. 2023;7(1):1–11. https://doi.org/10.1038/s41698-023-00357-0.
77. Sakai T, Toguchida J, Ohtani N, Yandell DW, Rapaport JM, Dryja TP. Allele-specific hypermethylation of the retinoblastoma tumor-suppressor gene. *Am J Hum Genet*. 1991;48(5):880–8.
78. Stirzaker C, Millar DS, Paul CL, et al. Extensive DNA methylation spanning the Rb promoter in retinoblastoma tumors. *Cancer Res*. 1997;57(11):2229–37.
79. Greger V, Debus N, Lohmann D, Höpping W, Passarge E, Horsthemke B. Frequency and parental origin of hypermethylated RB1 alleles in retinoblastoma. *Hum Genet*. 1994;94(5):491–6. https://doi.org/10.1007/BF00211013.
80. Li HT, Xu L, Weisenberger DJ, et al. Characterizing DNA methylation signatures of retinoblastoma using aqueous humor liquid biopsy. *Nat Commun*. 2022;13(1):5523. https://doi.org/10.1038/s41467-022-33248-2.
81. Liu J, Ottaviani D, Sefta M, et al. A high-risk retinoblastoma subtype with stemness features, dedifferentiated cone states and neuronal/ganglion cell gene expression. *Nat Commun*. 2021;12(1):5578. https://doi.org/10.1038/s41467-021-25792-0.
82. van Niel G, D'Angelo G, Raposo G. Shedding light on the cell biology of extracellular vesicles. *Nat Rev Mol Cell Biol*. 2018;19(4):213–28. https://doi.org/10.1038/nrm.2017.125.
83. Yuana Y, Sturk A, Nieuwland R. Extracellular vesicles in physiological and pathological conditions. *Blood Rev*. 2013;27(1):31–9. https://doi.org/10.1016/j.blre.2012.12.002.
84. Pitt JM, Kroemer G, Zitvogel L. Extracellular vesicles: masters of intercellular communication and potential clinical interventions. *J Clin Invest*. 126(4):1139–43. https://doi.org/10.1172/JCI87316.
85. Becker A, Thakur BK, Weiss JM, Kim HS, Peinado H, Lyden D. Extracellular Vesicles in Cancer: Cell-to-Cell Mediators of Metastasis. *Cancer Cell*. 2016;30(6):836–48. https://doi.org/10.1016/j.ccell.2016.10.009.
86. Nakata R, Shimada H, Fernandez GE, et al. Contribution of neuroblastoma-derived exosomes to the production of pro-tumorigenic signals by bone marrow mesenchymal stromal cells. *J Extracell Vesicles*. 2017;6(1):1332941. https://doi.org/10.1080/20013078.2017.1332941.
87. Tamura T, Yoshioka Y, Sakamoto S, Ichikawa T, Ochiya T. Extracellular Vesicles in Bone Metastasis: Key Players in the Tumor Microenvironment and Promising Therapeutic Targets. *Int J Mol Sci*. 2020;21(18):6680. https://doi.org/10.3390/ijms21186680.
88. Jackson HK, Mitoko C, Linke F, Macarthur D, Kerr ID, Coyle B. Extracellular Vesicles Potentiate Medulloblastoma Metastasis in an EMMPRIN and MMP-2 Dependent Manner. *Cancers (Basel)*. 2023;15(9):2601. https://doi.org/10.3390/cancers15092601.
89. Jankovičová J, Sečová P, Michalková K, Antalíková J. Tetraspanins, More than Markers of Extracellular Vesicles in Reproduction. *Int J Mol Sci*. 2020;21(20):7568. https://doi.org/10.3390/ijms21207568.
90. Berry JL, Xu L, Polski A, et al. Aqueous Humor Is Superior to Blood as a Liquid Biopsy for Retinoblastoma. *Ophthalmology*. 2020;127(4):552–4. https://doi.org/10.1016/j.ophtha.2019.10.026.
91. Wong, Elyssa. Inter-eye genomic heterogeneity in bilateral retinoblastoma via aqueous humor liquid biopsy. In: ; 2021. https://arvo2021.arvo.org/meetings/virtual/Eug3i6N2bDQ4kbXsJ

Care Innovation in Pediatric Radiology

Steven L. Blumer

Origins of Pediatric Radiology and Imaging Modalities

Physicians were not able to see pictures of what was happening inside a patient's body until November 8, 1895. On that day, German mechanical engineer and physicist Wilhelm Conrad Röntgen (also spelled Roentgen) discovered what he termed "X-rays" while experimenting with a Crookes cathode ray tube [1]. This revelation was and would become so monumental that it continues to alter the course of human disease today while earning him global recognition then; among the countless accolades was the 1901 receipt of the inaugural Nobel Prize in physics [2]. As X-ray absorption depends upon the densities of structures, X-ray absorption by bones, but less so by soft tissues and even less by fatty tissue, can create a clear picture [1]. The first and most well-known X-ray image in history was of the hand of Röntgen's wife, Bertha [1].

The Crookes tube used by Röntgen was promptly replicated in other labs throughout the world. Shortly after Conrad's discovery, the device was used to examine bone fractures [1]. On February 3, 1896, a 14-year-old boy who was believed to have fractured his wrist two weeks earlier while skating on the Connecticut River was medically evaluated and his wrist radiographed. The exposure took about twenty minutes and the resulting image on the emulsion-coated glass plate was able to demonstrate a fracture in his ulna. This image was not only apparently the first clinical radiographic examination in America, but this event also marked the birth of pediatric radiology in the Western Hemisphere [3]. A short time later, the Hospital for Sick Children in Toronto, Ontario, became the first children's hospital in North America to acquire X-ray equipment [3].

S. L. Blumer (✉)
UPMC Children's Hospital of Pittsburgh, Pittsburgh, PA, USA

University of Pittsburgh School of Medicine, Pittsburgh, PA, USA
e-mail: blumersl@upmc.edu

J. L. Wells (ed.), *A Clinical Lens on Pediatric Engineering*,
https://doi.org/10.1007/978-3-031-84655-7_8

Shortly after the discovery of X-ray technology, an Italian radiologist, Alessandro Vallebona, invented conventional tomography early in the twentieth century [1]. Conventional tomography is created by simultaneously moving the X-ray source and X-ray detector at the same time to maintain an object of interest in the center point of the scan plane and blur out all other objects [1]. Although this technique evolved, it proved to be largely ineffective in imaging soft tissues and large areas of the body [1]. **Soft tissue** is a term used to describe all the supporting tissues in the body that are not calcified or ossified; therefore, soft tissue does not include bone and teeth. Soft tissues include skin, fat, muscles, tendons, ligaments, fibrous tissues and vasculature. They generally exclude the internal body organs [4].

The development of computed tomography (CT) was able to make up for the shortcomings of conventional tomography. This technology utilizes radiographic projections from multiple angles and, then, builds a two-dimensional image using a mathematical model that incorporates all the projected data [1]. Sir Godfrey Hounsfield invented the first CT scanner at EMI research laboratories in 1967 and the first patient was scanned using this technology in 1971 [1]. Alan M. Cormack (who created the early mathematical models used in CT) and Hounsfield both received a Nobel Prize in 1979 for the development of "computer-assisted tomography [1]." CT has been hailed as one of the most important medical innovations in human history [1]. Images display soft tissue contrasted with anatomic detail, allowing for unprecedented diagnostic accuracy. Two-dimensional, cross-sectional images are created that can be positioned in space to generate a volume [1] (Fig. 1).

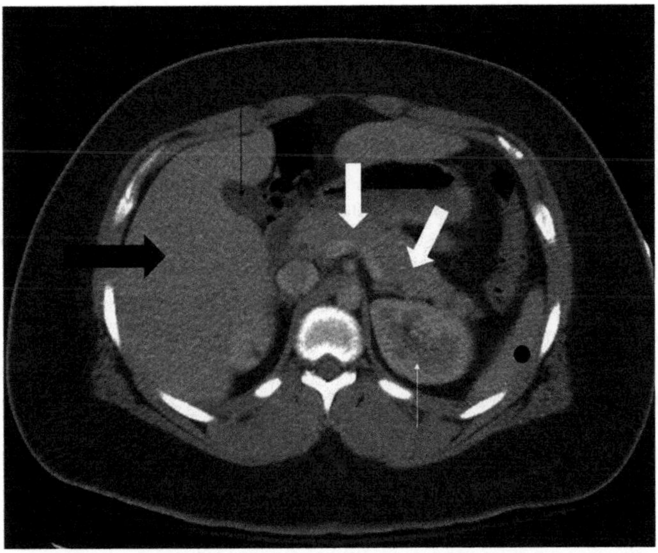

Fig. 1 Axial image from a pediatric CT scan of the abdomen and pelvis performed using intravenous contrast. Multiple organs are seen, including the liver (large black arrow), gallbladder (thin black arrow), pancreas (large white arrows), left kidney (thin white arrow), and spleen (black dot)

Intravenous iodine-based contrast media (IBCM) are utilized in CT, as well as an x-ray related modality known as fluoroscopy, for soft-tissue contrast and vessel imaging [5]. Acute allergic-like and physiological reactions are seen with IBCM. Fortunately, acute allergic-like reactions are very uncommon in pediatric patients, occurring in 0.18–0.46% of iodinated contrast studies, with most (80–82%) being mild reactions [5]. A 2009 study by Callahan et al. noted patient age as a factor in the incidence of contrast reactions as no contrast reactions were seen in children younger than 1 year and no severe reactions were seen in children below 7 years of age. The highest incidence of contrast reactions was seen in adolescents between 16–18 years of age [5, 6]. Usually, the onset of symptoms begins within the first 20 minutes after contrast administration. However, delayed reactions that take place between 1 hour and 7 days after IBCM administration have been seen, most commonly as cutaneous manifestations [5]. Allergies to shellfish or povidone-iodine do not create a significant risk of adverse reaction and, therefore, do not require premedication [5].

Contrast-induced acute kidney injury (CI-AKI) is directly caused by intravenous iodinated contrast and occurs within 48 hours following administration [5]. The pathophysiology of CI-AKI is poorly understood but is believed to be multifactorial and may be related to cytotoxicity and vasoconstriction caused by the contrast agent, worsened by patient risk factors that make certain patients more susceptible to kidney injury. These risk factors include hypotension, hypovolemia, pre-existing acute or chronic kidney injury, or concurrent use of other nephrotoxic agents [5]. The true risk of CI-AKI in pediatric patients is difficult to define, and there is no universal standard for measuring AKI in children [5]. There is a dearth of clear CI-AKI preventative guidelines in pediatric patients, with current practices often extrapolated from adult studies [5]. Hydration with intravenous saline is the main intervention to prevent CI-AKI in children at risk of developing this condition. However, care must be taken when administering fluids to avoid aggravating underlying conditions such as heart failure, cardiac arrhythmia and electrolyte imbalance. In children at high risk of developing CI-AKI, the benefits and risks of administering intravenous fluids should be discussed with the referring provider, who can determine the volume of intravenous fluid the child can receive safely [5].

A single dose of intravenous IBCM may contain up to 150 times the daily recommended iodine intake for young children [5]. Thyroid dysfunction, most commonly manifesting as hypothyroidism, has been associated with IBCM exposure in children and is believed to be related to decreased thyroid gland hormone production due to excess iodine exposure. This phenomenon is known as the Wolff–Chaikoff effect [5]. The incidence, duration of this effect, and its clinical significance in children in this setting are not well understood. More robust data is required to identify which children might specifically be vulnerable to thyroid dysfunction following IBCM administration. Just as important will be gaining clarity around which providers are responsible for ordering the thyroid function tests and following up on the results needed to appropriately manage these patients [5].

Children have also been imaged via nuclear medicine studies. Nuclear medicine utilizes radioactive tracers to create functional imaging. Unique sensors detect the radioactive emission of these tracers. Single-photon emission computed tomography (SPECT) and positron emission tomography (PET) both use co-registration of a morphological scan (commonly CT) with a functional nuclear medicine scan [1]. Functional imaging, also known as physiological imaging, is a technique of detecting or measuring changes in metabolism, blood flow, regional chemical composition, and absorption using medical imaging [7]. Functional imaging focuses on demonstrating physiological activities in a certain tissue or organ by using imaging modalities that often utilize tracers to reflect the spatial distribution of these tracers in the body. These tracers are usually analogous to some chemical compounds, like glucose, within the body. To achieve this, isotopes are used because they have similar chemical and biological characteristics. By appropriate proportionality, nuclear medicine physicians can determine the true intensity of certain substances in the body to determine the risk or danger of developing some diseases [7].

SPECT scans measure gamma rays, while PET scans utilize a radiotracer, usually F-18 fluorodeoxyglucose (FDG) that accumulates in body tissues and is metabolized, after which positrons are emitted. When positrons interact with electrons, they annihilate each other and emit a pair of gamma rays at the same time. Special detectors note the simultaneous arrival of the pair of gamma rays and measure the number of such arrivals. The data are used to map the metabolic activity of the organs in the body [1]. Tumor cells have a high metabolic rate; hence, this compound is also metabolized by tumor cells. FDG is metabolized to FDG-6-phosphate, a substance that cannot be further metabolized by tumor cells, which accumulates and concentrates in tumor cells (Fig. 2). This accumulation is detected and quantified [8]. James Robertson built the first single-plane PET scanner at the Brookhaven National Laboratory in 1961 [1].

PET scans have been shown to provide useful information in evaluating a wide variety of malignant pediatric tumors including lymphomas, sarcomas, neuroblastoma and Wilms tumor [9]. It has also been useful to evaluate benign pediatric tumors such as Langerhans cell histiocytosis (LCH) and Neurofibromatosis type I (NF1). Nonmalignant conditions, including infections and other inflammatory processes, are associated with hypermetabolic glucose metabolism. Thus, PET has the potential to contribute important information. Preliminary studies suggest a role for PET in the care of patients with chronic diseases as well as unidentified sources of infection [9].

Functional and morphological co-registration (i.e., the first hybrid design) was originally proposed for a PET-CT scanner by David Townsend and Ronald Nutt [1]. A prototype hybrid clinical PET-CT scanner in a single gantry was used clinically for the first time in 1998 [1]. Functional metabolic imaging is now routinely combined with high spatial resolution scans in clinical practice using both PET-CT as well as PET-Magnetic Resonance Imaging (PET-MRI). [1]. Interpretation of dedicated PET scans is limited by the difficulty in determining the anatomic region of increased activity within the body. The addition of contemporaneous CT images has several advantages, allowing for more precise anatomic location of lesions and the acquisition of diagnostic CT images along with PET scans in one effort [10]. PET-MRI scanners have been hindered in their development and are just entering clinical use. Their efficacy and clinical applications are unclear. However, it can be expected

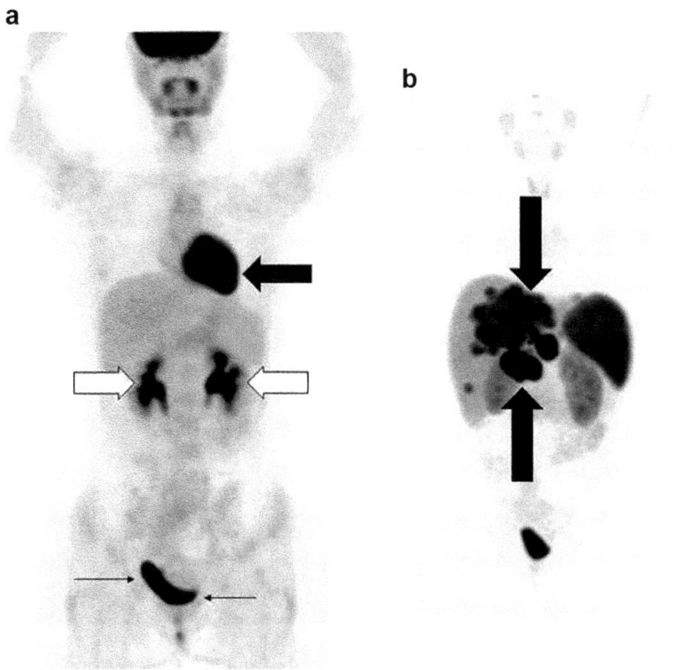

Fig. 2 (**a**) Coronal image from a normal pediatric PET scan shows normal intense uptake of FDG by the heart (large black arrow). FDG is normally excreted by the kidneys (white arrows) and bladder (thin black arrows). (**b**) Coronal image from a pediatric PET scan shows intense FDG uptake in a patient with a pancreatic tumor metastatic disease to the liver (arrows)

that they will be utilized to evaluate brain pathology (stroke and tumors) as well as for whole-body oncologic applications such as detecting metastatic disease. The main benefits of PET-MRI are the reduced radiation dose compared with PET-CT, better soft tissue contrast, and the ability to evaluate tissue chemistry [10].

The above modalities utilize ionizing radiation. Prior research has shown that children are at a greater risk than adults to develop cancer after being exposed to radiation from A-bomb explosions, nuclear power plant explosions, and medical procedures [11]. This association will be explored in greater detail later in this chapter. Fortunately, other imaging modalities that do not utilize ionizing radiation, namely ultrasound and MR, have been developed through the years and proven useful in pediatric radiology.

Ultrasonography (US) is one of the most used imaging modalities in the world. It offers a safe, reliable, noninvasive, and portable means to screen, diagnose, follow, and treat disease. This technology is driven by the science of sound waves. Images are created from the deflected echoes of inaudible high-frequency sound waves transmitted into an object of interest. US imaging became a viable method for medical imaging in the early 1900s and has developed into a very important diagnostic and therapeutic tool in modern medicine [12]—especially in pediatric radiology as it does not require the use of ionizing radiation (Fig. 3).

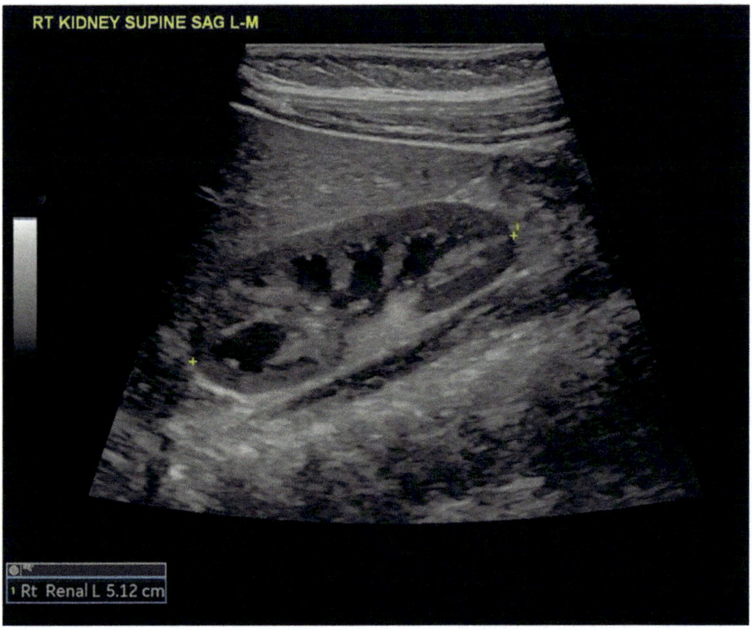

Fig. 3 Sagittal or long-axis view of the right kidney from an ultrasound performed on a pediatric patient. The yellow calipers seen measure the length of the kidney, which in this case was 5.12 cm

Magnetic resonance imaging (MRI) or (MR) is a powerful diagnostic imaging tool in both children and adults. This modality utilizes low-energy, non-ionizing radio waves and, therefore, is especially well suited for pediatric and longitudinal imaging exams. MRI takes advantage of a wide variety of intrinsic tissue-specific properties to produce a vast spectrum of tissue contrast, thus yielding detailed information about anatomy, physiology, and function. In clinical practice, the acquisition techniques utilized and the specific acquisition parameters selected are manipulated for each exam to generate the type of image contrast most relevant to answer the underlying medical question. In contrast to MRI, ultrasound is limited by the restriction of air and bony anatomy, and radiography and computed tomography both utilize ionizing radiation. Compared with other diagnostic imaging modalities, MRI has been found to be superior at imaging soft tissues. In addition, MRI allows for images to be obtained in any anatomic plane and can provide true volumetric data acquisition [13].

MRI, when carried out under the right conditions and guidelines, presents no apparent safety risks, and millions of MRI examinations are performed each year without incident. However, when not appropriately performed, safety concerns exist from the following: each of the three electromagnetic fields used, an external contrast agent (usually gadolinium-based), associated acoustic noise, and other issues such as sedation. Notably, most MR-related injuries and the small number of fatalities that have occurred primarily were a result of the failure to follow MR safety

Table 1 Common implanted medical devices in children

CSF shunt valves
Cochlear implants
Vagal nerve stimulator (VNS)
Intracranial neurostimulators
MAGEC growing spinal rods
Implantable continuous Doppler monitoring devices
Inspire UAS system
Pancreatobiliary stents
Endoscopic clips (endoclips)
NAVA catheter
Epicardial pacing wires/leads
MR conditional pacemakers and ICDs
PDA closure devices

Adapted from: Betz LH, Dillman JR, Jones BV, Tkach JA. MRI safety screening of children with implants: updates and challenges. Pediatr Radiol. 2023 Jun;53(7):1454-1468

guidelines for the MRI environment, or they happened because inaccurate or outdated MR safety-related information was utilized for biomedical implants and devices [13]. Some common implanted medical devices in children can be found in Table 1 [14].

Gadolinium-based contrast media (GBCM) are the only intravenous MRI agents approved for use in the United States [5]. There are four types of GBCM that are characterized by the chelating moiety as well as the ionicity of the chelate [5]. The chelating moiety is either macrocyclic, where Gd^{3+} is located in the cavity of the ligand, or linear. The chelating moiety can be ionic (mono-ionic or di-ionic) or non-ionic, where the metal and chelate both have equal charges. The most stable of the GBCM types in descending order are macrocyclic ionic, followed by macrocyclic non-ionic, linear ionic and linear non-ionic [5]. Nine intravascular GBCMs have been approved by the U.S. Food and Drug Administration (FDA) and are sold commercially in the United States. Seven of these agents are approved for pediatric patients [5]. Of these seven agents, four are approved for children of all ages, while the remaining three are approved for use in children 2 years of age or older [5]. The GBCM that is most kinetically stable is gadoteric acid (Dotarem; Guerbet LLC, Princeton, NJ) [5]. Dotarem is the only macrocyclic ionic type of GBCM that is available [5]. According to a 2016 survey of U.S. children's hospitals by Mithal et al., 80% of respondents reported using macrocyclic agents, and 58% of respondents had switched their main GBCM in the prior year. Most of those hospitals that did change most commonly switched to Dotarem due to concern over gadolinium deposition [5, 15].

Acute allergic-like reactions to intravenous GBCM occur and typically take place within an hour after contrast injection and vary from mild reactions requiring no medical intervention to severe reactions that can be life-threatening [5]. Reactions to contrast media can happen following the first exposure and do not require a sensitizing exposure as is usually the case with classic allergic reactions [5].

Allergic-like reactions to GBCM have been reported to be rare in the pediatric population, with a rate of approximately 0.04–0.05%, which is lower than the rates observed in adults between the ages of 18–65 (0.2%) [5]. Of the reactions seen in both the children and adult patients, most allergic-like reactions are mild, ranging between 74 and 96% [5] of all such reactions. Moderate (2–19% of reactions) and severe (2–7%) reactions happen less frequently [5]. Many mild reactions are limited cutaneous reactions including urticaria or rashes [5]. One study of adverse events following GBCM injections that were reported to the U.S. FDA Adverse Event Reporting System (AERS) estimates that the mortality rate from severe reactions to GBCM in all pediatric and adult patients is less than 1 per million [5, 16]. Treatment of acute allergic-like reactions for both GBCM and IBCM is similar and is based on the severity of patient symptoms [5].

To prevent acute allergic-like reactions, a premedication regimen can be prescribed involving the administration of corticosteroids and antihistamines at set intervals before injection of intravenous (IV) contrast [5]. At the institution of the author of a recent review article [5, 17], premedication is only given to children with a history of contrast reaction, of any severity. However, in the case of active asthma symptoms or moderate to severe food or medication allergies, it might be preferable to perform the exam at a location where a rapid response team can be utilized if necessary [5, 17]. For children with a history of severe reactions to intravenous contrast agents, the inciting contrast agent should be avoided unless deemed to be clinically necessary. If the contrast agent is clinically necessary, premedication is strongly advised, and an alternative class of contrast material should be utilized [5, 17]. As reported in the 2022 American College of Radiology (ACR) Manual on Contrast Media, there have not been studies that have evaluated the effectiveness of premedication prior to injection of gadolinium-based contrast agents in high-risk patients [5, 18]. Acute allergic-type reactions have taken place despite premedication, so physicians and care providers should be available to respond rapidly if a "breakthrough" reaction occurs [5, 19].

Physiological reactions can also happen after GBCM administration [5]. These reactions are non-allergic type reactions related to the molecular properties of the agent that give rise to direct chemotoxicity, osmotoxicity, or molecular binding [5, 18]. These types of reactions usually involve nausea, vomiting, anxiety as well as warmth at the site of injection. Usually, no treatment is necessary, and symptoms resolve without treatment. Vasovagal reactions with hypotension and bradycardia are more severe physiological reactions and, at times, require treatment with atropine and IV fluids if symptoms do not resolve after using conservative measures such as elevating the legs 60° and administering supplemental oxygen. It is important to recognize physiological reactions because, as opposed to acute-allergic type reactions, these types of reactions do not necessitate the use of premedication [5, 20]. It is important to note that physiological reactions should not be documented in the child's electronic medical record as acute allergic-like reactions because this might result in the child receiving unnecessary premedication before administration of contrast media in the future [5].

Nephrogenic systemic fibrosis (NSF) is a rare systemic fibrotic disorder seen in patients with decreased kidney function who characteristically present with scleroderma-like skin lesions [21]. This condition was first described in the literature in 2000 [21, 22]. NSF has attracted the attention of both the medical community as well as the public due to its high morbidity and mortality rates and possible association with GBCMs, which are extensively used [21]. As of 2010, there have been more than 500 cases reported to the U.S. Food and Drug Administration and more than 360 cases in the Yale International NSF Registry [21, 23]. Although NSF is possibly underrecognized and underreported, its incidence has been on a decline [21].

Only 23 reports of NSF have been recorded in children, with documented gadolinium exposure in 17 of those children [5]. As per the available documentation, all of the children had renal insufficiency, and 71% received gadodiamide, a linear non-ionic agent. There have been no reported cases of NSF in children younger than 6 years [5, 24]. No pediatric-specific guidelines are in place for GBCM administration; therefore, the ACR recommends following the adult guidelines for children with renal insufficiency [5]. At one major children's hospital, gadolinium agents are usually completely avoided for children with eGFR <30 mL/min/1.73 m^2, and either an alternative imaging modality or MRI without contrast is performed instead. However, based on 2021 consensus statements issued by the ACR and National Kidney Foundation, it would be reasonable to utilize a macrocyclic agent after appropriate informed consent has been obtained as well as communication between the radiologist and referring provider, given how rare NSF is with these agents [5]. For children with chronic kidney disease (CKD) on hemodialysis, scheduling GBCM administration just before regularly scheduled dialysis could be considered as hemodialysis has been found to remove GBCM from the body [5, 25]. As there is no evidence that hemodialysis prevents NSF, this therapy should not be initiated in children who do not need dialysis at baseline [5].

Despite evidence of gadolinium deposition in tissue, there is no scientific evidence that it leads to chronic adverse reactions in children who have normal renal function [5]. There is also no known literature supporting deposition results in histological tissue changes such as neurotoxicity [5, 18]. However, there have been patient reports of clinical symptoms after GBCM administration, which include headache, joint pain and stiffness, muscle spasm, fatigue, skin thickening and tightness, mental clouding and neuropathic pain. The ACR Committee on Drugs and Contrast Media suggested the term "symptoms associated with gadolinium exposure," or SAGE, to describe these symptoms [5, 26]. SAGE replaces old terminology including "gadolinium deposition disease" because no direct link has been found between gadolinium deposition in tissues and these reported symptoms [5, 26]. There is no known way to treat or stop gadolinium deposition in tissue from occurring [5]. The consensus in the radiology community is that the benefits of administering gadolinium in appropriate clinical scenarios outweigh the potential risks of gadolinium deposition, especially in life-threatening situations. However, without much longitudinal data about the clinical effects of deposition, radiologists

Table 2 Modalities commonly used to image children

Plain radiographs[a]
Ultrasound
Nuclear medicine[a]
MRI
CT[a]

[a] Use of ionizing radiation

should continue to use GBCM carefully and alter practice as needed as more information becomes available [5] (Table 2).

Differences Between Adult and Pediatric Radiology

While adults and children are often imaged with the same modalities, the actual practices of pediatric and adult radiology differ significantly, especially regarding age-related changes in physiology, anatomy, and diagnoses and differences in image acquisition parameters [27]. For example, pediatric patients' body size varies widely and runs the gamut from tiny premature neonates to adolescent bariatric patients [27]. Normal changes in children are seen in the size, shape, and appearance of multiple organs as well as the skeletal system along with differing patterns of injury and disease [27]. One example is changes in the appearance of the size of the thymus on multiple imaging modalities [27]. The thymus is a lymphatic organ located in the anterior mediastinum that plays an important role in the development and maturation of the immune system during childhood [28]. On frontal chest radiographs in infants and young children, the thymus is strikingly large and challenging to differentiate from the heart [28] (Fig. 4). The thymus usually has smooth borders and can be seen on radiographs through the age of 3 years [28]. As children mature and their immune systems develop, the thymus undergoes physiological involution and is ultimately replaced by fat [28].

In addition, the appearance of bone marrow on MRI changes over development [27, 29]. Specific types of fractures as well as skeletal fracture patterns on imaging studies often vary with age [27, 30]. Changes in physiology, such as differences in cerebral metabolism, organ perfusion, and functional immaturity, are seen [27, 31]. Variations in vascular lumen size, cardiac output, and circulation time as well as end-organ perfusion state necessitate different doses, concentrations, and administration techniques of intravascular contrast agents to yield the necessary tissue contrast enhancement [27, 32]. An additional challenge of imaging pediatric patients is the frequent requirement of the use of sedation or anesthesia in this population, which also has implications for observed tissue perfusion [27]. Other physiologic parameters such as respiratory rate and cardiac output require age consideration for imaging techniques such as respiratory and cardiac gating during image acquisition and accurate image interpretation [27].

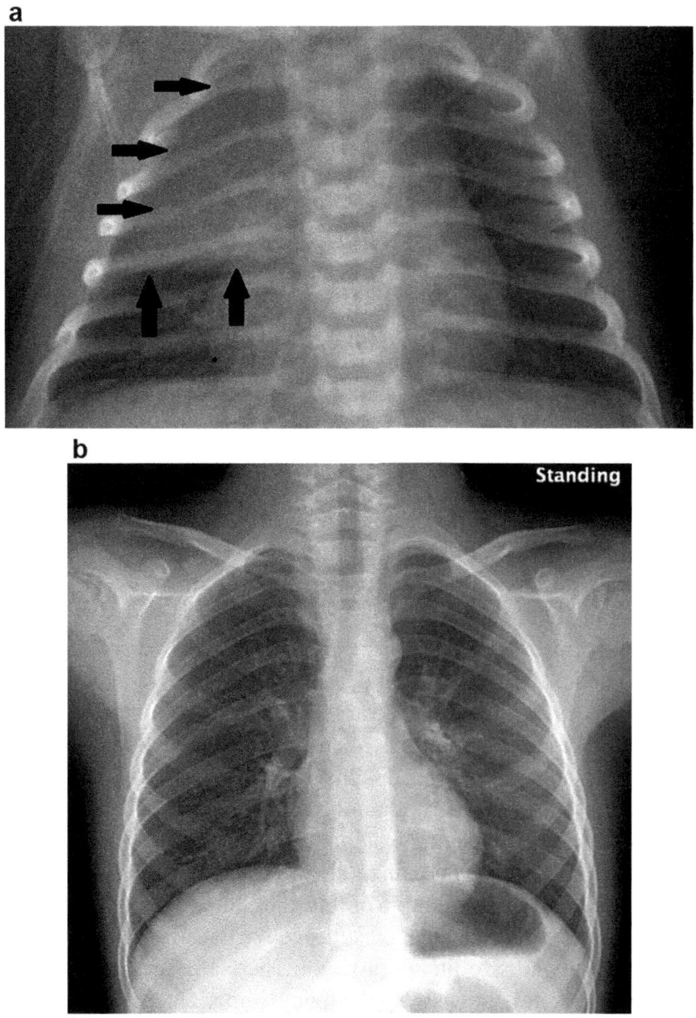

Fig. 4 (**a**) Chest radiograph in a young infant showing a prominent thymus, which is normal in this age group. This structure is often seen on chest radiographs until the age of 3 years. The smaller arrows show the lateral border of the thymus. The larger arrows show the lower border of the thymus. Note that the left side of the film is the patient's right side. (**b**) Chest radiograph in an older child. The thymus is no longer visualized and has involuted

As stated above, it is recognized that children are more susceptible than adults to similar doses of ionizing radiation [27]. In addition, children and adolescents are potentially exposed to more lifetime risk from either early or ongoing exposures and have a longer lifetime during which they can develop cancers from radiation exposures [33]. Due to concerns over the risks to children from exposure to ionizing radiation, a group of concerned pediatric radiologists started the "Image Gently" campaign [33]. This campaign is an educational, awareness, and advocacy campaign of the Alliance for Radiation Safety in Pediatric Imaging. The Alliance is a coalition of healthcare and regulatory organizations that wish to provide safe, high-quality pediatric imaging worldwide [33].

Image Gently is coordinated and run by volunteers with small financial contributions from the original four founding organizations (the American Association of Physicists in Medicine, the American College of Radiology, the American Society of Radiologic Technologists, and the Society for Pediatric Radiology) [33]. This campaign is composed of more than 60 healthcare organizations and agencies and hopes to include all stakeholders involved in pediatric imaging, including parents and families [33]. The stakeholders include more than 700,000 radiologists, radiology technologists, medical physicists, and pediatricians [33]. All educational and promotional materials for the Image Gently campaign are provided at no cost [33, 34].

One of the most important goals of Image Gently is to raise the awareness of all stakeholders about the three key pillars of radiation protection: justification, optimization, and dose limits [33]. Justification centers on appropriateness. Pediatric healthcare providers need to ask themselves whether the benefits outweigh the risks of performing an imaging test on an individual patient [33]. These risks include potential harm, including the induction of future cancers from exposure to ionizing radiation [33]. In addition, there are times when other diagnostic tests that do not utilize ionizing radiation, such as ultrasound or magnetic resonance imaging, can yield the same information as CT scans. Initiatives of the Image Gently campaign include creating and distributing parent education brochures, educating the medical community about which imaging tests are best suited for common clinical indications and when alternative modalities such as magnetic resonance imaging and ultrasound may be substituted for exams that use ionizing radiation [33].

The second important pillar of radiation protection is optimization [33]. Image Gently hopes to ensure that imaging examinations that utilize ionizing radiation are done under ideal conditions to maximize the information gained while minimizing radiation exposure [33]. As an example of this, the campaign worked with members to create CT exam protocols that can be adjusted for the size of the infant and child rather than a uniform one-size-fits-all protocol. There are free training modules for CT technologists to learn how to apply many important physics and technical parameters when performing imaging exams on children so that they use the lowest possible radiation dose to achieve the necessary exam quality. Image Gently also provides members with updates on the newest educational modules and information from cooperating organizations, arranges summits, and works with regulatory bodies to create and promote appropriate dose standardization and dose limits, which is the third key pillar of radiation protection [33]. Such efforts include working with equipment vendors and regulators to arrange summits and discuss standard dose metrics across vendors, explain the importance of accurate metrics for infants and children, and have all stakeholders comprehend the consequences of radiation exposure in children. Furthermore, the campaign educates patients and caregivers and promotes a collaborative decision-making environment in health care, allowing for the improvement of healthcare outcomes [33]. It should be noted that a similar campaign known as Image Wisely was also started in adults [35].

Significant progress has been made in reducing radiation exposure to pediatric patients since the launch of the Image Gently and Image Wisely campaigns. There have been more than 50,000 pledges to the Image Gently website and a similar amount to Image Wisely by the end of 2016 (with 20,000 new pledges in January 2017 alone) [35]. Starting in 2007–2008, a decreasing trend in the amount of CT scans performed on pediatric patients has been noted [35]. Further, improvements in CT hardware and software, some of which are discussed later in this chapter, show promise in lowering radiation exposure, in some cases by as much as 90 percent in the submillisievert range, which is similar to plain film exposures [35].

Advances in Alleviating Unique Psychosocial Issues in Pediatric Radiology

There are also other issues unique to the imaging of children as compared to adults. Many of these are psychosocial issues that involve the relationship between pediatric radiologists and patient families, professionalism, and effective communication, as well as how to approach imaging children who are not able to cooperate during certain exams [36].

In the practice of both pediatric and adult radiology, there are often patient family members with whom imagers must interact. However, in pediatric radiology, there are many unique features in the relationship among the radiologist, patient, and their family [36]. In the pediatric setting, communication more commonly takes place between the radiologist and parent than between the imager and the patient. When interacting with children, age-appropriate communication is crucial to success. Furthermore, the amount of interaction between the radiologist and the child-parent unit may be greater in pediatric settings than in adult settings because of associated issues such as the possible need for sedation, the need for consent from the parent instead of the patient (if the patient is a minor), as well as the need for an explanation of the procedure in a way that can be understood by the child as well as the parent. Many people also tend to be significantly more inquisitive and protective concerning events that involve their children. Due to this, descriptions of what to expect during the visit to the radiology area may need to be more detailed and nuanced when working with pediatric patients and their parents or guardians. The stress levels of parents when their child is or might be ill may be significant and such stress can bring out both the good and bad in people. Due to the intense bonds between most parents or guardians and their children, the relationship between the radiologist and caregivers is more successful when the radiologist exercises marked empathy, patience, professionalism, and effective communication [36].

Infants and young children are often not able to comply with the requirements for imaging studies that are usually met by adults [36]. For example, children often cannot sit still, remain in a certain position, concentrate for more than a short time or hold their breath when needed. Children of different ages have differing limitations [36]. As an example of this, infants and toddlers may not be able to hold still, whereas a four-year-old may be more likely to refuse to cooperate with instructions. These limitations affect exams performed on children utilizing almost all of the imaging modalities. There are several potential solutions that can be helpful in these situations. Commonly employed techniques include distracting the child, providing child-friendly surroundings, immobilization and sedation [36].

While sedation and general anesthesia are relatively safe in pediatric patients, with low reported rates of serious adverse events in institutions with well-organized sedation services, the risk of mild and moderate adverse events at the time of anesthesia remains substantial [37, 38]. In addition, the potential for long-term effects of anesthetic exposure on later neurodevelopmental outcomes based on preclinical models and some retrospective clinical data has also been raised recently [37]. These concerns caused the FDA to issue a warning in December 2016 to caution the use of repeated and prolonged (i.e., more than 3 hours) general anesthesia and sedation in children below 3 years of age [37, 39]. Although neurocognition in the developing brain is difficult to measure due to its complexity and the multiple contributing factors involved [37], a recent multi-national, multi-institutional, randomized controlled trial with follow-up at 2 years and 5 years of age has revealed that the use of single and brief exposures (i.e., less than 1 hour) of sedation/general anesthesia in this young pediatric cohort is considered to be safe [37]. Considering this, it is also important to note that not performing an MRI examination or obtaining suboptimal MR image quality in an attempt to avoid sedation/general anesthesia could be detrimental to patient care, and the final decision of its use should be based on a risk-benefit assessment [37].

Serious adverse events related to pediatric sedation and general anesthesia outside of the operating room are rare. The Pediatric Sedation Research Consortium (PSRC) has a large database of consecutive pediatric sedation/anesthesia administrations at more than 30 centers. Among almost 50,000 sedation/anesthesia encounters, no deaths occurred, and only two cardiac arrests were noted, with both patients recovering quickly [37, 38]. Airway-related events (stridor, laryngospasm, airway obstruction, wheezing or central apnea) happened in around 1 in 65 children undergoing sedation/anesthesia [37, 38]. One in 70 administrations of propofol necessitated airway and ventilation interventions of various severity [37, 38]. A subsequent analysis from the PSRC of around 140,000 sedation encounters revealed 0 deaths, 10 aspirations and 75 major complications [37, 40]. This analysis specifically sought to determine the relationship of nil per os (NPO, or pre-study fasting) status and aspiration and found that NPO status was not an independent risk predictor of aspiration or other major complications in this large dataset [37, 40].

Young infants, especially those who were born premature, are at higher risk of postoperative apnea related to sedation/anesthesia, which has long been thought to be worse in the setting of general anesthesia [37]. A multi-institutional prospective trial that included 722 infants (younger than 60 weeks of postmenstrual age) randomized infants to receive regional anesthesia or general anesthesia [37, 41]. Overall, the rate of apnea events was very low and occurred in 3% of participants. There was no evidence that regional anesthesia decreased the overall risk of apnea.

However, regional anesthesia was associated with a decreased risk of early postoperative apnea, the degree of postoperative desaturations, and the level of intervention for apnea [37, 41].

Propofol is a common agent for pediatric sedation/anesthesia in radiology procedures [37]. A retrospective review of 304 infants less than 6 months of age undergoing propofol administration for imaging studies revealed that 4.3% (13/304) had severe adverse events, with 18.4% (56/304) needing interventions for airway-related events [37, 42]. The PSRC has also shown that infants below 6 months of age who undergo propofol sedation/anesthesia are at a heightened risk of adverse pulmonary events [37, 43]. In addition to airway-related complications, sedation and general anesthesia is a significant risk for the development of hypoglycemia in neonates and infants [37].

Due to some of these risks, other techniques with fewer risks for children are often used instead of sedation when possible [37]. Distracting children with something other than the exam or procedure is usually an easy method to use [36]. Talking to older children about school and other activities has also been a useful technique [36]. Working with child life specialists in radiology can decrease the necessity of pharmacological sedation [44]. Certified child life specialists are often very helpful in coaching and distracting children, which can help them complete imaging exams without sedation. These specialists often utilize rattles and noise with very young children [36]. The use of video goggles allows some children younger than 8 years who would otherwise need sedation to have a non-sedated MRI [44]. In many instances, the use of therapy dogs has helped decrease MRI anxiety in children [37]. Utilizing a combined program including the use of a child life specialist, a combination of the other techniques described earlier to calm infants and young children, as well as a culture that avoids sedation whenever possible was able to decrease the frequency of the need for sedation in children less than 7 years of age by 34.6% for MRI and 44.9% for CT [36].

Imaging pediatric patients in child-friendly surroundings may help lessen a young child's anxiety and improve cooperation [36]. Paintings on the walls and equipment and cartoonish pictures in exam rooms can be helpful [36]. Eliminating or decreasing painful parts of the examination can be helpful in keeping a child's cooperation [36]. Placement of an intravenous line is sometimes needed for radiology exams such as CT or MRI scans that utilize intravenous contrast. Having these lines placed often causes a great amount of anxiety in children and causes children to become uncooperative. The use of topical analgesia to minimize the pain of line placement oftentimes makes the process less traumatic for pediatric patients [36]. It is also beneficial to schedule the sequencing of different examinations performed on the same day so that the most difficult exam is performed following the least difficult exam. For example, performing a renal ultrasound prior to a voiding cystourethrogram (VCUG), which is an exam that requires catheterization and evaluates for vesicoureteral reflux, would be preferable to the reverse sequence of exams [36].

Immobilization can also be a helpful technique. A feed-and-swaddle technique is sometimes employed for neonates and infants up to 6 months of age. This technique refers to the use of feeding and swaddling to induce natural sleep in infants because they tend to sleep when warmed and recently fed [37]. This technique can be combined with low-intensity lighting and minimal noise. Using this technique can mean the difference between requiring sedation use or not requiring its use for an imaging study. There are several commercially available immobilization devices that are helpful when performing imaging studies on pediatric patients. Institutions and practices that routinely image children should consider having this equipment available for use [36].

In some circumstances, distraction and immobilization may be unsuccessful and the use of sedation or general anesthesia will be needed. Because of the prolonged nature of the exam and the requirement for patients to remain still, many children below six years of age will need sedation or anesthesia for MRI examinations [36]. However, newer MRI techniques have been developed, which can shorten exam times and decrease the need for sedation and anesthesia in these patients [37]. The rapid nature of image acquisition of newer CT scanners and the use of sedation reduction programs mentioned above have allowed for sedation to be used less frequently than in the past for children undergoing a CT scan [36]. Other imaging exams that may necessitate sedation include a handful of nuclear medicine studies and most interventional radiology procedures [36].

Standards of care for sedation and anesthesia are a requirement of the Center for Medicare and Medicaid Services (CMS) and The Joint Commission [36]. Any radiology department planning to sedate children must have a defined sedation program that aligns with these guidelines [36]. Such programs must have protocols for presedation preparation, sedative agents used, monitoring during sedation and during postsedation recovery, in addition to discharge criteria [36]. At most pediatric healthcare institutions, sedation and anesthesia programs are not run by radiologists but instead are overseen by anesthesiologists, emergency physicians, or intensivists. At many of these institutions, such physicians have access to sedatives that are better suited for sedation during imaging studies, such as propofol or dexmedetomidine [36].

Advances in Pediatric MR Imaging

Advances in the technology used to image children as well as the techniques listed above to increase the ability of children to tolerate imaging examinations have allowed pediatric radiologists to image pediatric patients in different ways than in the past. Many of these new techniques include new technology and protocols allowing for exams that utilize ionizing radiation but at lower doses as well as new applications of the modalities that do not utilize ionizing radiation such as ultrasound and MRI.

It is interesting to note that some patients now have had an MRI exam performed on them before they are even born. While ultrasound is still the primary method for antenatal fetal evaluation, fetal MRI is now a valuable adjunct to ultrasound in confirming or excluding suspected abnormalities and in the detection of additional abnormalities, thus changing the outcome of pregnancy and optimizing perinatal management [45]. In contrast to ultrasound, MRI visualization of the fetus is not significantly limited by maternal obesity, fetal position, or oligohydramnios and visualization of the brain is not limited by the ossified skull [46] (Fig. 5). Due to its superior soft tissue contrast resolution, MRI is able to distinguish specific fetal structures such as lung, liver, kidney, and bowel [46]. In addition, MRI allows the fetus to be imaged in different planes as well as a large field of view, allowing for evaluation of fetuses with large or complex anomalies and visualization of the lesion within the context of the entire fetal body [46]. However, fetal MRI studies may give limited diagnostic information in early pregnancies due to the small size of the fetus as well as fetal movement [46].

MRI was first used in 1983 for examination of the placenta and fetus [45, 47]. The main drawback of fetal MRI was the motion of the fetus, which was overcome in the 1990s when ultrafast MRI sequences were developed [45, 48]. According to the Safety Committee of the Society for MRI, no known biological risks associated with fetal MRI have been proven to date [45]. Acoustic noise and biological effects are the primary safety concerns with fetal MRI. The noise intensity produced in fetal MRI by gradients can reach as high as 120 decibels (dB) [45]. Fetal hearing damage, which is a potential hazard, has not yet been confirmed in practice [45].

Fetal MRI is indicated in pregnant women when other non-ionizing diagnostic imaging techniques are not adequate or when the study provides important information that would otherwise necessitate exposure to ionizing radiation [45]. The American College of Radiology (ACR) white paper on MR safety states that

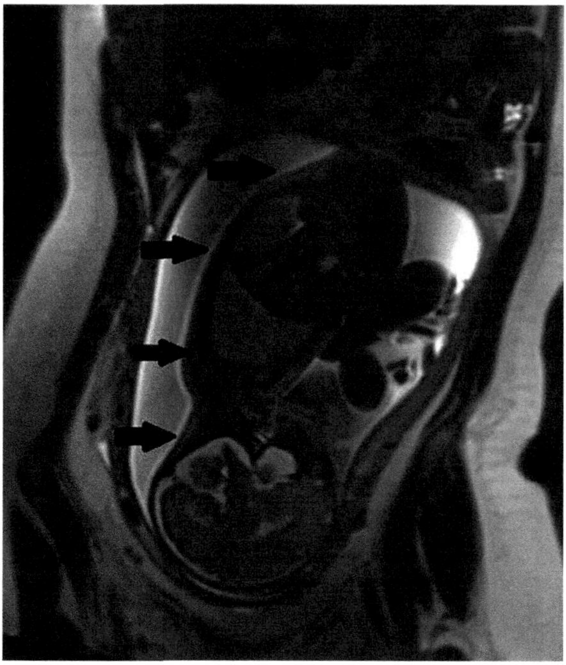

Fig. 5 Image from a fetal MRI shows the fetus (arrows) within the mother's uterus. The bright material surrounding the fetus is amniotic fluid, which is contained by the uterus

pregnant patients can be allowed to undergo MR imaging at any stage of pregnancy if the risk-benefit ratio to the patient warrants that the examination be performed and only if other diagnostic imaging methods that do not utilize ionizing radiation are inadequate [46, 49]. However, it is prudent to wait until the 17–18th weeks of gestation before performing fetal MRI due to the potential risks to the developing fetus as well as the current limitations of fetal MRI created by the small size and excessive motion of younger fetuses [46, 50]. Written informed consent is usually needed from the pregnant woman before the fetal MRI [46].

Shortly after birth, it is sometimes necessary to image critically ill neonates with MRI. In the past, this has proven to be a challenge. Unfortunately, neonates and unstable pediatric patients' accessibility to MRI is often limited due to the instability of these patients and equipment not being feasible for MRI [51]. This is especially important for neuroimaging as MRI is the preferred neuroimaging technique in pediatric patients [51]. Neonatal ultrasound is the standard technique in preterm and term neonates to evaluate for common brain pathologies including intraventricular hemorrhage, hypoxic-ischemic brain injury and stroke [51]. However, the success of ultrasound depends on an open fontanelle and deep brain structures can often not be seen on this modality [51]. The anterior or frontal fontanelle is a diamond-shaped soft membranous gap at the junction of the coronal and sagittal sutures of the calvarium. It remains open till approximately 18–24 months after birth, after which it is known as the bregma. It is the largest of the fontanelles and is the main sonographic window for sonographers to utilize when performing ultrasound examinations of the brain [52].

In addition to the aforementioned, the quality of the ultrasound exam is operator-dependent [51]. CT is a valuable imaging technique that is widely available [51]. CT often still requires transport of the patient to the CT scanner and some of these patients may be unstable. However, CT image acquisition is fast, and acute conditions can be diagnosed rapidly [51]. Due to the radiation exposure to children undergoing CT scans, it is advantageous to avoid utilizing this modality if possible. These reasons explain why MRI is preferred for neuroimaging in pediatrics. However, its routine accessibility is often a limiting factor. In patients undergoing invasive procedures in the intensive care unit, as well as in extremely premature infants, MRI is often not available and patient transport is not possible due to the severity of the underlying illness [51].

Low-field portable MRI (pMRI) has been reported to be a feasible and accurate technique to image the brain at the bedside [51, 53]. One first study in neonates and pediatric patients has shown that large intracranial hemorrhages and regions of ischemia detected by pMRI images correlate well with pathologies detected on conventional MRI [51, 54]. However, the study had not included severely sick neonates or children on extracorporeal life support (ECLS). Another subsequent study was performed on four patients receiving bedside high-quality MRI during ECLS treatment [51]. Although this study had a small sample size of only four patients, the authors concluded that it is safe and feasible to perform bedside MRI in this patient population. This opens the possibility of additional treatment decisions and may guide optimal treatment in these patients [51].

Recent advances in MRI have also allowed children to be imaged with this modality instead of the ways that they were traditionally imaged using modalities with ionizing radiation. For example, whole-body imaging in children was typically performed with radiography and positron-emission tomography, either with or without CT [55]. However, these imaging methods have the disadvantage of exposure to ionizing radiation, especially for children who are more susceptible to carcinogenic effects, even with low ionizing radiation doses. Development of MRI techniques, improvement of MRI coils, and introduction of new software have reduced exam time and allowed the use of MRl in whole-body evaluation [55]. In pediatrics, whole-body MRI (WBMRI) was originally utilized for staging lymphomas, but it is now utilized for the evaluation of other systemic diseases. In addition to assessment of cancer in children, WBMRI has also been utilized to evaluate inflammatory and/or infectious processes, osteonecrosis, soft tissue lesions such as myositis, neurofibromatosis, vascular malformations and multiple angiomatoses, non-accidental traumas, body fat composition, and perform virtual autopsy [55].

Imaging of the urinary system in children has been mainly based on ultrasound (US), plain film, and nuclear medicine scintigraphy [56]. All of these methods have their advantages as well as their limitations. Ultrasound is an optimal tool for evaluating the kidneys and their collecting systems and is safe and readily available [56]. However, this method does not provide information about renal function [57]. The plain film following intravenous contrast injection gives enough proper morphological evaluation of the collecting systems and indirect information about renal function. However, this method also uses ionizing radiation as well as intravenous contrast [56], which is required to opacify the kidneys.

Another method used for imaging the urinary tract is computed tomography (CT). When the acquisition of images became faster, this modality increased in popularity. Unfortunately, it is still limited due to radiation exposure and requires the use of intravenous contrast [56].

The gold standard in imaging renal function and determining renal split function or the relative function of one kidney relative to the other has been nuclear medicine scintigraphy imaging using the radionuclide Technetium-99m-mercaptoacetyltriglycine (99mTc-MAG3) [56]. This method also has drawbacks including low spatial and contrast resolution which can result in inadequate anatomic information. In addition, this modality also makes use of radionuclides which emit ionizing radiation [56, 58].

Like contrast-enhanced CT or CT urography, magnetic resonance urography (MRU) allows for the evaluation of the renal parenchyma, collecting systems, ureters, and bladder [56]. The better signal, excellent contrast resolution, and lack of ionizing radiation are characteristics that make MRU a promising technique for non-invasive assessment of the urinary tract. Nonetheless, it still necessitates the use of intravenous contrast, does not surpass CT in spatial resolution and has a lower sensitivity than CT for imaging renal stones [56].

Functional analysis of the MRU scan requires external post-processing with software that is freely available [57]. However, this can be a limiting factor in the full routine implementation of functional analysis of the MRU and the generation of functional parameters like those obtained in nuclear medicine studies [56]. Furthermore, one study determined that split renal functions obtained by functional MRI (fMRU) studies show a lack of accuracy in comparison with the clinical standard MAG3 scintigraphy. This finding was more common in patients with severely diminished function of one kidney [58]. However, the study also found that fMRU allows for an adequate assessment of urinary tract obstruction and a high-resolution morphologic evaluation of the whole urinary tract. Thus, the authors concluded that fMRU is suitable to add diagnostic value, especially as a complementary examination for complex individual cases in the presurgical state [58].

fMRU allows for quantification of the perfusion of the renal parenchyma and glomerular filtration as well as visualization of kidney excretory functions and urination [56]. All of these advantages make fMRU ideally suited for a comprehensive assessment of the kidneys and urinary tract in children [57]. Unfortunately, the requirement for anesthesia is still a main limitation of the study although sedation is primarily used for younger patients undergoing this examination [56].

MR enterography is a technique for assessment of the bowel on MRI [59]. It is increasingly being used for inflammatory bowel disease and has almost replaced barium follow-through examinations in many institutions [59]. Its lack of radiation in comparison to barium follow-through examinations makes it a good technique for bowel evaluation in children. It has been proven to be highly sensitive in detecting Crohn's disease in both adults and children [59]. It is also superior to barium studies in demonstrating extra-enteric findings and complications including fistulas and abscesses [59]. Even though, at present, it is almost solely used for the evaluation of inflammatory bowel disease, this technique has the potential to be used in other pathologies affecting the bowel [59].

Ultrasound (US) and computed tomography (CT) are the initial modalities for examining the pediatric pancreaticobiliary ductal system [60]. If they are unable to provide an accurate diagnosis, endoscopic retrograde cholangiopancreatography (ERCP) is usually performed [60]. However, ERCP is a difficult study in young children and infants as it requires special equipment and expertise that are not available at many institutions [60]. In addition, ERCP is an invasive procedure with possible complications, and it is both more difficult and more hazardous in the pediatric population than when carried out in adults [60].

Magnetic resonance cholangiopancreatography (MRCP) is an innovative technique for evaluating the biliary tree and pancreatic duct, which was originally used in adults at the beginning of the 1990s [60]. It is a reliable diagnostic exam to evaluate biliary pathologies including primary sclerosing cholangitis (PSC), Caroli disease, and choledochal cysts [60, 61]. The advantages of MRCP are that it is noninvasive, does not require contrast material or ionizing radiation, and can be done on an outpatient basis [60, 62]. The safety is deemed to be like that of the US [60]. Due to these factors, MRCP is increasingly replacing ERCP or percutaneous cholangiogram for imaging many pancreaticobiliary diseases [60]. Given its success in adults, MRCP has begun to be used in children for indications including suspected biliary atresia, choledochal cyst, cholelithiasis, choledocholithiasis, bile plug syndrome, pancreatitis, and liver transplantation [60, 63].

Echocardiography is the primary method for cardiovascular imaging in children. It is inexpensive, quick, accessible, non-invasive and especially helpful in neonates and infants for whom it is possible to achieve good sonographic imaging windows. However, it is operator-dependent and provides only limited views of vascular structures outside of the heart [64, 65]. Cardiac catheterization provides helpful hemodynamic data and allows for simultaneous therapeutic intervention if needed. However, it is invasive with rare complications that are potentially fatal, involves exposure to ionizing radiation and requires the use of iodinated contrast agents [64].

Cardiac MRI exams are also used to evaluate cardiac structure and function in pediatric patients [64]. However, cardiac MRI (CMR) scanners are not often mobile, and availability is limited in comparison to echocardiography. Even in institutions that offer conventional MRI, CMR requires significant software, training and expertise [64, 66]. The data acquisition time is long, usually 20–50 minutes, depending on what information is needed, and the space within MRI magnets is limited. It can be claustrophobic, and for children who cannot cooperate with the scanning procedure, general anesthesia may be necessary [64]. It is also less suitable for clinically unstable patients that need intensive monitoring, and in the event of a cardiopulmonary arrest, the patient has to be removed from the magnet of the CMR scanner before advanced life support can be started. All monitoring equipment utilized during a CMR scan needs to be MRI-compatible, requiring a switch to compatible pumps before the patient enters the scanner [64].

Computerized tomography (CT) is frequently utilized as an alternative to CMR. This modality permits the acquisition of a high-resolution data set in a much shorter time and is, therefore, useful for children and unstable patients who are unable to tolerate a prolonged CMR exam [64]. In the pediatric population, it is considered superior to CMR when evaluating the anatomy of the airway in cases of vascular rings, when assessing pulmonary vasculature (especially in thromboembolic disease of the pulmonary arteries, when breath-holding capability is often compromised), and when assessing for the presence or absence of any major aortopulmonary collateral arteries (MAPCAs) [64]. CT can also be utilized as an alternative when CMR is contraindicated due to the presence of implanted devices, foreign bodies or claustrophobia. However, it is less detailed in terms of tissue characterization and exposes patients to high doses of ionizing radiation, especially in cases involving serial imaging when the cumulative dose of radiation is significant [64, 67].

Despite its disadvantages, CMR is still a versatile tool with distinct advantages over other modalities. However, outcomes are most successful when it is used in combination with other imaging technologies in a manner to obtain an answer to a specific clinical question [64].

Advances in Pediatric Ultrasound Imaging

There has also been innovation in ultrasound techniques that have allowed children to be imaged with this modality in new ways. Point-of-care ultrasound (POCUS) is a non-invasive bedside technique with many pediatric uses [68]. It is portable and increasingly affordable with many pediatric applications supported by prior research [68, 69]. In the pediatric emergency medicine literature, POCUS has been demonstrated to increase diagnostic accuracy [68, 70], assist in procedural guidance [68, 71], and shorten the length of stay [68, 72]. The diagnostic information provided by POCUS can be used to reduce the need for other imaging modalities and other more invasive tests, assisting in radiological stewardship and high-value care [68]. Compared to adults, small children offer the additional advantage of usually being easier to scan, as their size allows for easy penetration by ultrasound waves with high-frequency transducers, giving superior resolution images [68].

Ultrasound is usually the initial imaging modality for evaluating abnormalities in children. Abnormalities are usually first detected by radiography or ultrasound, after which the patient may need to undergo contrast-enhanced CT or MRI for further evaluation [73]. Contrast-enhanced ultrasound (CEUS) is increasingly changing this approach as it can increase the sensitivity and specificity of ultrasound, negating the need for further imaging [73]. Increased interest in CEUS has resulted in a rapid expansion of this exam for many approved and off-label use cases in children [73].

Ultrasound contrast agents (UCAs) are microbubbles made up of inert gas within a phospholipid and/or protein shell. The shell is cleared by the hepatobiliary system and the gas is exhaled during respiration [73]. Due to the lack of renal excretion of UCAs, CEUS may be performed in children with poor renal function, pediatric patients being treated with other medications that may harm the kidneys, as well as neonates with potentially immature renal function [73]. Microbubbles are like red blood cells (RBCs) in their ability to travel through the capillary bed allowing visualization of both venous and arterial circulation [73]. RBCs, also called erythrocytes, are cells that carry oxygen using hemoglobin and form part of the cellular portion of blood as it travels throughout the body. These cells are numerous and typically produced in the bone marrow in a process known as erythropoiesis. Normal red blood cells live around 120 days before they are destroyed. However, certain diseases will change the survival time of these cells [74].

The safety profile of UCAs is excellent, and unlike CT, CEUS does not utilize ionizing radiation. UCAs do not deposit in soft tissues, unlike gadolinium-based contrast agents [73]. Because CEUS can be carried out as a portable examination, it is especially suitable for patients who are too ill to be transported safely. Its portability also allows for easy patient monitoring and maintenance of patient body temperature [73]. CEUS can be repeated more easily and is cheaper than CT and MRI [73].

Renal or hepatic function tests are not necessary to obtain prior to the administration of UCAs. The lack of a need for additional blood is beneficial to neonates with small total blood volume and expedites study performance [73]. CEUS can often be scheduled more quickly than CT or MRI, especially in patients who require sedation or general anesthesia. Quick turnover is important to decrease patient and family anxiety due to an unknown diagnosis [73].

Although CEUS is advantageous in multiple clinical scenarios, a disadvantage is the need for UCA administration and/or study oversight by a radiologist, preventing the radiologist from interpreting other studies when performing this examination [73]. In addition, artifacts that are seen on conventional ultrasound exams as well as those specific to CEUS have been described [73, 75]. An artifact in radiology usually refers to something seen on an image that is not present but appears due to a quirk of the modality itself. The term is also used to describe findings that are due to things external to the patient that may distort or obscure the image. Examples of things external that may cause artifacts include clothing, cardiac monitor leads, or body parts of the caregiver [76].

Appropriate use of CEUS is also important. If a diagnosis can be made using a conventional grayscale and Doppler ultrasound or if a patient will need to have a CT or MRI examination regardless of the CEUS result, then CEUS should not be performed to prevent a delay in care [73, 77]. One significant advantage of CEUS is its quantification capability. However, most of the software utilized for quantification is not available in the United States for clinical or human research use [73].

Evaluation of focal liver lesions is the most common indication for CEUS. CEUS should be the test of choice for evaluating a focal liver lesion in pediatric patients because of its many advantages [73]. It is becoming a possible alternative to CT in hemodynamically stable children who have suffered blunt abdominal trauma and improves the diagnostic performance of ultrasound in adults as well as children [73]. CEUS may be helpful for imaging complications of pediatric pneumonia, including complications that involve consolidated lung and the pleural space and may become an acceptable alternative to contrast-enhanced CT in unstable patients [73, 78]. Solid renal masses show varied appearances on CEUS [73]. CEUS has demonstrated the ability to distinguish solid renal masses from pseudotumors (including a hypertrophied column of Bertin), demonstrate renal infarction, and diagnose pyelonephritis and renal abscess [73, 79]. There is also mounting evidence of its usefulness in monitoring patients with inflammatory bowel disease [73]. Another area for further study is its role in assessing infants with necrotizing enterocolitis [73]. Additional other potential applications in children include evaluating pancreatic and splenic masses [73]. Furthermore, several off-label applications of brain CEUS are currently being investigated and interest in using this technique for brain imaging is increasing [73].

Voiding urosonography (VUS) refers to the examination of the urinary tract following the administration of ultrasound contrast into the urinary bladder to evaluate for vesicoureteric reflux (VUR) [80]. VUR is defined as backward flow of urine from the urinary bladder into the ureters [81]. Urinary tract infections (UTIs) are the most common source of bacterial infections among young children who are febrile [82]. VUR occurs in less than 0.5% of asymptomatic children but is noted to occur in as much as 50% of children with a UTI [81]. The importance of VUR is its association with renal scarring as it is present in almost every child with severe renal scarring [81]. Furthermore, there is an association between the grade or severity of VUR and the prevalence of renal scarring [81]. Other complications of VUR include pyelonephritis, interference with normal renal growth and the development of renal arterial hypertension [81].

Due to the above, diagnostic imaging for VUR is common in children. At present, there are three modalities available to diagnose VUR. These include voiding cystourethrography (VCUG), radionuclide cystography (RNC) and voiding urosonography (VUS) [80]. VUS is performed with ultrasound and allows the use of ionizing radiation to be avoided, which is not true of RNC and VCUG. However, in VCUG a significant decrease in radiation dose has been achieved with the introduction of digital pulsed fluoroscopy [80]. Nevertheless, one author recently concluded that VUS achieves similar or better diagnostic results when compared with VCUG for the evaluation of VUR [83]. In addition, they state that it should be used as the primary and sole imaging study in children with suspected VUR, especially when performed by an experienced examiner [83]. However, the authors also stated that if the assessment of either an ectopic insertion of a ureter or the male urethra is a concern, especially for inexperienced sonographers, first-time studies in boys can be carried out by VCUG, with all future exams performed by VUS [83].

Noninvasive tests for the evaluation of liver fibrosis are especially helpful in children to avoid general anesthesia and potential complications of testing by invasive means. Ultrasound elastography is increasingly being utilized to estimate liver fibrosis in the pediatric population [84]. This technique appraises tissue elasticity by measuring tissue response after the application of an external stimulus [85]. Elasticity is defined as the tissue's tendency to resist deformation or to recover its shape after the external force has been removed [85]. Differences in elasticity can be due to physiological or pathological factors [85]. To compare tissue elasticity, the displacement speed with which the region of interest responds to the applied stimulus is used [85]. Tissue movement is subsequently detected utilizing a variety of manufacturer-dependent methods [85].

The technique is continuously improving, and a sensitivity and specificity of 81% and 91%, respectively, were recently demonstrated for liver fibrosis in pediatric patients [84, 86]. The gold standard for liver fibrosis evaluation is the liver biopsy, which has disadvantages such as its invasive nature, the need for general anesthesia and the potential for sampling errors and clinical complications [84, 87]. In contrast, ultrasound elastography is noninvasive and allows for the evaluation of the entire organ with minimal discomfort for the patient, although it still lacks the liver biopsy's ability to help determine the etiology of the fibrosis [84]. In addition to ultrasound elastography, MR elastography (MRE) is now routinely performed in clinical practice to evaluate liver fibrosis in children [88]. Like ultrasound elastography, this noninvasive method is increasingly being used to replace liver biopsy for the diagnosis, quantitative staging, and treatment monitoring of patients with chronic liver disease [88]. Although ultrasound has the advantages of portability and lower equipment cost, available evidence suggests that MRI may have better reliability and accuracy in detecting liver fibrosis [88].

Ultrasound has also proven to be beneficial in imaging and guiding therapeutic interventions in children in the emergency room setting. This is especially true regarding children who present with suspected appendicitis, hypertrophic pyloric stenosis and intussusception. In each of these conditions, the use of ultrasound offers the option to image children while avoiding exposure to ionizing radiation.

Appendicitis is the most common indication for emergent abdominal surgery in children, and the duration of symptoms prior to diagnosis and treatment correlates with the likelihood of a perforated appendix and associated complications [89]. Clinical diagnosis remains difficult as symptoms, physical examination findings, and laboratory studies are inconsistent and nonspecific [89]. A recent meta-analysis involving 8605 pediatric patients from 21 studies concluded that "once acute appendicitis is suspected, no single history, physical examination, laboratory finding, or score attained on [the pediatric appendicitis score] can eliminate the need for imaging studies [70, 89]."

Because of this, a stepwise approach of an initial ultrasound followed by CT or MRI when necessary is now the preferred imaging approach [89]. Ideally, radiologists should perform or closely supervise the US examination, while considering that there is no evidence-based age threshold above which ultrasound has been demonstrated to be less effective and also acknowledging that the diagnostic yield of ultrasound decreases with increasing weight or body mass index [89]. Structured reporting is ideal for an ultrasound of the appendix, and a high-quality US examination that is unable to visualize the appendix but does not demonstrate secondary signs of inflammation should be considered negative and a distinct reporting category from otherwise indeterminate or equivocal studies [89].

If CT is felt to be the next best imaging modality at an institution, as is likely to be the case in many, if not most, hospitals, radiologists should optimize protocols to decrease radiation doses as much as possible while maintaining diagnostic accuracy [89]. If an institution decides to incorporate MRI into their pediatric appendicitis imaging pathway, it can be done without administration of IV contrast material or sedation in patients as young as 4 years old. However, in some scenarios the addition of IV contrast material is useful for problem-solving [89] (Fig. 6).

Midgut malrotation or simply "malrotation" is a developmental rotation and fixation abnormality of the embryonic bowel, which results in a narrow attachment of the mesenteric root [90]. Malrotation may be associated with bowel obstruction from Ladd bands or midgut volvulus, or simply "volvulus," and can present at any age although it usually occurs in early infancy. It occasionally presents in older patients with chronic symptoms [90]. In volvulus, the bowel twists around the superior mesenteric artery (SMA) and narrow mesenteric root, causing bowel obstruction and possible infarction or bowel death that results in substantial morbidity and mortality if this condition is not emergently diagnosed and surgically corrected [90].

For several decades, the fluoroscopic upper gastrointestinal (UGI) series has been the preferred study to diagnose malrotation and volvulus [90]. UGI necessitates contrast administered by mouth or via an enteric tube and exposes patients to ionizing radiation [90]. More recently, a growing body of research supports the use of ultrasound (US) to supplement or replace the UGI series to evaluate for both conditions [90].

Infantile hypertrophic pyloric stenosis (IHPS) is an entity that affects young infants, in which the region of the stomach known as the pylorus becomes abnormally thickened and manifests as obstruction of contents from the stomach emptying into the duodenum or first part of the small bowel [91]. Infants with IHPS are usually clinically normal at birth. However, during the first few weeks after birth, they develop forceful nonbilious vomiting described as "projectile [91]." This is in distinction to patients with malrotation and midgut volvulus who present with bilious vomiting. In IHPS, gastric outlet obstruction leads to emaciation and, if left

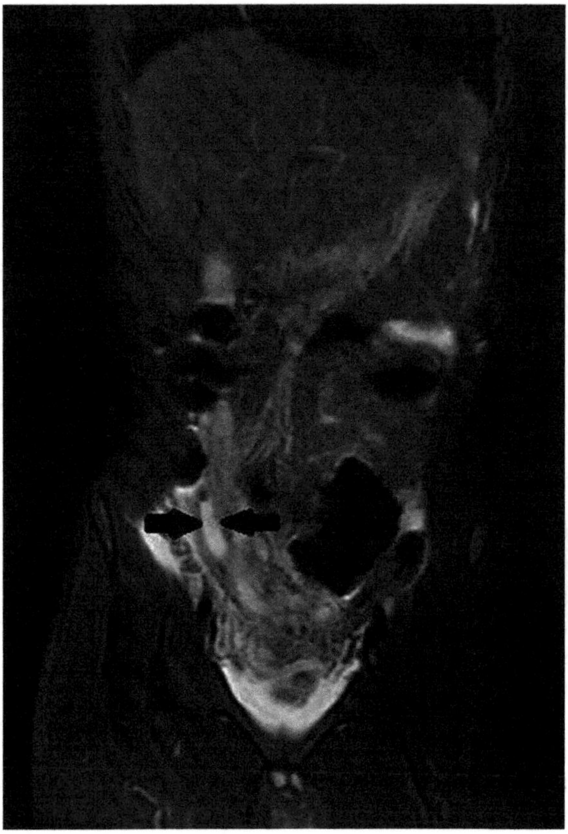

Fig. 6 MRI examination in a child performed to evaluate for appendicitis. The study demonstrates a dilated blind ending tubular structure in the patient's right lower quadrant filled with bright fluid (arrows) compatible with appendicitis

untreated, can be fatal. Surgery cures this condition [91]. The clinical diagnosis is based on palpation of the thickened pylorus, or "olive [91]." Abdominal palpation is accurate but not always successful and depends on factors including the experience of the examiner, the presence of gastric distention, and a calm infant [91].

In 1977, a paper was published showing five cases in which IHPS was diagnosed correctly by using ultrasound [91, 92]. Prior to this paper, the UGI was the main modality used to diagnose it. However, after this article was published, many additional articles were written on the sonographic diagnosis of IHPS. These articles, along with the introduction of real-time ultrasound capability and improvements in equipment resolution, aspired to define the measurements of the abnormal pyloric canal. However, these articles generated controversy regarding normal and abnormal values and suggestions regarding various indices and signs to help make the diagnosis [91]. In 1988, the author of an editorial suggested that the diagnosis should be made clinically by repeated attempts to palpate the olive in lieu of imaging, which the author felt to be time-consuming and often not accurate [91, 93]. The authors of subsequent papers addressed the issue of cost-effectiveness and suggested that the use of sonography does not justify the cost when compared with the cost of UGI [91]. In 1994, De Backer et al. proposed endoscopy as the most expedient and accurate method to diagnose IHPS [91, 94]. Despite the past debate about the ideal modality to image suspected cases of IHPS, ultrasound is currently the modality of choice to diagnose this condition [95].

Ultrasound is also used as a screening tool for hip dysplasia [96]. Developmental dysplasia of the hip (DDH) is one of the most frequent causes of disability in pediatric patients [96, 97]. DDH refers to a wide spectrum of pathologic conditions, ranging from fine acetabular dysplasia to hip dislocation that cannot be reduced [96, 97]. The prior term congenital dysplasia of the hip has been replaced by developmental dysplasia of the hip, because many of the clinical manifestations of DDH cannot be detected at birth and are recognized at a later age [96, 97]. Ultrasound is the preferred modality for examining the hips in infants below 6 months of age. This modality allows for dynamic evaluation of the hips with stress maneuvering, as well as direct imaging of the cartilaginous portions of the hip that cannot be visualized on plain radiographs [96, 98]. Hip US has become the most frequently utilized diagnostic modality for DDH during early infancy because the early and accurate diagnosis of DDH is the most important factor contributing to appropriate treatment [96, 99].

In neonates and infants with suspected spinal and paraspinal anomalies, MRI was and remains the preferred modality for imaging [100]. However, ultrasonography has recently seen extensive improvement in image quality with the introduction of a new generation of high-frequency ultrasound scanners that have brought its diagnostic value to the same level as that of MRI in certain conditions [100]. Relative advantages of sonography over MRI include wide and cheap availability, no requirement for sedation or general anesthesia, and lack of vulnerability to artifacts from patient movement, cerebrospinal fluid (CSF) pulsation, and vascular flow, which can negatively affect the quality of MR images [100].

Ultrasound is an ideal, effective, noninvasive modality to evaluate the spinal cord in neonates and young infants due to the lack of ossification of the posterior elements of the spine [100]. It is a well-established technique to evaluate the spinal canal, cord, and meningeal coverings and for characterizing almost all spinal anomalies with high geometric resolution in neonates and infants [100].

Advances in Pediatric CT Imaging

As the use of CT in children has increased significantly during the past decades, so, too, has concern about the long-term risks of exposure to ionizing radiation from diagnostic imaging in childhood [101]. As already mentioned, children are more prone to the deleterious effects of ionizing radiation as they have a longer life expectancy, during which radiation-related effects may accumulate [101]. In response to these concerns, initiatives such as the Image Gently campaign were initiated [33, 34] and resulted in a profound cultural shift toward the more selective utilization of CT in children [35, 101].

Despite this change in practice patterns, CT is still a very important diagnostic tool for many pediatric pathologies due to its speed, high spatial resolution, ability to image children in different anatomical planes, and widespread accessibility [101]. Recent advances in CT technology have decreased radiation doses from CT scans in adults and children while maintaining diagnostic image quality. One such technique is dual-energy CT, which allows for the concurrent acquisition of CT images at two different X-ray energy spectra [102]. By obtaining high- and low-energy spectral data, this technique can produce unique qualitative and quantitative information regarding tissue composition, allowing differentiation of multiple materials, including iodinated contrast agents [101, 102]. Clinical applications include evaluation of the arterial system, lung perfusion, neoplasms, bowel diseases, renal calculi, tumor response to therapy, and metal implants [103]. It is important to note that the radiation exposure level of dual-energy CT is equivalent to or less than that of conventional single-energy CT [103]. These technical advances are especially important for vulnerable children who are likely to need multiple CT scans at a young age [101].

Some pediatric CT examinations can be replaced with newer MRI techniques, some of which were previously discussed. Although these exams do not use ionizing radiation, the longer exam times and the enclosed environment that come with MRI imaging studies often necessitate the use of general anesthesia for infants and young children [101]. Recent concerns about the potential immediate and long-term risks associated with anesthesia in infants and young children have already been discussed. However, anesthesia for imaging studies also has a significant impact on cost and utilization of resources [101, 105]. Furthermore, the findings from recent studies have demonstrated tissue deposition of gadolinium-based contrast material after multiple contrast-enhanced MRI exams in patients with normal renal function. This finding is of uncertain clinical significance but has nonetheless prompted recommendations for more selective utilization of gadolinium-based contrast agents in pediatric patients [101]. Newer quicker CT scanning techniques may obviate the need for sedation or exposure to gadolinium-based contrast agents in many children and allow for fast, conscious, free-breathing studies with low radiation doses [101].

Iterative reconstruction has emerged as an important technique in the last few decades. Iterative reconstruction is a method used in CT imaging to improve image quality [104]. Conventional CT scans use a technique known as filtered back projection, which can occasionally result in images with noise or artifacts. However, iterative reconstruction utilizes a more sophisticated method. It involves using the data from the initial image and repeatedly refining it through a series of mathematical iterations. Each iteration aims to decrease errors or inconsistencies in the image, resulting in a clearer, more accurate representation of the examined area. Iterative reconstruction has evolved with different techniques, which have allowed for significant decreases in radiation doses without compromising image quality [104].

CT is an important diagnostic modality in the management of critically ill children, especially those patients with neurosurgical conditions such as traumatic brain injury. Previously, these scans have required patients to be transferred to the radiology department at times of severe physiological instability, such as incipient cerebral herniation. This exposes children with actual, or potential, spinal injuries to the risks of transfer. Transporting children from pediatric intensive care (PICU), often overnight, also requires senior staff. Portable CT (PCT) scanning offers a solution to this problem [106]. One study found that PCT scanning is safe for unstable neurosurgical patients who require emergent diagnostic head CT scans and, in addition, decreases the risks that come with transfer and the utilization of staff needed to provide care for other children in the PICU. While this study did not specifically address image quality, all images were noted to be diagnostic for the study indication [106].

Advances in Pediatric X-ray Technology

Patients with different types of scoliosis and lower-extremity anomalies or malalignment often need repeated radiographic imaging for diagnosis, monitoring, surgical planning and postoperative follow-up [107]. As mentioned previously, repeated radiographic exposure is potentially problematic due to the harmful effects attributed to ionizing radiation, especially among pediatric patients. A newer radiologic device is now utilized to image children and adolescents for the indications listed above. This device is manufactured by EOS Imaging (formerly Biospace Med), which is in Paris, France [107]. This device uses biplanar digital slot-scanning technology, which allows concurrent orthogonal radiograph images to be acquired. EOS scans the patient mechanically in both planes simultaneously while the patient stands upright [107, 108]. It can produce high-quality images while reducing radiation dosages [107, 108].

The exposure time for an examination of the spine is around 15 seconds, during which patients are asked to remain still and hold their breath [107]. When younger, smaller children or children who find it difficult to remain still are scanned, X-ray technologists can coach them before the exam (instructing them to "play statue" or "freeze") and can position them facing towards the technologist and imaging source to keep eye contact. Unfortunately, this is not always successful. Recovery options for digital slot-scanning include rescanning after additional coaching or altering the imaging parameters. These changes often increase radiation dosages and, therefore, more conventional radiographic techniques might be used preferentially in these patients. Advantages of digital slot-scanning, aside from lower radiation doses, include better total spine visualization, especially in larger patients, and the ability to construct 3-D images [107, 108]. One drawback of this technology is that multiple artifacts have been described appearing on these images that may lead radiologists to misinterpret the findings [107, 108].

Artificial Intelligence and Pediatric Radiology

The prominence of artificial intelligence (AI) in medicine, and especially in radiology, is becoming increasingly noticeable [109]. The term artificial intelligence (AI) includes machine learning and deep learning. AI is defined as "the theory and development of computer systems and algorithms able to perform tasks normally requiring human intelligence [110]." This definition covers a wide variety of tools including, but not limited to, visual perception, speech recognition and decision-making [110]. AI works by "a set of advanced computational algorithms that learn patterns in data to help make predictions on unseen data sets [110]."

Machine learning (ML) is a type of artificial intelligence (AI) that focuses on creating systems that "learn and improve as they consume more data [111]." In ML, algorithms are procedures that are run on data to create a machine learning model. Machine learning models are the outputs of a machine learning algorithm run on data [112]. Deep learning is a type of machine learning that consists of a neural network with three or more layers. Data enters the neural network through the input layer. Hidden layers process and send data to other layers. The result or prediction is made in the output layer. Neural networks try to model human learning by processing and analyzing very large amounts of information, also called training data. They perform a given task with that data, repeatedly, and improve in accuracy each time. It is similar to the way that humans study and practice to improve skills [113]. AI can be subdivided into artificial narrow intelligence, where a computer can carry out a very specific task as well as or better than people, and artificial general (generalist) intelligence, where a computer goes beyond specific tasks to perform higher-order reasoning, reproducing human thought processes [114].

AI has received widespread and increasing interest in healthcare as a means to save time and cost and improve efficiencies [115]. The high-performance statistics and diagnostic accuracies reported by using AI algorithms (when compared to predefined reference standards), especially from image pattern recognition studies, have resulted in many applications proposed for clinical radiology, especially for improved image interpretation [115]. While certain radiology subspecialties, such as those relating to cancer screening, have received a great deal of attention in the media and scientific community, pediatric imaging has been neglected [115].

According to a recent paper from 2023, there were > 200 AI software as a medical device (SaMD) cleared by the FDA that were listed in the online American College of Radiology (ACR) Data Science Institute catalog and only 6 (or 3%) indicated that they were intended for use in children [27]. While the number of FDA cleared algorithms have likely increased since then, the relative paucity of pediatric AI likely has not changed. The paucity of commercially available AI algorithms for pediatric patients has multiple causes. Almost all of the publicly available data sets of imaging studies used to train AI software have exclusively included adult patients. Furthermore, a long-standing shortage of pediatric radiology subspecialists decreases access to expertise for the necessary labeling of images to help develop AI software for pediatric patients. These challenges reduce feasibility and increase development costs for vendors, impeding SaMD development for children. In addition, AI pricing and licensing are usually volume-based; therefore, the lower volume of pediatric studies in comparison to adult studies may result in decreased revenue for pediatric AI developers [27]. Finally, there is increased heterogeneity in data because of the wide variations of normal findings in children at different stages of their development [115].

A few studies investigating the use of adult AI in children demonstrated challenges with the generalizability of AI in pediatric patients. For example, studies using AI developed to detect vertebral body fractures trained on adult patient data were neither reliable nor sensitive for the detection of vertebral body fractures in children [27, 116, 117]. In another study, the application in children of commercially available AI designed for adult chest radiographs noted an overall accuracy of 87.5% in the analysis of 2273 pediatric chest radiographs [27, 118]. Accuracy improved to 96.9% when patients below 2 years of age and those patients with cardiomegaly, corresponding to 33.3% of the original data set, were excluded from the study [27, 118]. Another recent investigation evaluated the ability of adult AI software to differentiate between benign and malignant thyroid nodules in pediatric patients. The specificity for pediatric malignant thyroid nodule classification was reported as only 36.1% [27, 119]. Therefore, the small number of available studies investigating the use of adult AI software on pediatric imaging exams indicates a lack of straightforward generalizability to the pediatric patient population [27].

Furthermore, there are potential pitfalls of using adult AI algorithms in children. There is a wide range of body sizes in children compared to adults, along with different growth and development patterns. Normal anatomy for children varies at different ages in comparison to adults. Certain diseases such as idiopathic ileocolic intussusception and hypertrophic pyloric stenosis are frequently seen in children and are not often seen in adults. Finally, pediatric patients are often imaged utilizing varying parameters such as different contrast and radiation dosages [27].

The lack of pediatric AI can be considered a health equity issue [27]. To raise awareness of this problem, a pediatric radiology AI (PAI) workgroup was created by the ACR Commission on Informatics to educate the radiology community about this issue and offer solutions to address the inadequacies in pediatric AI development [27]. In short, the design, training, validation, and safe implementation of AI in children requires careful and specific approaches that can be different from those used for adults. Furthermore, the committee proposed several solutions to help ensure that AI becomes accessible, safe, effective, and reliable for pediatric patients [27].

Like the "Image Gently" campaign, the PAI workgroup is sponsoring the "Image IntelliGently" campaign to ensure equal access to high-quality and safe AI for pediatric patients. Individuals and organizations who sign this pledge aim to promote the safe, equitable and reliable use of AI for medical imaging in children [120]. Also, members of the pediatrics panel of the ACR Data Science Institute (DSI), which consists of several pediatric radiologists, have published on the ACR DSI website a list of several potential use cases for AI in pediatric radiology [121]. These use cases could possibly be developed and brought to market by vendors in the future.

However, pediatric radiology has not completely been shut out of the benefits of using AI. Artificial intelligence, as it applies to radiology, can be divided into interpretive and non-interpretive applications. Interpretive applications all carry out specific tasks by using computer vision, an evaluation of the same images (and perhaps other data) that radiologists utilize, often following manual or automated segmentation of images into smaller regions of interest. The use cases for interpretive

Table 3 Interpretive use cases of AI in pediatric radiology

Orthopedic use cases	Nonorthopedic use cases
Bone Age	Differentiation of head and neck soft-tissue tumor subtypes
Anatomic labeling of radiographs	Automated assessment of neck radiograph adequacy
Scoliosis quantification	Attention deficit–hyperactivity disorder prediction with functional magnetic resonance imaging
Leg length discrepancy measurement	Ovarian torsion diagnosis
Supracondylar fracture detection	Cystic fibrosis scoring on chest radiographs
Elbow injury detection	Categorization of congenital renal anomalies
	Liver stiffness classification using magnetic resonance imaging and clinical data
	Intussusception detection on radiographs
	Fetal magnetic resonance imaging evaluation

Adapted from: Otjen JP, Moore MM, Romberg EK, Perez FA, Iyer RS. The current and future roles of artificial intelligence in pediatric radiology. Pediatr Radiol. 2022 Oct;52(11):2065-2073

applications in pediatric radiology can be roughly divided into orthopedic and non-orthopedic applications and are listed in Table 3 [122]. While mature pediatric specific models for some of these orthopedic use cases have been developed, most of them have not been cleared by the FDA and are not broadly in use [122]. Most of those that are FDA-cleared include both adult and pediatric data training sets and are not specifically designed for use in children. The inclusion of adult training data may bias the algorithms and can possibly affect their accuracy in pediatric patients. Non-interpretive areas include applications that can assist in protocoling exams, image acquisition such as image reconstruction on MRI or CT, correction of artifacts, and reducing radiation doses on imaging exams [122]. These applications can often be used in adult as well as pediatric imaging studies [122]. There are also hybrid systems that are not solely designed to interpret images but instead to incorporate image evaluation, such as triaging exams for radiologist interpretation by detecting acute pathologies such as acute abdomen on body CT or intracranial hemorrhage on head CT [122]. Most of these types of models have also not been trained on pediatric specific data.

The most well-known example of AI application to pediatric radiology was generated by the 2017 Radiological Society of North America (RSNA)'s Machine Learning Challenge, in which a compilation of 14,236 bone-age hand radiographs and associated data for each study was made freely available to train machine-learning algorithms. Greater than 100 AI applications were entered, and the winning entry deviated from ground truth data by just 4.3 months. In comparison, the average difference between radiologists on the same dataset was 7.3 months. In addition, many of the top-performing entries performed better than human radiologists in this regard, despite widely differing designs in algorithms [122]. The big, well-annotated dataset with a well-defined ground-truth and quantitative output make bone-age exams especially attractive for evaluation with AI [122].

A convolutional neural network (CNN) is a kind of deep learning network that is mainly used to identify and classify images and to recognize objects within images [123]. A CNN bone age model described by Pan et al. [122, 124] based on 15,129 trauma hand radiographs with a test set of 214 radiographs from a different independent institution also demonstrated good results, with an average absolute error of 11.1 months when compared with chronological age. This was slightly better than two radiologists. Pan et al. [122, 125] also showed an improvement in bone age estimation by combining multiple techniques that had been submitted as a part of the RSNA bone age challenge, which was able to bring the average error down to 3.8 months using an ensemble of four models.

Other fully mature applications created using machine learning to some degree are in use, including the BoneXpert application [122, 126] although this model has not yet been cleared by the FDA. This product was trained with 1559 hand radiographs of Danish children and tested on 122 images and has been validated in and calibrated for different patient populations [122, 127]. AI for multiple other interpretive non-orthopedic pediatric radiology use cases has been described and is listed in Table 3 [122]. Unfortunately, very few, if any, of these interpretive AI applications have been cleared for use by the FDA, although these can still be utilized experimentally in the United States [122].

The current generation of interpretive AI models in radiology can perform only a small number of interpretation tasks, and they rely extensively on data that have been specifically labeled and categorized. Although focusing on the image as an isolated model input has some value, it does not reflect the true cognitive work of radiology, which involves comprehensive interpretation of medical imaging studies, comparing current and prior exams, and correlating this information with clinical contextual data to generate diagnostic and patient management recommendations [128].

Nevertheless, the rapid development of newer AI models may help solve some of the factors that have been implicated in the shortfall of pediatric radiology AI. Self-supervised models can learn from medical data without the need for annotations. These models are useful when labeled data are scarce [128], which has been the case for pediatric imaging studies [27]. Multimodal models can comprehend and synthesize different types of medical data (e.g., medical images and electronic health records (EHR)). These models are useful for tasks that require a detailed understanding of the patient, such as diagnosis and individualized treatment planning for patients with complex medical histories [128] that are often seen at children's hospitals. Foundation models function as a starting point for creating more specific AI models. These models are trained on vast quantities of data and can be fine-tuned for specific applications, such as segmenting anatomic structures or lesion detection [128]. As these models can segment structures by themselves, they may help overcome the shortage of pediatric radiologists that, in the past, have been needed to label anatomic structures to help train older AI models.

Another newer type of model worthy of discussion is known as Generative Adversarial Networks (GANs). This type of model actually consists of two separate models called a generator and a discriminator. The generator is trained to produce fake images while the discriminator is responsible for determining whether the image created by the generator is fake or real. The training of the model is considered complete when the discriminator is unable to realize that the generator's output images are fake, and the generator becomes able to produce high-quality fake images that are close to the real ones [129]. GANs can thus be used to create synthetic images which can help overcome the lack of datasets needed to train pediatric radiology AI models [27]. In addition to creating synthetic images, a recent review showed that GANs can be applied to pediatric MRI, X-ray, CT, ultrasound, and PET for image translation, segmentation, reconstruction, quality assessment, data augmentation, and disease diagnosis [129].

These newer models are helping to pave the way for a new generation of general medical AI models with the potential to perform the entire task of radiologic image interpretation. These models will be able to accurately generate a full radiologic report by interpreting a wide range of findings with degrees of uncertainty and specificity based on the image, by combining the clinical context with the imaging data, and by utilizing previous imaging in the decision of the model. This comprehensive approach more closely replicates the overall cognitive work in radiology [128]. Early investigations of these models have shown that they can detect many diseases on images at an expert level without needing further annotation. This capability is known as zero-shot learning [128]. Zero-shot learning in AI refers to the ability of a model to correctly infer or make decisions about data classes it has not been trained on or previously seen. The model is trained on a set of classes and then, subsequently, tested on a different, unique set of classes. The goal is to enable the model to generalize from its training data to different unseen classes utilizing high-level attributes or descriptions [130]. Such a technique would likely also prove helpful in developing more pediatric radiology AI applications due to the current shortage of publicly available pediatric datasets needed to train algorithms.

There is a great deal of potential for artificial intelligence to advance pediatric healthcare, and there will likely be many exciting developments in the coming years. Telemedicine is one of the areas where AI is expected to have a major impact [131]. Telemedicine is especially applicable in the field of radiology, where the practice of teleradiology is prevalent. This is of particular importance for pediatric patients residing in rural or remote areas where the shortage of pediatric subspecialists as well as pediatric radiologists is often worse than in more urban settings. In addition to improving access to care and lowering healthcare costs, AI-based tools can help physicians, including radiologists, diagnose and treat children from remote settings [131].

To help facilitate the use of remote AI tools to help treat and diagnose children remotely, a pediatric moonshot initiative was recently implemented. This group hopes to connect to approximately 1 million healthcare machines, such as ultrasound machines as well as CT and MRI scanners in radiology, at the world's approximately 500 childrens' hospitals [132]. This would allow access to large amounts of patient data and imaging studies [132]. The use of edge computing would allow all protected health information to remain within the confines of an institution's network [132]. The goal of this project would be to create privacy-preserving real-time AI applications to help provide remote healthcare to children living in underserved areas and, therefore, help reduce healthcare inequity [132].

Conclusion

The field of pediatric radiology has come a long way since what was believed to be the first radiograph on a child was performed in 1896. Pediatric radiology continued to develop as a subspecialty of radiology with the advent of other imaging modalities such as CT, MRI, ultrasound and nuclear medicine scintigraphy. Although the modalities used to image adults and children are similar, the practice of pediatric and adult radiology has significant differences. This is due to differing growth and development of children and varying anatomy as well as differing sensitivities to ionizing radiation and psychosocial needs that children have in comparison to adults. More recent advances in the main imaging modalities have allowed pediatric radiologists to diagnose pathologies that they were previously unable to and, in some instances, without exposing children to the potential dangers of ionizing radiation. Artificial intelligence has recently garnered significant attention due to its potential to revolutionize the practice of medicine. Unfortunately, the development of AI in children has lagged far behind the development in adults, causing a health inequity between the two populations. However, campaigns are underway to help alleviate this problem and, in addition, a new class of AI models have been developed that do not have the limitations of prior models and have the potential to increase the development of AI in children.

Competing Interests The author is employed full-time by Bayer as Americas Director, Radiology Digital Medical Affairs in addition to his clinical work. Bayer owns the Calantic and Blackford Radiology AI marketplaces. In addition, the author has stock equity in Rad AI.

References

1. Bercovich E, Javitt MC. Medical imaging: from Roentgen to the digital revolution, and beyond. Rambam Maimonides Med J. 2018 Oct 4;9(4):e0034.
2. The Noble Prize. "Wilhelm Conrad Röntgen Biographical." https://www.nobelprize.org/prizes/physics/1901/rontgen/biographical/. Accessed February 10, 2024.
3. Griscom NT. History of pediatric radiology in the United States and Canada: images and trends. Radiographics. 1995 Nov;15(6):1399-422.
4. Radiopedia. "Soft tissue." https://radiopaedia.org/articles/soft-tissue?lang=us. Accessed February 10, 2024
5. Cheeney SHE, Maloney E, Iyer RS. Safety considerations related to intravenous contrast agents in pediatric imaging. Pediatr Radiol. 2023
6. Callahan MJ, Poznauskis L, Zurakowski D, Taylor GA. Nonionic iodinated intravenous contrast material-related reactions: incidence in large urban children's hospital-retrospective analysis of data in 12,494 patients. Radiology. 2009 Mar;250(3):674-81.
7. Wikipedia. "Functional imaging." https://en.wikipedia.org/wiki/Functional_imaging#:~:text=Functional%20imaging%20(or%20physiological%20imaging,regional%20chemical%20composition%2C%20and%20absorption. Accessed February 17, 2024.

8. Radiopedia. "Positron emission tomography." https://radiopaedia.org/articles/positron-emission-tomography?lang=us#:~:text=Tumor%20cells%20have%20a%20high,accumulation%20is%20detected%20and%20quantified. Accessed February 17, 2024.
9. Kaste SC. PET-CT in children: where is it appropriate? Pediatr Radiol. 2011 Sep;41 Suppl 2(0 2):509-13.
10. Mettler Jr FA, Guibertau MJ. Essentials of nuclear medicine 7th ed. Philadelphia (US): Elsevier; 2019. Chapter 2, Instrumentation and quality control; p.19-59.
11. Kutanzi KR, Lumen A, Koturbash I, Miousse IR. Pediatric exposures to ionizing radiation: carcinogenic considerations. Int J Environ Res Public Health. 2016 Oct 28;13(11)
12. Kaproth-Joslin KA, Nicola R, Dogra VS. The history of us: from bats and boats to the bedside and beyond: RSNA Centennial Article. Radiographics. 2015 May-Jun;35(3):960-70.
13. Krishnamurthy R, Tkach JA. Chapter 2, Magnetic resonance safety." In: Coley BJ, editor. Caffey's Pediatric Diagnostic Imaging. 13th ed. vol. 1. Philadelphia (US): Elsevier; 2019. p. 43-54.
14. Betz LH, Dillman JR, Jones BV, Tkach JA. MRI safety screening of children with implants: updates and challenges. Pediatr Radiol. 2023 Jun;53(7):1454-1468.
15. Mithal LB, Patel PS, Mithal D, Palac HL, Rozenfeld MN. Use of gadolinium-based magnetic resonance imaging contrast agents and awareness of brain gadolinium deposition among pediatric providers in North America. Pediatr Radiol. 2017 May;47(6):657-664.
16. Prince MR, Zhang H, Zou Z, Staron RB, Brill PW. Incidence of immediate gadolinium contrast media reactions. AJR Am J Roentgenol. 2011 Feb;196(2):W138-43.
17. Maloney E, Iyer RS, Phillips GS, Menon S, Lee JJ, Callahan MJ. Practical administration of intravenous contrast media in children: screening, prophylaxis, administration and treatment of adverse reactions. Pediatr Radiol. 2019 Apr;49(4):433-447.
18. ACR Committee on Drugs and Contrast Media. ACR manual on contrast media (version 2022).
19. Dillman JR, Ellis JH, Cohan RH, Strouse PJ, Jan SC. Allergic-like breakthrough reactions to gadolinium contrast agents after corticosteroid and antihistamine premedication. AJR Am J Roentgenol. 2008 Jan;190(1):187-90.
20. Beckett KR, Moriarity AK, Langer JM. Safe use of contrast media: what the radiologist needs to know. Radiographics. 2015 Oct;35(6):1738-1750.
21. Kaewlai R, Abujudeh H. Nephrogenic systemic fibrosis. AJR Am J Roentgenol. 2012 Jul;199(1):W17-23.
22. Cowper SE, Robin HS, Steinberg SM, Su LD, Gupta S, LeBoit PE. Scleromyxoedema-like cutaneous diseases in renal-dialysis patients. Lancet. 2000 Sep 16;356(9234):1000-1.
23. Braverman IM, Cowper S. Nephrogenic systemic fibrosis. F1000 Med Rep. 2010 Nov 24;2:84.
24. Nardone B, Saddleton E, Laumann AE, Edwards BJ, Raisch DW, McKoy JM, Belknap SM, Bull C, Haryani A, Cowper SE, Abu-Alfa AK, Miller FH, Godinez-Puig V, Dharnidharka VR, West DP. Pediatric nephrogenic systemic fibrosis is rarely reported: a RADAR report. Pediatr Radiol. 2014 Feb;44(2):173-80.
25. Weinreb JC, Rodby RA, Yee J, Wang CL, Fine D, McDonald RJ, Perazella MA, Dillman JR, Davenport MS. Use of intravenous gadolinium-based contrast media in patients with kidney disease: consensus statements from the American College of Radiology and the National Kidney Foundation. Radiology. 2021 Jan;298(1):28-35.
26. McDonald RJ, Weinreb JC, Davenport MS. Symptoms associated with gadolinium exposure (SAGE): a suggested term. Radiology. 2022 Feb;302(2):270-273.
27. Sammer MBK, Akbari YS, Barth RA, Blumer SL, Dillman JR, Farmakis SG, Frush DP, Gokli A, Halabi SS, Iyer R, Joshi A, Kwon JK, Otero HJ, Sher AC, Sotardi ST, Taragin BH, Towbin AJ, Wald C. Use of artificial intelligence in radiology: impact on pediatric patients, a white paper from the ACR pediatric AI workgroup. J Am Coll Radiol. 2023 Aug;20(8):730-737.
28. Nasseri F, Eftekhari F. Clinical and radiologic review of the normal and abnormal thymus: pearls and pitfalls. Radiographics. 2010 Mar;30(2):413-28.

29. Chan BY, Gill KG, Rebsamen SL, Nguyen JC. MR imaging of pediatric bone marrow. Radiographics. 2016 Oct;36(6):1911-1930.
30. Nguyen JC, Markhardt BK, Merrow AC, Dwek JR. Imaging of pediatric growth plate disturbances. Radiographics. 2017 Oct;37(6):1791-1812.
31. Figaji AA. Anatomical and physiological differences between children and adults relevant to traumatic brain injury and the implications for clinical assessment and care. Front Neurol. 2017 Dec 14;8:685.
32. Strauss KJ. Pediatric interventional radiography equipment: safety considerations. Pediatr Radiol. 2006 Sep;36 Suppl 2(Suppl 2):126-135.
33. Applegate KE, Cost NG. Image gently: a campaign to reduce children's and adolescents' risk for cancer during adulthood. J Adolesc Health. 2013 May;52(5 Suppl):S93-7.
34. The Image Gently Alliance. "Image Gently." http://www.imagegently.org. Accessed April 8, 2024.
35. Image Wisey. "Pediatric CT and Image Gently." https://www.imagewisely.org/Imaging-Modalities/Computed-Tomography/Pediatric-CT-and-Image-Gently. Accessed April 8, 2024.
36. Donnelly LF. Special considerations in pediatric imaging. In: Donnelly LF editor. Fundamentals of pediatric imaging. 3rd ed. Philadephia (US): Elsevier; 2022. p.1-8.
37. Artunduaga M, Liu CA, Morin CE, Serai SD, Udayasankar U, Greer MC, Gee MS. Safety challenges related to the use of sedation and general anesthesia in pediatric patients undergoing magnetic resonance imaging examinations. Pediatr Radiol. 2021 May;51(5):724-735.
38. Cravero JP, Blike GT, Beach M, Gallagher SM, Hertzog JH, Havidich JE, Gelman B; Pediatric Sedation Research Consortium. Incidence and nature of adverse events during pediatric sedation/anesthesia for procedures outside the operating room: report from the Pediatric Sedation Research Consortium. Pediatrics. 2006 Sep;118(3):1087-96.
39. United States Food and Drug Administration. FDA drug safety communication: FDA review results in new warnings about using general anesthetics and sedation drugs in young children and pregnant women. Online document. https://www.fda.gov/drugs/drug-safety-and-availability/fda-drug-safety-communication-fda-review-results-new-warnings-about-using-general-anesthetics-and. Accessed March 16, 2025.
40. Beach ML, Cohen DM, Gallagher SM, Cravero JP. Major adverse events and relationship to nil per os status in pediatric sedation/anesthesia outside the operating room: a report of the Pediatric Sedation Research Consortium. Anesthesiology. 2016 Jan;124(1):80-8.
41. Davidson AJ, Morton NS, Arnup SJ, de Graaff JC, Disma N, Withington DE, Frawley G, Hunt RW, Hardy P, Khotcholava M, von Ungern Sternberg BS, Wilton N, Tuo P, Salvo I, Ormond G, Stargatt R, Locatelli BG, McCann ME; General anesthesia compared to spinal anesthesia (gas) consortium. apnea after awake regional and general anesthesia in infants: the general anesthesia compared to spinal anesthesia study-comparing apnea and neurodevelopmental outcomes, a randomized controlled trial. Anesthesiology. 2015 Jul;123(1):38-54.
42. Jenkins E, Hebbar KB, Karaga KK, Hirsh DA, Fortenberry JD, McCracken CE, Simoneaux SF, Mallory MD, Kamat PP. Experience with the use of propofol for radiologic imaging in infants younger than 6 months of age. Pediatr Radiol. 2017 Jul;47(8):974-983.
43. Cravero JP, Beach ML, Blike GT, Gallagher SM, Hertzog JH; Pediatric Sedation Research Consortium. the incidence and nature of adverse events during pediatric sedation/anesthesia with propofol for procedures outside the operating room: a report from the Pediatric Sedation Research Consortium. Anesth Analg. 2009 Mar;108(3):795-804.
44. Kadom N. Engaging patients and families in pediatric radiology. Pediatr Radiol. 2020 Oct;50(11):1492-1498.
45. Rathee S, Joshi P, Kelkar A, Seth N. Fetal MRI: a pictorial essay. Indian J Radiol Imaging. 2016 Jan-Mar;26(1):52-62.
46. Saleem SN. Fetal MRI: an approach to practice: a review. J Adv Res. 2014 Sep;5(5):507-23.
47. Smith FW, Adam AH, Phillips WD. NMR imaging in pregnancy. Lancet. 1983 Jan 1;1(8314-5):61-2.

48. Levine D, Barnes PD, Sher S, Semelka RC, Li W, McArdle CR, Worawattanakul S, Edelman RR. Fetal fast MR imaging: reproducibility, technical quality, and conspicuity of anatomy. Radiology. 1998 Feb;206(2):549-54.
49. Kanal E, Borgstede JP, Barkovich AJ, Bell C, Bradley WG, Felmlee JP, Froelich JW, Kaminski EM, Keeler EK, Lester JW, Scoumis EA, Zaremba LA, Zinninger MD; American College of Radiology. American College of Radiology white paper on MR safety. AJR Am J Roentgenol. 2002 Jun;178(6):1335-47.
50. Glenn OA, Barkovich AJ. Magnetic resonance imaging of the fetal brain and spine: an increasingly important tool in prenatal diagnosis, part 1. AJNR Am J Neuroradiol. 2006 Sep;27(8):1604-11.
51. Sabir H, Kipfmueller F, Bagci S, Dresbach T, Grass T, Nitsch-Felsecker P, Pantazis C, Schmitt J, Schroeder L, Mueller A. Feasibility of bedside portable MRI in neonates and children during ECLS. Crit Care. 2023 Apr 4;27(1):134.
52. Radiopedia. "Anterior fontanelle." https://radiopaedia.org/articles/anterior-fontanelle?lang=us. Accessed April 13, 2024.
53. Beekman R, Crawford A, Mazurek MH, Prabhat AM, Chavva IR, Parasuram N, Kim N, Kim JA, Petersen N, de Havenon A, Khosla A, Honiden S, Miller PE, Wira C, Daley J, Payabvash S, Greer DM, Gilmore EJ, Taylor Kimberly W, Sheth KN. Bedside monitoring of hypoxic ischemic brain injury using low-field, portable brain magnetic resonance imaging after cardiac arrest. Resuscitation. 2022 Jul;176:150-158.
54. Sien ME, Robinson AL, Hu HH, Nitkin CR, Hall AS, Files MG, Artz NS, Pitts JT, Chan SS. Feasibility of and experience using a portable MRI scanner in the neonatal intensive care unit. Arch Dis Child Fetal Neonatal Ed. 2023 Jan;108(1):45-50.
55. Teixeira SR, Elias Junior J, Nogueira-Barbosa MH, Guimarães MD, Marchiori E, Santos MK. Whole-body magnetic resonance imaging in children: state of the art. Radiol Bras. 2015 Mar-Apr;48(2):111-20.
56. Grzywińska M, Świętoń D, Sabisz A, Piskunowicz M. Functional magnetic resonance urography in children-tips and pitfalls. Diagnostics (Basel). 2023 May 18;13(10):1786.
57. Dickerson EC, Dillman JR, Smith EA, DiPietro MA, Lebowitz RL, Darge K. Pediatric MR urography: indications, techniques, and approach to review. Radiographics. 2015 Jul-Aug;35(4):1208-30.
58. Rodigas J, Kirsch H, John U, Seifert P, Winkens T, Stenzel M, Mentzel HJ. Static and functional MR urography to assess congenital anomalies of the kidney and urinary tract in infants and children: comparison with MAG3 renal scintigraphy and sonography. AJR Am J Roentgenol. 2018 Jul;211(1):193-203.
59. Chavhan GB, Babyn PS, Walters T. MR enterography in children: principles, technique, and clinical applications. Indian J Radiol Imaging. 2013 Apr;23(2):173-8.
60. Huang CT, Lee HC, Chen WT, Jiang CB, Shih SL, Yeung CY. Usefulness of magnetic resonance cholangiopancreatography in pancreatobiliary abnormalities in pediatric patients. Pediatr Neonatol. 2011 Dec;52(6):332-6.
61. Fernández-Esparrach G, Ginès A, Sánchez M, Pagés M, Pellisé M, Fernández-Cruz L, López-Boado MA, Quintó L, Navarro S, Sendino O, Cárdenas A, Ayuso C, Bordas JM, Llach J, Castells A. Comparison of endoscopic ultrasonography and magnetic resonance cholangiopancreatography in the diagnosis of pancreatobiliary diseases: a prospective study. Am J Gastroenterol. 2007 Aug;102(8):1632-9.
62. Almehdar A, Chavhan GB. MR cholangiopancreatography at 3.0 T in children: diagnostic quality and ability in assessment of common paediatric pancreatobiliary pathology. Br J Radiol. 2013 May;86(1025):20130036.
63. Metreweli C, So NM, Chu WC, Lam WW. Magnetic resonance cholangiography in children. Br J Radiol. 2004 Dec;77(924):1059-64.
64. Mitchell FM, Prasad SK, Greil GF, Drivas P, Vassiliou VS, Raphael CE. Cardiovascular magnetic resonance: diagnostic utility and specific considerations in the pediatric population. World J Clin Pediatr. 2016 Feb 8;5(1):1-15.

65. Lai WW, Geva T, Shirali GS, Frommelt PC, Humes RA, Brook MM, Pignatelli RH, Rychik J; Task force of the pediatric council of the American Society of Echocardiography; Pediatric Council of the American Society of Echocardiography. Guidelines and standards for performance of a pediatric echocardiogram: a report from the task force of the Pediatric Council of the American Society of Echocardiography. J Am Soc Echocardiogr. 2006 Dec;19(12):1413-30.
66. Fratz S, Chung T, Greil GF, Samyn MM, Taylor AM, Valsangiacomo Buechel ER, Yoo SJ, Powell AJ. Guidelines and protocols for cardiovascular magnetic resonance in children and adults with congenital heart disease: SCMR expert consensus group on congenital heart disease. J Cardiovasc Magn Reson. 2013 Jun 13;15(1):51.
67. Brenner D, Elliston C, Hall E, Berdon W. Estimated risks of radiation-induced fatal cancer from pediatric CT. AJR Am J Roentgenol. 2001 Feb;176(2):289-96.
68. McGinness A, Lin-Martore M, Addo N, Shaahinfar A. The unmet demand for point-of-care ultrasound among general pediatricians: a cross-sectional survey. BMC Med Educ. 2022 Jan 3;22(1):7.
69. Conlon TW, Nishisaki A, Singh Y, Bhombal S, De Luca D, Kessler DO, Su ER, Chen AE, Fraga MV. Moving beyond the stethoscope: diagnostic point-of-care ultrasound in pediatric practice. Pediatrics. 2019 Oct;144(4):e20191402.
70. Benabbas R, Hanna M, Shah J, Sinert R. Diagnostic accuracy of history, physical examination, laboratory tests, and point-of-care ultrasound for pediatric acute appendicitis in the emergency department: a systematic review and meta-analysis. Acad Emerg Med. 2017 May;24(5):523-551.
71. Doniger SJ, Ishimine P, Fox JC, Kanegaye JT. Randomized controlled trial of ultrasound-guided peripheral intravenous catheter placement versus traditional techniques in difficult-access pediatric patients. Pediatr Emerg Care. 2009 Mar;25(3):154-9.
72. Harel-Sterling M, Diallo M, Santhirakumaran S, Maxim T, Tessaro M. Emergency department resource use in pediatric pneumonia: point-of-care lung ultrasonography versus chest radiography. J Ultrasound Med. 2019 Feb;38(2):407-414.
73. Squires JH, McCarville MB. Contrast-enhanced ultrasound in children: implementation and key diagnostic applications. AJR Am J Roentgenol. 2021 Nov;217(5):1217-1231.
74. Radiopedia. “Red blood cells.” https://radiopaedia.org/articles/red-blood-cells-1?lang=us. Accessed April 14, 2024.
75. Fetzer DT, Rafailidis V, Peterson C, Grant EG, Sidhu P, Barr RG. Artifacts in contrast-enhanced ultrasound: a pictorial essay. Abdom Radiol (NY). 2018 Apr;43(4):977-997.
76. Radiopedia. “Radiological image artifact.” https://radiopaedia.org/articles/radiological-image-artifact?lang=us Accessed April 14, 2024.
77. Schooler GR, Squires JH, Alazraki A, Chavhan GB, Chernyak V, Davis JT, Khanna G, Krishnamurthy R, Lungren MP, Masand PM, Podberesky DJ, Sirlin CB, Towbin AJ. Pediatric hepatoblastoma, hepatocellular carcinoma, and other hepatic neoplasms: consensus imaging recommendations from American College of Radiology pediatric Liver Reporting and Data System (LI-RADS) working group. Radiology. 2020 Sep;296(3):493-497.
78. Deganello A, Rafailidis V, Sellars ME, Ntoulia A, Kalogerakou K, Ruiz G, Cosgrove DO, Sidhu PS. Intravenous and intracavitary use of contrast-enhanced ultrasound in the evaluation and management of complicated pediatric pneumonia. J Ultrasound Med. 2017 Sep;36(9):1943-1954.
79. Kapur J, Oscar H. Contrast-enhanced ultrasound of kidneys in children with renal failure. Journal of Medical Ultrasound. 2015 Jun 1;23(2):86-97.
80. Darge K. Voiding urosonography with ultrasound contrast agents for the diagnosis of vesicoureteric reflux in children. I. Procedure. Pediatr Radiol. 2008 Jan;38(1):40-53.
81. Navarro OM. Chapter 6: genitourinary. In: Donnelly LF, editor. Fundamentals of pediatric imaging. 3rd ed. London (GB): Elsevier; 2022. p. 139-174.

82. Leroy S, Vantalon S, Larakeb A, Ducou-Le-Pointe H, Bensman A. Vesicoureteral reflux in children with urinary tract infection: comparison of diagnostic accuracy of renal US criteria. Radiology. 2010 Jun;255(3):890-8.
83. Back SJ, Ntoulia A. Imaging for vesicoureteral reflux: point-contrast-enhanced voiding urosonography is a sensitive and radiation-free imaging method that improves patient comfort. AJR Am J Roentgenol. 2023 Jan;220(1):24-25.
84. Mjelle AB, Mulabecirovic A, Havre RF, Rosendahl K, Juliusson PB, Olafsdottir E, Gilja OH, Vesterhus M. Normal liver stiffness values in children: a comparison of three different elastography methods. J Pediatr Gastroenterol Nutr. 2019 May;68(5):706-712.
85. Močnik M, Marčun Varda N. Ultrasound elastography in children. Children (Basel). 2023 Jul 27;10(8):1296.
86. Kim JR, Suh CH, Yoon HM, Lee JS, Cho YA, Jung AY. The diagnostic performance of shear-wave elastography for liver fibrosis in children and adolescents: a systematic review and diagnostic meta-analysis. Eur Radiol. 2018 Mar;28(3):1175-1186.
87. Potter C, Hogan MJ, Henry-Kendjorsky K, Balint J, Barnard JA. Safety of pediatric percutaneous liver biopsy performed by interventional radiologists. J Pediatr Gastroenterol Nutr. 2011 Aug;53(2):202-6.
88. Serai SD, Franchi-Abella S, Syed AB, Tkach JA, Toso S, Ferraioli G. MR and ultrasound elastography for fibrosis assessment in children: practical implementation and supporting evidence-AJR expert panel narrative review. AJR Am J Roentgenol. 2024 Jan 3.
89. Swenson DW, Ayyala RS, Sams C, Lee EY. Practical imaging strategies for acute appendicitis in children. AJR Am J Roentgenol. 2018 Oct;211(4):901-909.
90. Nguyen HN, Navarro OM, Bloom DA, Feinstein KA, Guillerman RP, Munden MM, Sammer MBK, Silva CT. Ultrasound for midgut malrotation and midgut volvulus: AJR expert panel narrative review. AJR Am J Roentgenol. 2022 Jun;218(6):931-939.
91. Hernanz-Schulman M. Infantile hypertrophic pyloric stenosis. Radiology. 2003 May;227(2):319-31.
92. Teele RL, Smith EH. Ultrasound in the diagnosis of idiopathic hypertrophic pyloric stenosis. N Engl J Med. 1977 May 19;296(20):1149-50.
93. Is ultrasound really necessary for the diagnosis of hypertrophic pyloric stenosis? Lancet. 1988 May 21;1(8595):1146.
94. De Backer A, Bové T, Vandenplas Y, Peeters S, Deconinck P. Contribution of endoscopy to early diagnosis of hypertrophic pyloric stenosis. J Pediatr Gastroenterol Nutr. 1994 Jan;18(1):78-81.
95. Nasser MP. Chapter 5: gastrointestinal. In: Donnelly LF, editor Fundamentals of pediatric imaging. 3rd ed. London (GB): Elsevier; 2022. p. 95-138.
96. Kang YR, Koo J. Ultrasonography of the pediatric hip and spine. Ultrasonography. 2017 Jul;36(3):239-251.
97. Alsaleem M, Set KK, Saadeh L. Developmental dysplasia of hip: a review. Clin Pediatr (Phila). 2015 Sep;54(10):921-8.
98. Matrawy KA, Nouh MR. Ultrasound screening for developmental dysplasia of the hip and its socioeconomic impact: experience of tertiary care health level. Alexandria Journal of Medicine. 2014 Mar 1;50(1):25-9.
99. Omeroğlu H. Use of ultrasonography in developmental dysplasia of the hip. J Child Orthop. 2014 Mar;8(2):105-13.
100. Nair N, Sreenivas M, Gupta AK, Kandasamy D, Jana M. Neonatal and infantile spinal sonography: A useful investigation often underutilized. Indian J Radiol Imaging. 2016 Oct-Dec;26(4):493-501.
101. Gottumukkala RV, Kalra MK, Tabari A, Otrakji A, Gee MS. Advanced ct techniques for decreasing radiation dose, reducing sedation requirements, and optimizing Image quality in children. Radiographics. 2019 May-Jun;39(3):709-726.
102. McCollough CH, Leng S, Yu L, Fletcher JG. Dual-and multi-energy CT: principles, technical approaches, and clinical applications. Radiology. 2015 Sep;276(3):637-53.

103. Siegel MJ, Ramirez-Giraldo JC. Dual-energy CT in children: imaging algorithms and clinical applications. Radiology. 2019 May;291(2):286-297.
104. Mese I, Altintas Mese C, Demirsoy U, Anik Y. Innovative advances in pediatric radiology: computed tomography reconstruction techniques, photon-counting detector computed tomography, and beyond. Pediatr Radiol. 2024 Jan;54(1):1-11.
105. Vanderby SA, Babyn PS, Carter MW, Jewell SM, McKeever PD. Effect of anesthesia and sedation on pediatric MR imaging patient flow. Radiology. 2010 Jul;256(1):229-37.
106. Agrawal S, Hulme SL, Hayward R, Brierley J. A portable CT scanner in the pediatric intensive care unit decreases transfer-associated adverse events and staff disruption. Eur J Trauma Emerg Surg. 2010 Aug;36(4):346-52.
107. Blumer SL, Dinan D, Grissom LE. Benefits and unexpected artifacts of biplanar digital slot-scanning imaging in children. Pediatr Radiol. 2014 Jul;44(7):871-82.
108. Wybier M, Bossard P. Musculoskeletal imaging in progress: the EOS imaging system. Joint Bone Spine. 2013 May;80(3):238-43. https://doi.org/10.1016/j.jbspin.2012.09.018.
109. Barreiro-Ares A, Morales-Santiago A, Sendra-Portero F, Souto-Bayarri M. Impact of the rise of artificial intelligence in radiology: what do students think? Int J Environ Res Public Health. 2023 Jan 16;20(2):1589.
110. Laborie LB, Naidoo J, Pace E, Ciet P, Eade C, Wagner MW, Huisman TAGM, Shelmerdine SC. European society of paediatric radiology artificial intelligence taskforce: a new taskforce for the digital age. Pediatr Radiol. 2023 Apr;53(4):576-580.
111. OCI. "What is Machine Learning?" https://www.oracle.com/artificial-intelligence/machine-learning/what-is-machine-learning/ Accessed April 20, 2024.
112. Brownlee J. "Difference between algorithm and model in machine learning." August 19, 2020. https://machinelearningmastery.com/difference-between-algorithm-and-model-in-machine-learning/ Accessed March 18, 2025.
113. Coursera. "What is Deep Learning? Definition, Examples and Careers," March 26, 2024, https://www.coursera.org/articles/what-is-deep-learning Accessed April 20, 2024.
114. European Society of Radiology (ESR). What the radiologist should know about artificial intelligence - an ESR white paper. Insights Imaging. 2019 Apr 4;10(1):44.
115. Davendralingam N, Sebire NJ, Arthurs OJ, Shelmerdine SC. Artificial intelligence in paediatric radiology: future opportunities. Br J Radiol. 2021 Jan 1;94(1117):20200975.
116. Alqahtani FF, Messina F, Kruger E, Gill H, Ellis M, Lang I, Broadley P, Offiah AC. Evaluation of a semi-automated software program for the identification of vertebral fractures in children. Clin Radiol. 2017 Oct;72(10):904.e11-904.e20.
117. Alqahtani FF, Crabtree NJ, Bromiley PA, Cootes T, Broadley P, Lang I, Offiah AC. Diagnostic performance of morphometric vertebral fracture analysis (MXA) in children using a 33-point software program. Bone. 2020 Apr;133:115249.
118. Shin HJ, Son NH, Kim MJ, Kim EK. Diagnostic performance of artificial intelligence approved for adults for the interpretation of pediatric chest radiographs. Scientific reports. 2022 Jun 17;12(1):10215.
119. Yang J, Page LC, Wagner L, Wildman-Tobriner B, Bisset L, Frush D, Mazurowski MA. Thyroid nodules on ultrasound in children and young adults: comparison of diagnostic performance of radiologists' impressions, ACR TI-RADS, and a deep learning algorithm. AJR Am J Roentgenol. 2023 Mar;220(3):408-417.
120. American College of Radiology. "Image IntelliGently Pledge." https://www.acr.org/Practice--Management-Quality-Informatics/Informatics/Pediatric-Radiology-AI-Resources/Image-IntelliGently-Campaign/Pledge Accessed April 20, 2024.
121. American College of Radiology Data Science Institute. "Define-AI Use Case Directory." https://www.acrdsi.org/DSI-Services/Define-AI Accessed April 20, 2024.
122. Otjen JP, Moore MM, Romberg EK, Perez FA, Iyer RS. The current and future roles of artificial intelligence in pediatric radiology. Pediatr Radiol. 2022 Oct;52(11):2065-2073.
123. Nvidia. "Convolutional Neural Network." https://www.nvidia.com/en-us/glossary/convolutional-neural-network/ Accessed April 20, 2024.

124. Pan I, Baird GL, Mutasa S, Merck D, Ruzal-Shapiro C, Swenson DW, Ayyala RS. Rethinking Greulich and Pyle: a deep learning approach to pediatric bone age assessment using pediatric trauma hand radiographs. Radiol Artif Intell. 2020 Jul 29;2(4):e190198.
125. Pan I, Thodberg HH, Halabi SS, Kalpathy-Cramer J, Larson DB. Improving automated pediatric bone age estimation using ensembles of models from the 2017 RSNA machine learning challenge. Radiol Artif Intell. 2019 Nov 20;1(6):e190053.
126. BoneXpert. https://bonexpert.com Accessed April 20, 2024.
127. Thodberg HH, Kreiborg S, Juul A, Pedersen KD. The BoneXpert method for automated determination of skeletal maturity. IEEE Trans Med Imaging. 2009 Jan;28(1):52-66.
128. Rajpurkar P, Lungren MP. The current and future state of AI interpretation of medical images. N Engl J Med. 2023 May 25;388(21):1981-1990.
129. Ng CKC. Generative adversarial network (generative artificial intelligence) in pediatric radiology: a systematic review. Children. 2023; 10(8):1372.
130. Ted AI San Francisco. "What is zero-shot learning in AI?" https://ai-event.ted.com/glossary/zero-shot-learning Accessed April 20, 2024.
131. Medium.com. "How AI is Benefitting Child Healthcare: The Latest Advances and Future Possibilities." May 23, 2023. https://cogitotech.medium.com/how-ai-is-benefitting-child-healthcare-the-latest-advances-and-future-possibilities-8fa4aed0f33a Accessed April 20, 2024.
132. Pediatric Moonshot. https://pediatricmoonshot.com Accessed April 20, 2024.

Advancements in Pediatric Liver Transplantation: Innovations and Future Directions

Jennifer Vittorio and Adam Griesemer

Overview of Pediatric Liver Transplantation

Pediatric liver transplantation is a lifesaving procedure for select patients with both acute and chronic liver failure. More than 500 pediatric liver transplants occur each year, accounting for nearly 18,000 liver transplants performed on recipients <18 years of age as of December 2021 [1]. Advancements and innovations in surgical techniques, postoperative management, and immunosuppression have resulted in improvements in both short- and long-term outcomes. While pediatric liver transplants account for <10% of all liver transplants performed annually in the United States, sufficient differences in disease pathophysiology and patient management warrant further discussion.

Disease categories suitable for referral to a pediatric liver transplant program include acute liver failure, cholestasis, autoimmune, oncologic, vascular, infectious, and underlying metabolic or genetic diseases. The most common indications for pediatric liver transplantation based on Organ Procurement and Transplantation Network/Scientific Registry of Transplant Recipients (OPTN/SRTR) data from 2019 through 2021 are illustrated in Fig. 1.

J. Vittorio (✉) · A. Griesemer
New York University (NYU) Transplant Institute, NYU Langone Health, New York, NY, USA
e-mail: Jennifer.Vittorio@nyulangone.org; Adam.Griesemer@nyulangone.org

J. L. Wells (ed.), *A Clinical Lens on Pediatric Engineering*,
https://doi.org/10.1007/978-3-031-84655-7_9

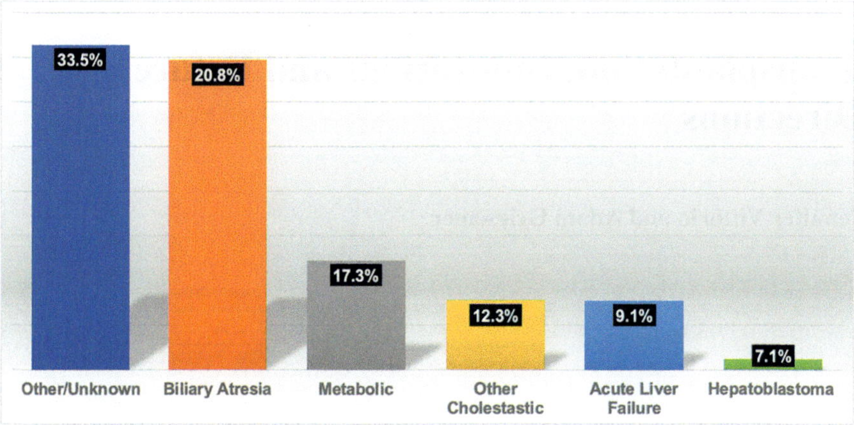

Organ Procurement and Transplantation Network (OPTN) and Scientific Registry of Transplant Recipients (SRTR). OPTN/SRTR 2021 Annual Data Report. Rockville, MD: Department of Health and Human Services, Health Resources and Services Administration; 2023.

Fig. 1 Indications for pediatric liver transplant

Historically, biliary atresia has been the most common indication for pediatric liver transplantation, accounting for nearly 40–50% of cases each year in the United States [2]. A summary of common pediatric liver disorders is detailed in Table 1. Early referral for transplant evaluation should be anticipated for children with chronic liver disease and evidence of deteriorating liver function characterized by poor weight gain, growth failure, variceal hemorrhage, intractable ascites, recurrent cholangitis or episodes of spontaneous bacterial peritonitis, pruritus, advancing encephalopathy, and/or uncorrectable coagulopathy [3].

Innovations in pediatric liver transplant have significantly transformed the field of pediatric hepatology and transplantation, offering numerous benefits and advancements in patient care. Included among these are improved survival rates, reduced waitlist mortality, expansion of the donor pool, advancements in organ preservation, tailored immunosuppression, and innovations in diagnosis and monitoring.

Table 1 Common pediatric liver disorders

Condition	Incidence and heritability	Clinical features	Extrahepatic manifestations	Treatment	Monitoring	Prognosis
Biliary atresia	1 in 10,000–20,000. Not inherited	Progressive, idiopathic, fibro-obliterative disease of extrahepatic biliary tree that presents with biliary obstruction exclusively in the neonatal period	Biliary atresia splenic malformation (BASM) 15% • Situs inversus • Asplenia or polysplenia • Malrotation • Interrupted IVC • Cardiac anomalies Other congenital anomalies 10% • Intestinal atresia • Imperforate anus • Kidney anomalies • Cardiac malformations	Kasai HPE Liver transplant	Portal HTN HCC Cholangitis FSV deficiency Progression to ESLD with indications for liver transplant	90% survival to adulthood 20–40% survival with native liver at 20 years

(continued)

Table 1 (continued)

Condition	Incidence and heritability	Clinical features	Extrahepatic manifestations	Treatment	Monitoring	Prognosis
Alagille syndrome	1 in 30,000–50,000. Autosomal dominant with variable penetrance. JAG1 > 90% NOTCH2	Characterized by paucity of intrahepatic bile ducts Multisystemic disease with potential for cardiac and vascular anomalies, renal involvement, pancreatic insufficiency, bone health	Cardiac anomalies Butterfly vertebrae Posterior embryotoxon Dysmorphic facies (broad nasal bridge, triangular facies, deep-set eyes) Renal involvement Short stature Cerebral and systemic vascular malformations (intracranial abnormalities may predispose to hemorrhage/stroke)	Largely supportive and directed at symptoms of pruritus and malabsorption Biliary diversion Liver transplant	Portal HTN HCC Nutrition FSV deficiency Pruritus Brain MRI–low threshold for imaging in setting of head trauma or symptoms Multidisciplinary care: Cardiology, nephrology, gastroenterology, ophthalmology, endocrinology, neurology	~25% reach adulthood with native liver
Progressive familial intrahepatic cholestasis (PFIC)	1/50,000–100, 000 Autosomal recessive ATP8B1 ABCB11 ABCB4 TJP2 NR1H4 MYO5B	Chronic liver disease with hallmarks of cholestasis and pruritus due to defects in bile acid transporters Increased risk for HCC and hepatic decompensation	PFIC 1 is associated with extrahepatic manifestations—Sensorineural hearing loss, diarrhea, pancreatic insufficiency, short stature. PFIC 1 hepatic steatosis, worsening diarrhea noted following transplant	Largely supportive and directed at symptoms of pruritus and malabsorption Biliary diversion Liver transplant	Portal HTN HCC (esp. PFIC 2) FSV deficiency Pruritus Counseling: Importance of nutrition, OCP contraindicated, cholestasis in pregnancy	44% reach adulthood with native liver 75% of those alive with native liver underwent SBD by 18 years of age

Condition	Incidence and heritability	Clinical features	Extrahepatic manifestations	Treatment	Monitoring	Prognosis
Caroli disease	1 in 100,000 Autosomal recessive Associated with ARPKD	Multifocal, segmental dilatation of the large intrahepatic bile ducts Caroli disease • Ectasia of the large bile ducts Caroli syndrome • Bile duct dilation with CHF	Patients with renal involvement may have enlarged palpable kidneys	Largely supportive • Choleretic agents • Antibiotics Surgical portosystemic shunt in the setting of portal HTN	Portal HTN Cholangitis Cholangiocarcinoma	Variable prognosis depending upon disease severity and renal involvement
Alpha-1 antitrypsin deficiency	1/80,000–100,000 Autosomal recessive SERPINA1	Chronic liver disease with variable phenotype Clinical presentation ranges from abnormal LFTs to cirrhosis and HCC	Pulmonary involvement typically beginning in adulthood	Supportive Liver transplant ERT available to prevent progression of lung disease	Portal HTN HCC Pulmonary function testing Counseling: Avoidance of alcohol, smoking, vaping and secondhand smoke	Variable
Autoimmune liver disease	Familial patterns	Chronic liver disease, variable phenotype with increased risk of relapsing disease	Dependent on underlying disease, may have systemic involvement	Immunosuppression Liver transplant	Portal HTN HCC Disease relapse Treatment adherence Side effects of immunosuppression Bone health Indications for liver transplant	Variable

(continued)

Table 1 (continued)

Condition	Incidence and heritability	Clinical features	Extrahepatic manifestations	Treatment	Monitoring	Prognosis
Wilson's disease	1/10,000–30,000 Autosomal recessive ATP7B	Impairment of copper transport resulting in accumulation in liver, brain and cornea	Neurologic and psychiatric manifestations	Low copper diet Zinc Chelation therapy Liver transplantation	Portal HTN HCC Dietary and treatment adherence Side effects of treatment Indications for liver transplant Neurocognitive testing	Variable Excellent for those who achieve copper detoxification before developing cirrhosis
Mitochondrial hepatopathy	Typically autosomal recessive	Present with multisystemic or neuromuscular involvement in a patient with acute or chronic liver disease, presence of lactic acidosis and hepatic steatosis	Dependent on underlying disorder	Supportive Liver transplant may be considered in certain conditions without significant neurologic or cardiac involvement	Growth and nutrition Neurologic development	Dependent on underlying disorder
Urea cycle defect	1 in 8200 Autosomal recessive with the exception of OTC deficiency which is X-linked	Typically present in first 24–48 h of life with seizures, hyperammonemia, metabolic decompensation leading to severe neurologic sequelae	Somnolence, poor feeding, lethargy, emesis, coma and seizures. Can result in significant neurologic insults. Patients with partial enzyme deficiency—Chronic emesis, developmental delay, seizures, psychiatric illness	Protein restriction ammonia scavengers Arginine Hemodialysis Liver transplantation	Dietary compliance Ammonia levels Seizures Neurologic sequelae	Dependent on underlying disorder and timing of diagnosis

Condition	Incidence and heritability	Clinical features	Extrahepatic manifestations	Treatment	Monitoring	Prognosis
Hepatoblastoma	~100 cases per year Most cases sporadic Inherited syndromes associated with HB: Beckwith-Wiedemann Trisomy 18 and 21 Li-Fraumeni Type 1a glycogen storage Familial adenomatous polyposis	90% occur in children <5 years Often presents as asymptomatic abdominal mass		Perioperative chemotherapy followed by resection Liver transplant for nonmetastatic, unresectable disease	Monitor for tumor recurrence	5-year event-free survival rate of 75–90% in children treated with chemotherapy and surgery

IVC inferior vena cava, *HPE* hepatoportoenterostomy, *HTN* hypertension, *HCC* hepatocellular carcinoma, *FSV* fat-soluble vitamin, *ESLD* end-stage liver disease, *LT* liver transplant, *MRI* magnetic resonance imaging, *OCP* oral contraceptive pill, *SBD* surgical biliary diversion, *LFTs* liver function tests, *ERT* enzyme replacement therapy, *HB* hepatoblastoma

Diagnostic and Therapeutic Updates in Pediatric Liver Disease

Ongoing research and innovations have deepened the understanding of pediatric liver disease, leading to more accurate diagnoses and improved treatment strategies. Widespread accessibility to improved diagnostic testing has allowed for early identification of potential complications of liver disease. This has direct implications for patient selection, timing of transplantation, and overall post-transplant care.

Neonatal cholestasis refers to a condition characterized by the impairment or obstruction of bile flow from the liver presenting in the first few months of life. As a result, bile acids and other substances accumulate in the liver, leading to symptoms such as jaundice (yellowing of the eyes and skin), dark urine, pale-colored stools, and pruritus (itching). Neonatal cholestasis can be caused by various conditions including congenital liver disorders, bile duct abnormalities, infections, and underlying metabolic or genetic disease. Prompt diagnosis and management are crucial to prevent complications and ensure healthy infant development.

Genetic disorders account for more than 25–50% of all cases of neonatal cholestasis [4, 5]. Identification of the precise etiology in the infant and child can be challenging due to the broad overlap of clinical presentations. Routine serologic and other diagnostic assessments are often insufficient to distinguish between these disorders. Historically, genetic testing was limited to the analysis of a single gene based on clinical observation and initial evaluation; however, given the diverse etiology and a substantial number of potential genetic causes of cholestasis, single-gene testing has become impractical, inefficient, and expensive. Next-generation sequencing (NGS), including multigene panels, whole exome sequencing (WES), and whole genome sequencing (WGS), now enables simultaneous, rapid testing of multiple genes with results available within days to weeks. This allows for the utilization of genetic testing early in the diagnostic process. A recent study using a 66 gene panel in 716 infants and older children with cholestasis or liver disease of unknown etiology reported a positive or likely positive molecular diagnosis in 11.7% and a single pathogenic or likely pathogenic variant in another 12.7% of patients [4]. The simultaneous analysis of multiple genes in a single panel reduces the time to identify, or eliminate, specific genes as a cause of cholestasis.

While the increased accessibility and affordability of NGS has allowed for rapid evaluation of these patients, widespread testing also presents the need for additional considerations. The ongoing discovery of new gene variants, single mutations in known pathogenic genes, unclear pathogenicity and variable phenotypes have made interpretation of these results challenging at times. The use of NGS without selection may give rise to unnecessary or uninterpretable information, which may complicate the diagnosis rather than clarify it [5]. With continued implementation, multigene panels may reshape the clinical algorithm of diagnosing cholestatic disorders, including the role and utility of invasive procedures such as liver biopsy and early referral for liver transplantation.

Timely diagnosis also allows for the early initiation of treatment modalities. The management of cholestasis is often aimed at the biochemical consequences of impaired bile flow. These include malnutrition, pruritus, xanthoma formation, and progressive fibrosis. Together, these complications can significantly impact the quality of life of both patients and caretakers. When medical therapies fail, complications of cholestasis have historically been managed with complex surgical procedures, including biliary diversion and liver transplantation.

Bile acids, often elevated in the setting of cholestasis, are reabsorbed from the ileum via specific transporters. Inhibitors of the apical sodium-dependent bile acid transporter (ASBT), also known as ileal bile acid transporter (IBAT) inhibitors, are novel therapies aimed at blocking entry of bile acids into enterocytes through selective binding of these transporters to ameliorate symptoms of pruritus. These minimally absorbed oral therapies interrupt the enterohepatic circulation of bile acids by increasing fecal excretion and reducing hepatic uptake. This also reduces activation of both the hepatic and ileal farnesoid X receptor (FXR), thereby increasing bile acid synthesis from cholesterol in the liver. Thus, IBAT inhibitors both reduce bile acid reabsorption while increasing bile acid synthesis and secretion in the hepatocytes. In pediatric trials of patients with Alagille syndrome and Progressive Familial Intrahepatic Cholestasis (PFIC), this resulted in the reduction of pruritus, serum bile acids and xanthomas. Clinically meaningful improvements in growth and quality of life were also observed [6–8].

There are currently two IBAT inhibitors approved for use in children. Odevixibat has been approved for the treatment of cholestatic pruritus in patients with PFIC >3 months of age and Alagille syndrome >12 months, while maralixibat has been approved for patients with PFIC >5 years of age and Alagille syndrome >3 months of age. Though there is emerging data to suggest that these therapies may delay the progression of liver disease and prolong transplant-free survival, long-term data and further exploration of these findings are necessary [9]. While these therapies were initially targeted for the treatment of pediatric cholestatic disorders, IBAT inhibitors are now undergoing investigation for use in the treatment of adult cholestatic liver diseases such as primary sclerosing cholangitis and primary biliary cholangitis.

Historical Perspective

The first liver transplant was performed in 1963 by Dr. Thomas Starzl on a 2-year-old child with biliary atresia, a universally fatal disease. The patient died in the operating room secondary to uncontrollable hemorrhage. For several years, the only technical option for pediatric liver transplantation was to await a whole liver from a size-matched donor. Given the low number of available pediatric donors, up to 50% of children died on the transplant wait list before they could receive a suitable offer [10]. The development of techniques, such as reduced-size and split-liver grafts, later allowed surgeons to transplant portions of livers from adult donors and changed the fate of liver transplantation for pediatric patients. The liver is a remarkable organ with the unique ability to regenerate. When portions of the liver are damaged or removed, the remaining healthy tissue can rapidly proliferate and restore liver function. This regenerative capacity allows partial sections of the liver to be utilized for

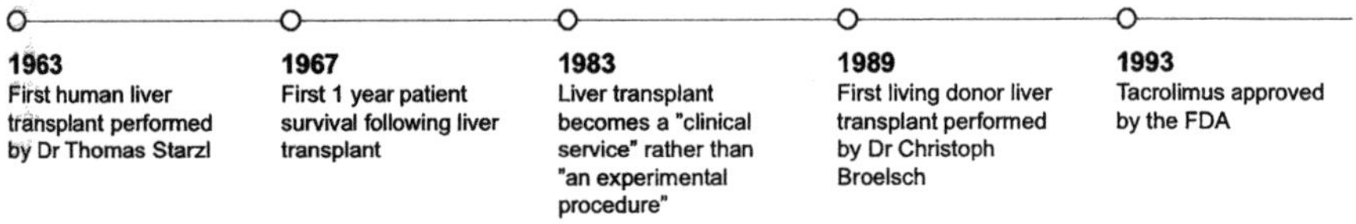

Fig. 2 Timeline of pediatric liver transplantation

transplantation, as the liver can grow back to its original size over time. The introduction of living donor liver transplant in the 1980s further revolutionized the field by addressing the shortage of suitable organs for children (Fig. 2).

Continuous refinements in organ preservation methods, diagnostic tools and multidisciplinary care have also contributed to increased success rates and better long-term outcomes for pediatric liver transplant recipients. The advent of calcineurin inhibitors such as cyclosporine and tacrolimus has led to decreased episodes of acute cellular rejection and improvements in long-term graft survival. Most transplant centers have adopted tacrolimus as the mainstay of immunosuppressive regimens given its superiority in the prevention of rejection and absence of adverse cosmetic effects.

Current Challenges in Pediatric Liver Transplantation

At present, there is a shortage of suitable donors for liver transplantation in the United States. Prolonged waitlist times not only result in increased waitlist mortality but also lead to ongoing concerns for poor growth and nutrition, cognitive and social development, increasing lifetime morbidity. Pretransplant mortality rates among pediatric patients on the liver transplant waitlist are 5.6 deaths per 100 patient-years and highest among children younger than 1 year of age at 21.7 deaths per 100 patient-years, according to the most recent OPTN/SRTR data [1].

In the early era of organ transplantation, waiting time and severity of illness were dictated by patient location (home, hospital, ICU). It was later recognized that wait time had no bearing on patient mortality outside of patients with acute liver failure and that an allocation policy based on objective medical criteria was needed. This led to the implementation of the MELD/PELD scores in 2002. The Model for End-stage Liver Disease (MELD) score is used for adults 12 years and older, and the PELD score is applied for children <12 years old. Data derived from the Studies of Pediatric Liver Transplantation research group was utilized to create a Pediatric End-stage Liver Disease (PELD) score, using bilirubin, INR, serum albumin, age >1 year, and growth failure to predict waitlist mortality [11]. While the adoption of the MELD/PELD scoring system has improved access and accountability of the allocation system, it has not been proven to be a successful predictor of outcome following transplantation. As a result, standardized exception scores were created to allocate additional points for those patients whose mortality risk was thought to be significantly unrepresented by the current scoring system. In addition, non-standardized exception points could also be awarded on a case-by-case basis. In

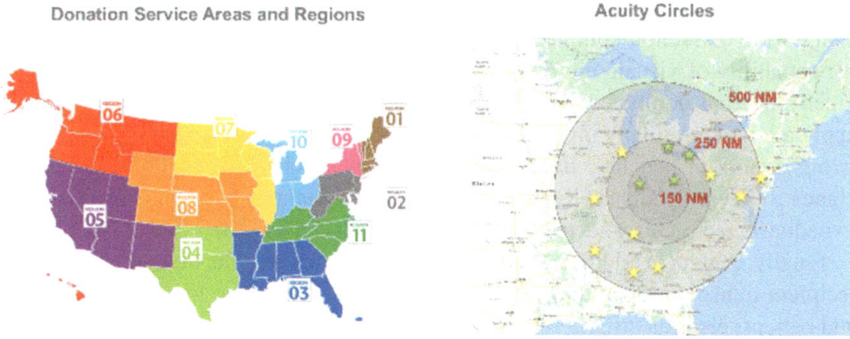

Fig. 3 Donation service areas (DSAs) (left) and acuity circles (right) reflecting the current organ allocation system. Acuity circles are concentric circles measured in nautical miles around the donor hospital. www.optn.transplant.hrsa.gov

2020, 61% of children were listed for transplant with the exception of MELD or PELD score (OPTN/SRTR 2020 Annual Report).

Pretransplant mortality improved for children following the implementation of acuity circle-based distribution in 2020, which replaced donor service areas (Fig. 3). This policy also outlined the prioritization of pediatric donors to pediatric recipients [1].

The most recent OPTN policy changes occurred in July 2023 when the new PELD Cr score was adopted. This new score adds creatinine to the score calculation and age-adjusted mortality risk to improve the representation of waitlist mortality. Given its recent implementation, the results of these changes on pediatric prioritization and waitlist mortality have not yet been evaluated.

Innovations in Surgical Techniques

The last few years have seen significant advancements in surgical innovation with increased application of minimally invasive surgical techniques in liver transplantation. The first notable achievement was the application of laparoscopic liver surgery in the donor for pediatric living donor liver transplantation [12]. The application of minimally invasive surgical techniques in kidney transplant led to a surge of live donor transplants for that organ due to decreased pain, postoperative hospital length of stay, and faster return to work for the donor, thus decreasing the burden of donation. For liver donation, the same benefits were found to apply for the donors of the left lateral segment when the surgery was performed laparoscopically. Laparoscopic hepatectomy for donation to children is associated with fewer complications for the donor compared to open liver donation or laparoscopic kidney donation [13, 14]. Critically, there are no differences in recipient outcomes when the donor's liver is removed laparoscopically.

Most recently, surgeons have begun performing the donor operation robotically, with equivalent donor outcomes compared to laparoscopic donation [15]. The robotic approach has the advantage of articulated instruments that increase precision during dissection compared to the laparoscopic approach. Additionally, it has the capability of visualizing resection planes and the bile ducts using indocyanine green injection combined with near-infrared cameras. It is hoped that the broad application of minimally invasive donor hepatectomies will increase the numbers of living donors in pediatric liver transplantation, especially since the survival outcomes following pediatric living donor liver transplantation are superior compared to pediatric recipients of deceased donor liver transplants [16].

Finally, centers in Saudi Arabia and the US have reported the first cases where the recipient transplant operation was performed robotically. The long-term outcomes and patient selection criteria for this approach have yet to be determined. It is likely there will be a delay between the widespread application of this technique in adult recipients and its routine use in smaller pediatric recipients where decreased space limits the use of multiple robotic arms and the ability to maneuver the instruments.

Expanding the Donor Pool

The most challenging issue in transplantation is obtaining a suitable donor liver to use for transplantation. In pediatric liver transplant, donor livers can be obtained from living donors who donate a part of their liver. Typically, this involves removing the left lateral segment of the liver (Fig. 4), which represents about 20% of the mass of the donor's liver. With the use of minimally invasive surgical techniques (see above), the donor stays in the hospital for 3–6 days and experiences a full recovery

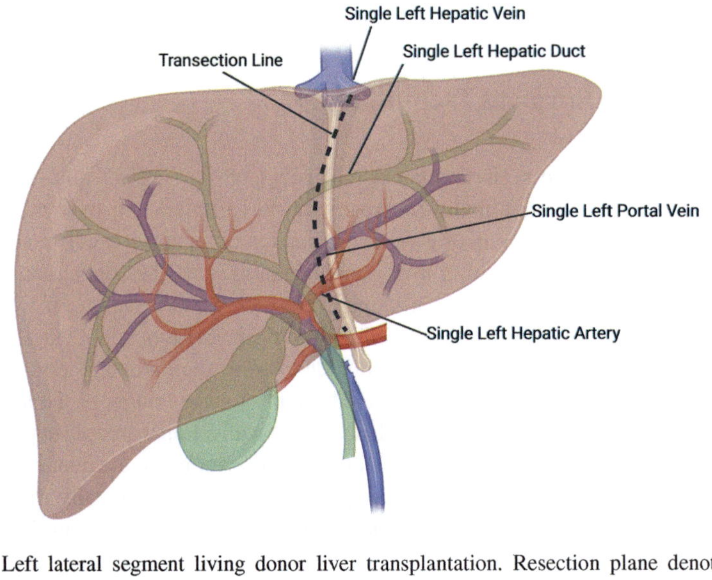

Fig. 4 Left lateral segment living donor liver transplantation. Resection plane denoted by a dotted line

in 3–4 weeks. The major limitations for this type of transplant include the need for a donor with a compatible blood type, the need for a care team for the donor and the recipient, and the ability of the donor to take time away from work during recovery. Unfortunately, not every recipient has a donor that meets these criteria.

In pediatrics, the donor organ must fit into the abdominal cavity of the recipient. For deceased donors, we can use size-matched pediatric livers or "split" adult livers where the whole liver is cut into two pieces, one of which is the appropriate size for the pediatric recipient while the other side is appropriate for an adult recipient. In each recipient, the transplanted lobe grows to become the appropriate size for the recipient [17]. Since the function of the left and right sides of the liver are the same, each lobe provides complete liver function in its recipient. Recent advances in surgical experience and techniques have led to equivalent survival outcomes for whole or split deceased donor liver transplants in pediatric recipients. In the US, the limiting factor for split liver transplantation is the allocation of the whole liver to the pediatric patient that then triggers the "splitting" of the liver. While in the US the allocation of adult organs is based entirely on illness severity ranking on the waiting list, some countries have adopted a "split first" allocation policy to prioritize allocation of whole livers from young donors to pediatric recipients prior to offering them to adult recipients. Countries with this allocation system have decreased pediatric waitlist mortality, shorter waiting time for transplantation, and no demonstrable effect on adult liver transplant outcomes [18].

Deceased donors can broadly be separated into two categories. The first are donors who have been declared brain dead and go to the operating room for controlled organ donation (donation after brain death, DBD). The second category is donors that have controlled withdrawal of life support and where the donation begins only after the cessation of the heartbeat for 5 min (donation after circulatory death, DCD). Generally, the donor's livers are better preserved after DBD than DCD donation since the withdrawal of life support and the lack of circulation leads to ischemic injury of the liver in DCD donation. The bile duct of the liver is especially sensitive to ischemia and livers from DCD donors have a higher incidence of biliary complications after transplantation. While studies examining the outcomes of DCD transplants in pediatric recipients have shown good outcomes, this type of organ is rarely used in pediatrics [19].

Several techniques have recently been developed to enhance organ preservation of donor livers during their transportation from the donor hospital until they are transplanted into the recipient. The traditional method of organ preservation is to flush the liver with a preservation solution and pack it in ice (approximately 4 °C) for transportation in cold storage. New perfusion devices have been developed to provide continuous perfusion of the liver with either a cold preservation solution [20] or a warm preservation solution [21]. These devices allow livers to be preserved for longer times, enabling them to travel further distances between donor and recipient hospitals. Additionally, the warm perfusion devices provide data regarding liver function prior to transplantation. This data allows the surgical team reassurance that the liver will function well after the operation. It also allows the surgeon to assess the degree of ischemic injury in a DCD donor and may lead to broader use of DCD livers in pediatric recipients.

Finally, xenotransplantation of organs from pigs to humans has recently been performed using pig kidneys and hearts [22, 23]. The application for liver xenotransplantation is likely limited to brief periods of support using the pig liver on an external perfusion circuit as a bridge to native liver recovery or to support a recipient while the team waits for a suitable human liver to use for transplantation [24]. This assessment is based on the incredible number of critical metabolic functions performed by the liver that may not be performed correctly by the pig liver. For example, in preclinical models of liver xenotransplantation, a continuous infusion of a human clotting factor is required to prevent bleeding and achieve recipient survival of 1 month, which is the longest reported survival of a liver xenograft recipient [25]. The recent report that a brain-dead human was connected to an externally perfused genetically modified pig liver demonstrates that short-term liver support with a pig liver is feasible, though the ability of the pig liver to perform the functions of the human liver still needs to be determined.

Immunosuppression

Adequate immunosuppression is needed to support graft function but must be balanced between potential medication side effects and risks of over immunosuppression. Immunosuppression protocols vary amongst transplant centers. Current practice includes calcineurin inhibitor (CNI)-based regimens. In the last several decades, immunosuppression for pediatric liver transplant patients has been reduced. The use of anti-lymphocyte antibodies has decreased, and tacrolimus has replaced cyclosporine as the primary CNI at most transplant centers. Most centers chose tacrolimus-based regimens to avoid the cosmetic side effects of gingival hyperplasia and hirsutism associated with cyclosporine. Augmentation with mycophenolate mofetil or mammalian target of rapamycin (mTOR) inhibitors may reduce reliance on steroids and high-dose CNIs. Steroids can often be withdrawn within 3–6 months [2]. A summary of common immunosuppressive agents is provided in Table 2. Monitoring for graft dysfunction and adequate immunosuppression includes liver indices such as alanine aminotransferase (ALT), aspartate aminotransferase (AST), gamma-glutamyl transpeptidase (GGT), and immunosuppression trough levels.

Tolerance enables the recipient's immune system to maintain the allograft without the need for long-term immunosuppressant medications, allowing patients to reduce or even discontinue medication while still ensuring the success of the transplant. In general, immunosuppression withdrawal in the context of liver transplantation refers to the gradual reduction or complete cessation of immunosuppressive medications given to transplant recipients. The goal is to assess whether some patients can maintain graft function without these drugs while minimizing potential side effects and complications associated with long-term immunosuppression. Two-thirds of late mortality is directly attributed to complications of immunosuppression such as infection and malignancy. A large, multicenter trial of immunosuppression withdrawal demonstrated that 37.5% of selected pediatric liver transplant recipients were operationally tolerant [26]. Withdrawal failure almost always occurred after a

Table 2 Common immunosuppressive agents utilized in pediatric liver transplant

Medication	Mechanism of action	Side effects
Corticosteroids	Inhibit cytokine transcription	Salt retention, hypertension, elevated blood sugars, osteopenia, cataracts, dyslipidemia, catabolic state
Cyclosporine Tacrolimus	Inhibits calcineurin Inhibits IL-2 synthesis	Nephrotoxicity, neurotoxicity, hypertension, diabetes
Mycophenolate mofetil (MMF)	Inhibits nucleotide synthesis, DNA synthesis, T-cell proliferation	Gastrointestinal upset (diarrhea), bone marrow suppression
Basiliximab	Blocks IL-2 receptors on T-helper cells Prevents T-cell proliferation	Infusion reaction, chills, fever, rash, leukopenia
Sirolimus	mTOR inhibitor Blocks IL-2 receptor signal transduction	Thrombocytopenia, leukopenia, anemia, hyperlipidemia, poor wound healing
Thymoglobulin	Depletion of T-cell, B-cell, NK-cell	Infusion reaction, fever, chills, headache, diarrhea, HTN, nausea, leukopenia, thrombocytopenia

IL-2 interleukin-2, *mTOR* mammalian target of rapamycin

substantial reduction of the baseline immunosuppressive dose, suggesting that some patients may have been receiving more immunosuppression than necessary. Dose reduction may be safe for nontolerant subjects. Future development of accurate, noninvasive biomarkers is necessary to help guide minimization efforts and personalized targets for immunosuppression following transplantation.

Long-Term Outcomes

In years prior, the main concern for children with end-stage liver disease was to find a suitable organ for transplant. As access to organs improves, the focus has shifted toward optimization of graft health and long-term survival. The current 5- and 10-year graft survival rates for pediatric liver transplant recipients are 84.1% and 75.2%, respectively [1]. Pediatric liver transplant recipients may face several long-term complications that require ongoing medical attention.

For most adult liver recipients, a 20- to 25-year patient and allograft survival equates with a full life. In contrast, 20–25 years brings a pediatric recipient to young adulthood. Routine laboratory monitoring is typically implemented to screen for rejection and ongoing liver health. Even in the setting of normal liver function tests, silent allograft damage is detected in almost 50% of stable, long-term pediatric liver transplant patients [2]. As liver tests are neither sensitive nor specific for determining allograft health, direct or indirect assessment by liver histology or vibration-controlled transient elastography (VCTE) is necessary for long-term surveillance. VCTE is a noninvasive imaging technique used to assess liver stiffness, which is an indicator of liver fibrosis. This information aids in diagnosing liver diseases, monitoring disease progression, and determining treatment effectiveness. It is quick, painless, and does not require an invasive procedure such as liver biopsy.

Post-transplant lymphoproliferative disorder (PTLD) is a potential complication that can occur following organ transplantation. PTLD comprises a spectrum of lymphoproliferative conditions that range from a benign abnormal proliferation of lymphoid cells to lymphoma. It is often associated with Epstein-Barr virus (EBV). Treatment of PTLD is based on immunological cell typing and clinical presentation. Documented PTLD requires an immediate decrease or withdrawal of immunosuppression, taking into account the increased risk of organ rejection. If a tumor expresses the B-cell marker CD20, the anti-CD20 monoclonal antibody rituximab has been successfully used. Patients who have aggressive monoclonal malignancies have a poor prognosis. Recently, autologous EBV-specific cytotoxic T-lymphocytes have proved effective in enhancing EBV-specific immune responses and reducing viral load in organ transplant recipients with active infection and have been successfully used as first-line treatment of EBV-related PTLD [27].

Long-term outcomes are largely the result of a complex interplay of medical, developmental, psychosocial, and systems-related factors. There is an increased prevalence of executive dysfunction, changes in neurocognition, mental health conditions and risk-taking behaviors in adolescents and young adults [28–30].

Adolescent and young adult transplant recipients demonstrate significantly higher levels of executive dysfunction compared to healthy controls. Executive dysfunction is a key developmental aspect of self-management and transition readiness. Increasing independence combined with incomplete brain maturity may result in increased barriers to adherence in this population. The relatively rapid development of the limbic system, which is responsible for reward-seeking behavior and emotions, paired with slower maturation of the prefrontal cortex, which is responsible for planning, impulse control, and organization, may lead young adults to choose actions that maximize short-term rewards over long-term outcomes. As caretakers step back, lapses in adherence may occur [31].

Young adult solid organ transplant recipients are at risk for neurocognitive deficits and executive dysfunction when compared to healthy peers. Many patients exhibit deficits in working memory, attention control, processing speed and metacognitive skills (the ability to organize, plan, self-monitor, and initiate tasks). These may be the result of factors present before, during and after transplant—including end-stage organ failure, poor nutrition, metabolic derangements, or recurrent infections [29, 32, 33]. Other contributing factors may include missed school days, and concern for the impact of chronic immunosuppression on neurocognitive development. While common, many deficits often go unrecognized. Early recognition of such deficits, through formal assessment of cognitive function, will allow appropriate support structures to be put into place.

These inherent findings impact adherence, transition readiness and long-term medical outcomes. Routine assessment of executive functioning and psychosocial well-being is essential to help guide intervention efforts, decrease barriers to adherence and promote developmentally appropriate levels of treatment responsibility. This requires an increased awareness at the level of both pediatric and adult providers.

Long-term outcomes must also look beyond patient and graft survival and gauge important measures such as adherence, health care costs, and health-related quality of life (HRQOL). Objective, reliable metrics exist to assess allograft health, adherence, and complications of immunosuppression. Attention should also be paid to establish such metrics for determinants of HRQOL. Proposed outcome measures and routine health maintenance following pediatric liver transplantation are summarized in Table 3.

Table 3 Health maintenance and outcome measures of successful pediatric liver transplantation

Outcome	Measures	Assessment tools
Medical outcomes		
Patient survival		SRTR data
Allograft survival		SRTR data
Allograft health	Liver function tests	Laboratory assessment
	Synthetic function	Physical exam
	Episodes of acute and chronic rejection	Liver biopsy
	Fibrosis stage	VCTE
Long-term complications of immunosuppression	Renal insufficiency Hypertension Hyperlipidemia Diabetes Cancer Osteopenia/osteoporosis Infection	GFR, urine protein/creatinine BP monitoring, lipid panel, ECHO, cardiac catheterization HgA1C Physical exam, imaging, histology DEXA scan, vitamin D levels Routine vaccinations
Adherence		
Medication	The standard deviation of immunosuppression levels	MLVI
Laboratory monitoring		Laboratory monitoring adherence
Office visits	LTFU designation	Office visit attendance
Health care utilization		
	Invasive procedures Emergency department visits Hospitalizations Hospital length of stay Retransplantation End-of-life care	
Health-related quality of life		
	Patient satisfaction Caretaker satisfaction	PedsQL, PedsQL TM, PeLTQL
	School/vocation	High school diploma or equivalent, employment
	Skills necessary for independent living	Housing, financial security
	Anxiety/depression/PTSD	PHQ-9, GAD-7
	Substance use disorder Unplanned pregnancy	NIDA quick screen

SRTR scientific registry of transplant recipients, *VCTE* vibration controlled transient elastography, *GFR* glomerular filtration rate, *DEXA* dual X-ray absorptiometry, *MLVI* medication level variability index, *LTFU* lost to follow-up, *PedsQL* pediatric quality of life inventory transplant module, *PeLTQL* pediatric liver transplantation quality of life, *PHQ-9* patient health questionnaire 9, *GAD-7* general anxiety disorder-7, *NIDA* National Institute on Drug Abuse

Health Care Transition and Transfer of Care

Pediatric liver transplantation involves a multidisciplinary approach to patient care, incorporating the expertise of transplant surgeons, hepatologists, advanced practice providers, nurses, nutritionists, and psychosocial support teams. Innovations in care models have facilitated collaborative efforts, ensuring comprehensive and holistic care for pediatric transplant recipients. Adaptations of these enhanced care models have been utilized to improve outcomes surrounding health care transition and transfer of care.

Advances in medical therapies and surgical techniques have resulted in more pediatric liver transplant recipients reaching young adulthood. Young adult liver transplant recipients transferring from pediatric to adult models of care have increased rates of rejection, graft loss, and medical complications [34]. Health care *transition* (HCT) is the process of changing from a pediatric to an adult model of health care. *Transfer* of care is an event, or the purposeful, planned movement of a patient from pediatric to adult healthcare facilities [35]. Barriers to successful HCT are multifaceted. The goal of a transition program is to optimize the likelihood of sustained well-being in young adults by fostering their ability to achieve physical, social, and psychosocial potential [36–38]. The means to achieve this goal requires an organized transition process to support youth in acquiring independent health care skills, preparing for an adult model of care, and transferring to new providers without disruption in treatment. This can only be achieved through a multidisciplinary approach to transition planning. Successful HCT requires a partnership between pediatric and adult providers across all disciplines to ensure a transition process that does not end at the time of transfer but continues throughout early adulthood.

Joint consensus guidelines from the North American Society of Pediatric Gastroenterology, Hepatology and Nutrition (NASPGHAN) in collaboration with the American Society of Transplantation (AST) were developed to facilitate the adoption of a structured, multidisciplinary approach to transition planning utilizing The Six Core Elements of Health Care Transition™ for use by both pediatric and adult specialists [37]. A structured transition program is associated with improved health, patient experience, and utilization of health care [39]. Got Transition©, a federally funded national resource center on HCT, provides guidelines for program development through The Six Core Elements of Health Care Transition™ [40]. These elements define a sequential clinical process with tools and templates for use by both pediatric and adult practices. An adaptation of the Six Core Elements pertaining to adolescents and young adults following pediatric liver transplantation is summarized in Table 4.

The ongoing development of evidenced-based multidisciplinary HCT programs will help to narrow disparities in long-term pediatric liver transplant outcomes, optimize health and assist youth in reaching their full potential. Achieving this goal requires an organized transition process to support young adults in acquiring

Table 4 Six core elements of health care transition

Core element	Pediatric providers	Adult providers
I	Transition policy • Describe the transition process • Review privacy/consent • Discuss the anticipated age of transfer	Transition and care policy • Outline an adult approach to patient care • Review practice expectations • Review privacy/consent
II	Tracking and monitoring • Track the progress of AYA through HCT	Tracking and monitoring • Create a young adult database • Track the progress/completion of core elements to the goal of independence
III	Transition readiness • Conduct regular transition readiness assessments • Assess adherence, self-management • Develop goals with the patient, caregiver	Orientation to adult practice • Conduct regular transition readiness assessments • Assess adherence, self-management, patient engagement
IV	Transition planning • Develop/update plan of care • Prepare transfer packet and checklist • Discuss the timing of transfer • Assist in finding an adult provider • Provide links to insurance resources	Integration into adult practice • Collaborate with pediatric team for transfer planning • Participate in virtual meet and greet with patient and pediatric team
V	Transfer of care • Send the completed transfer packet • Communicate with an adult team • Confirm the date of the first adult appointment	Continue transition of care/adult approach to care • Review the transfer packet prior to the appointment • Orient to adult practice • Establish a preferred method of communication • Complete self-care assessment
VI	Transfer completion • Contact patient 3–6 months after last pediatric visit, confirm transfer • Obtain feedback from AYA on the transition process • Obtain feedback from the adult team on transfer • Remain available as a consultant	Ongoing care • Confirm transfer completion with the pediatric team • Obtain feedback from AYA on the transfer process

AYA adolescent and young adult, *HCT* health care transition

independent health care skills and transferring to new providers without disruption in treatment [33]. This is often a labor and resource-intensive undertaking, which may not receive the necessary support from local institutions. Widespread implementation of HCT programs will require rigorous evidence and endorsement from professional societies and governing organizations to enable better financing for these resource-intensive programs [41].

Future Directions and Conclusions

In summary, innovations in pediatric liver transplantation have significantly improved patient outcomes by addressing various challenges in organ availability, surgical techniques, immunosuppression, and post-transplant care. These advancements continue to shape the landscape of pediatric hepatology, offering hope for a better quality of life and survival for children with liver disease. The future of pediatric liver transplant continues to hold promising innovations centered on reducing long-term immunosuppression and embracing precision medicine. Incorporating genetic and molecular profiling may lead to improved donor-recipient matching and personalized treatment plans. The continued expansion of living donor transplants, coupled with refinements in surgical techniques, aims to enhance outcomes for pediatric patients, addressing the challenges associated with organ shortage and transplant procedures.

Competing Interests There was no external funding for this manuscript.
Jennifer Vittorio has received speaker and consultant honoraria from Mirum Pharmaceuticals.
Adam Griesemer has received research support from United Therapeutics, Chioronex, and eGenesis.

References

1. Kwong, A.J., et al., *OPTN/SRTR 2021 Annual Data Report: Liver.* Am J Transplant, 2023. **23**(2 Suppl 1): p. S178–S263.
2. Kelly, D.A., et al., *Long-term medical management of the pediatric patient after liver transplantation: 2013 practice guideline by the American Association for the Study of Liver Diseases and the American Society of Transplantation.* Liver Transpl, 2013. **19**(8): p. 798–825.
3. Squires, R.H., et al., *Evaluation of the pediatric patient for liver transplantation: 2014 practice guideline by the American Association for the Study of Liver Diseases, American Society of Transplantation and the North American Society for Pediatric Gastroenterology, Hepatology, and Nutrition.* J Pediatr Gastroenterol Nutr, 2014. **59**(1): p. 112–31.
4. Karpen, S.J., et al., *Use of a Comprehensive 66-Gene Cholestasis Sequencing Panel in 2171 Cholestatic Infants, Children, and Young Adults.* J Pediatr Gastroenterol Nutr, 2021. **72**(5): p. 654–660.
5. Jeyaraj, R., et al., *The Genetics of Inherited Cholestatic Disorders in Neonates and Infants: Evolving Challenges.* Genes (Basel), 2021. **12**(11).
6. Gonzales, E., et al., *Efficacy and safety of maralixibat treatment in patients with Alagille syndrome and cholestatic pruritus (ICONIC): a randomised phase 2 study.* Lancet, 2021. **398**(10311): p. 1581–1592.
7. Shneider, B.L., et al., *Placebo-Controlled Randomized Trial of an Intestinal Bile Salt Transport Inhibitor for Pruritus in Alagille Syndrome.* Hepatol Commun, 2018. **2**(10): p. 1184–1198.
8. Thompson, R.J., et al., *Odevixibat treatment in progressive familial intrahepatic cholestasis: a randomised, placebo-controlled, phase 3 trial.* Lancet Gastroenterol Hepatol, 2022. **7**(9): p. 830–842.
9. Sokol, R.J., et al., *Predictors of 6-year event-free survival in alagille syndrome patients treated with maralixibat, an ileal bile acid transporter inhibitor.* Hepatology, 2023.

10. Emond, J.C., et al., *Transplantation of two patients with one liver. Analysis of a preliminary experience with 'split-liver' grafting.* Ann Surg, 1990. **212**(1): p. 14–22.
11. Freeman, R.B., Jr., et al., *Improving liver allocation: MELD and PELD.* Am J Transplant, 2004. **4 Suppl 9**: p. 114–131.
12. Samstein, B., et al., *Totally laparoscopic full left hepatectomy for living donor liver transplantation in adolescents and adults.* Am J Transplant, 2013. **13**(9): p. 2462–2466.
13. Broering, D.C., et al., *Pure Laparoscopic Living Donor Left Lateral Sectionectomy in Pediatric Transplantation: A Propensity Score Analysis on 220 Consecutive Patients.* Liver Transpl, 2018. **24**(8): p. 1019–1030.
14. Soubrane, O., et al., *Laparoscopic Living Donor Left Lateral Sectionectomy: A New Standard Practice for Donor Hepatectomy.* Ann Surg, 2015. **262**(5): p. 757–761; discussion 761–3.
15. Troisi, R.I., et al., *Safety and Feasibility Report of Robotic-assisted Left Lateral Sectionectomy for Pediatric Living Donor Liver Transplantation: A Comparative Analysis of Learning Curves and Mastery Achieved With the Laparoscopic Approach.* Transplantation, 2021. **105**(5): p. 1044–1051.
16. Mogul, D.B., et al., *Fifteen-Year Trends in Pediatric Liver Transplants: Split, Whole Deceased, and Living Donor Grafts.* J Pediatr, 2018. **196**: p. 148–153.e2.
17. Chaudhry, S.G., et al., *Growth of liver allografts over time in pediatric transplant recipients.* Pediatr Transplant, 2018. **22**(2).
18. Angelico, R., et al., *A national mandatory-split liver policy: A report from the Italian experience.* Am J Transplant, 2019. **19**(7): p. 2029–2043.
19. Sasaki, K., et al., *Should We Be Utilizing More Liver Grafts From Pediatric Donation After Circulatory Death Donors? A National Analysis of the SRTR from 2002 to 2017.* Transplantation, 2021. **105**(9): p. 1998–2006.
20. Guarrera, J.V., et al., *Hypothermic machine preservation in human liver transplantation: the first clinical series.* Am J Transplant, 2010. **10**(2): p. 372–381.
21. Martins, P.N., et al., *The role of normothermic machine perfusion in liver transplantation.* Int J Surg, 2020. **82s**: p. 52–60.
22. Montgomery, R.A., et al., *Results of Two Cases of Pig-to-Human Kidney Xenotransplantation.* N Engl J Med, 2022. **386**(20): p. 1889–1898.
23. Griffith, B.P., et al., *Genetically Modified Porcine-to-Human Cardiac Xenotransplantation.* N Engl J Med, 2022. **387**(1): p. 35–44.
24. Kim, J.I., A. Torres-Hernandez, and A. Griesemer, *Pigs or Pumps: A new strategy emerges for liver perfusion.* Hepatology, 2023. **78**(3): p. 694–696.
25. Shah, J.A., et al., *Prolonged Survival Following Pig-to-Primate Liver Xenotransplantation Utilizing Exogenous Coagulation Factors and Costimulation Blockade.* Am J Transplant, 2017. **17**(8): p. 2178–2185.
26. Feng, S., et al., *Efficacy and Safety of Immunosuppression Withdrawal in Pediatric Liver Transplant Recipients: Moving Toward Personalized Management.* Hepatology, 2021. **73**(5): p. 1985–2004.
27. Chiou, F.K., et al., *Cytotoxic T-lymphocyte therapy for post-transplant lymphoproliferative disorder after solid organ transplantation in children.* Pediatr Transplant, 2018. **22**(2).
28. Gogtay, N., et al., *Dynamic mapping of human cortical development during childhood through early adulthood.* Proc Natl Acad Sci U S A, 2004. **101**(21): p. 8174–8179.
29. Ohnemus, D., et al., *Health-Related Quality of Life and Cognitive Functioning in Pediatric Liver Transplant Recipients.* Liver Transpl, 2020. **26**(1): p. 45–56.
30. McCormick King, M.L., et al., *Emotional functioning, barriers, and medication adherence in pediatric transplant recipients.* J Pediatr Psychol, 2014. **39**(3): p. 283–293.
31. Fernandez, H.E. and B.J. Foster, *Long-Term Care of the Pediatric Kidney Transplant Recipient.* Clin J Am Soc Nephrol, 2022. **17**(2): p. 296–304.
32. Sorensen, L.G., et al., *Cognitive and academic outcomes after pediatric liver transplantation: Functional Outcomes Group (FOG) results.* Am J Transplant, 2011. **11**(2): p. 303–311.

33. Troen, A.M., et al., *Cognitive dysfunction and depression in adult kidney transplant recipients: baseline findings from the FAVORIT Ancillary Cognitive Trial (FACT).* J Ren Nutr, 2012. **22**(2): p. 268–276.e3.
34. Lawrence, Z.E., et al., *Adherence, Medical Outcomes, and Health Care Costs in Adolescents/Young Adults Following Pediatric Liver Transplantation.* J Pediatr Gastroenterol Nutr, 2020. **70**(2): p. 183–189.
35. Anthony, S.J., et al., *Perceptions of transitional care needs and experiences in pediatric heart transplant recipients.* Am J Transplant, 2009. **9**(3): p. 614–619.
36. Bell, L.E., et al., *Adolescent Transition to Adult Care in Solid Organ Transplantation: a consensus conference report.* Am J Transplant, 2008. **8**(11): p. 2230–2242.
37. Vittorio, J., et al., *Health Care Transition for Adolescents and Young Adults With Pediatric-Onset Liver Disease and Transplantation: A Position Paper by the North American Society of Pediatric Gastroenterology, Hepatology, and Nutrition.* J Pediatr Gastroenterol Nutr, 2023. **76**(1): p. 84–101.
38. Fredericks, E.M., et al., *Assessment of transition readiness skills and adherence in pediatric liver transplant recipients.* Pediatr Transplant, 2010. **14**(8): p. 944–953.
39. Schmidt, A., et al., *Outcomes of Pediatric to Adult Health Care Transition Interventions: An Updated Systematic Review.* J Pediatr Nurs, 2020. **51**: p. 92–107.
40. *White PH, Cooley WC; Transitions Clinical Report Authoring Group; American Academy of Pediatrics; American Academy of Family Physicians; American College of Physicians. Supporting the Health Care Transition From Adolescence to Adulthood in the Medical Home. Pediatrics. 2018;142(5):e20182587.* Pediatrics, 2019. **143**(2).
41. Vittorio, J., *Health Care Transition: A Time of Increased Vulnerability for Pediatric Liver Transplant Recipients.* J Pediatr Gastroenterol Nutr, 2020. **71**(6): p. 704–706.

Beyond User-Centered Design: Designing Digital Health Technologies for Substance Use Prevention and Treatment in Children and Youth

Kanu Priya Singh, Deepa R. Camenga, Kammarauche Aneni, and Kimberly Hieftje

K. P. Singh (✉)
XR Pediatrics at Yale, Yale School of Medicine, New Haven, CT, USA

Yale Center for Immersive Technologies in Pediatrics, Yale School of Medicine, New Haven, CT, USA

Yale Department of Pediatrics, Yale School of Medicine, New Haven, CT, USA

Current Affiliation: Department of Technical Communication & Interactive Design, Norman J. Radow College of Humanities & Social Sciences, Kennesaw State University, Kennesaw, GA, USA
e-mail: ksingh22@kennesaw.edu

D. R. Camenga
Yale Center for Immersive Technologies in Pediatrics, Yale School of Medicine, New Haven, CT, USA

Yale Department of Pediatrics, Yale School of Medicine, New Haven, CT, USA

Yale Department of Emergency Medicine, Yale School of Medicine, New Haven, CT, USA

Yale School of Public Health, New Haven, CT, USA
e-mail: deepa.camenga@yale.edu

K. Aneni
Yale Center for Immersive Technologies in Pediatrics, Yale School of Medicine, New Haven, CT, USA

The ACCESS Lab, Yale Child Study Center, Yale School of Medicine, New Haven, CT, USA

Yale Department of Biomedical Informatics and Data Science, Yale School of Medicine, New Haven, CT, USA
e-mail: uche.aneni@yale.edu

K. Hieftje
XR Pediatrics at Yale, Yale School of Medicine, New Haven, CT, USA

Yale Center for Immersive Technologies in Pediatrics, Yale School of Medicine, New Haven, CT, USA

Yale Department of Pediatrics, Yale School of Medicine, New Haven, CT, USA
e-mail: kimberly.hieftje@yale.edu

J. L. Wells (ed.), *A Clinical Lens on Pediatric Engineering*,
https://doi.org/10.1007/978-3-031-84655-7_10

Introduction

Digital health describes the use of technology to deliver services, support, or interventions to improve health outcomes or prevent disease. It can leverage various digital tools and platforms to provide personalized, timely, and accessible health information and services with high fidelity and reach. Typically, digital technologies reduce the overall workforce costs of health interventions through reductions in health professional training and time [2–4]. A recent review documented technologies used in digital health interventions for youth. These include web and mobile applications, serious games, and extended reality-based interventions (XR; virtual, augmented, and mixed reality) [4]. Newer modalities reported by recent studies include wearable technologies [5] and artificial intelligence (AI) [6].

A leading public health problem that digital technologies can help address is adolescent substance use [7, 8]. Adolescent substance use is associated with the top three causes of death in this age group—motor vehicle crashes, drug poisonings and overdoses, and gun violence [9]—as well as a heightened risk of future addiction and mental health disorders [10]. Unfortunately, access to prevention interventions has decreased over the past 20 years [11], and only ~20% of adolescents with substance use disorders receive treatment [12]. Substance use during adolescence is also linked with poorer academic performance, unhealthy peer relationships, legal involvement, risky sexual behaviors, and preventable injuries [13]. As youth transition to young adulthood, continued substance use is similarly linked with poorer job performance and educational attainment [13]. Given these concerning statistics, increasing access to evidence-based substance use prevention and treatment interventions is critical for child and youth well-being.

Traditionally, substance use prevention interventions have been delivered in schools and communities in classrooms or group-based formats [14, 15]. Similarly, treatment interventions have generally been delivered in clinical settings with individual, group, and family formats [14]. However, the provider workforce is not sufficient to meet the demand [16], and requires training of teachers and specialty providers, which is costly and takes time [15]. As such, there has been a growing interest in the use of digital interventions that can increase access to evidence-based interventions [15]. The World Health Organization (WHO) has identified digital health interventions as cost-effective, functional, and scalable solutions to address prevention and treatment gaps [17]. Despite the growing interest in digital health interventions, more evidence-based research is needed to support its use in substance use prevention and treatment in children and youth [4, 14, 15, 18, 19]. Although digital interventions, in general, employ a variety of frameworks, including user-centered design (UCD), human-centered design, or patient-centered design to guide the design process [20], a critical challenge with these frameworks is that they are primarily designed for adults and may miss a large population of children and youth who are at risk for or engage in unsafe substance use [12, 18, 21].

In this context, this chapter discusses a need to go beyond the conventional UCD approaches when designing digital technologies for the prevention and treatment of substance use in children and youth. We first review existing digital health interventions and describe their design processes. Then, we detail the benefits and limitations of existing methods and tools for designing digital interventions for children.

Finally, we propose a novel overarching framework, the Child-Centered Design Framework (CCD), to help guide the development of digital interventions for substance use in children and youth.

Overview of Existing Digital Technologies that Address the Prevention and Treatment of Substance Use in Children and Youth

Digital health interventions for substance use prevention and treatment can be delivered through various modalities, including interactive e-learning modules, computers, mobile devices (e.g., cell phones or tablets), or virtual, augmented, and mixed-reality headsets [4, 22]. According to recent reviews, the most common substance targeted by the existing digital interventions is alcohol [4, 19] which is also the most common substance used by youth [1, 12], followed by cannabis, tobacco, and other substances (e.g., opioids) [4, 19].

To date, most digital health technologies that address substance use are classified as web-based interventions or those that deliver health education and services via the Internet [4, 23–27]. Real Teen [28], Pop4Teens [29], and Climate School courses [30] are examples of effective web-based prevention interventions that have been tested in randomized-controlled clinical trials. Although these web-based interventions have reported successful prevention and reduction of substance use in adolescents, they generally are less interactive than interventions delivered through more interactive modalities, such as text messages or apps [22, 28]. Examples of apps that focus on reducing adolescent substance use include D-ARIANNA [Digital–Alcohol Risk Alertness Notifying Network for Adolescents and Young Adults]—an e-Health app designed to identify binge drinking behavior in adolescents [31, 32], and SmarTrek—a smartphone app with interactive components including text messages aimed to reduce risky alcohol use in adolescents [33].

Over time, the field has shifted towards using game-based approaches and extended reality technologies to create more immersive and engaging experiences around substance use prevention and treatment. Recent reviews have shown the effects of serious games—defined as video games with non-entertainment goals—in improving engagement, as well as knowledge, attitudes, and behaviors related to positive health (e.g., adherence to cancer medication, increased physical activity) and outcomes (e.g., reduced asthma attacks) [34–37]. For example, Take Control game—a partially immersive virtual reality-based serious game designed to allow users to actively combat substance cues by using physical movements—reported a statistically significant increase in self-efficacy, attitude, and behaviors [38].

Extended reality, including virtual reality, as an emerging videogame technology, has been found optimal for medical and health use [39, 40]. VR technologies offer various advantages, including structured learning, control, realism, immersion, and automated feedback [41] and are, therefore, used to enhance learning performance, motivation, and engagement [42]. A recent review of VR applications in the medical field demonstrated how VR is being used to facilitate quality healthcare services [40], while another review found VR effective in reducing substance use in adults, specifically [39]. Studies have also shown that immersive video games improve learning outcomes [43, 44]. For example, a virtual reality-based e-cigarette prevention game for adolescents called Invite Only VR: A Vaping Prevention Game reported significant improvement in knowledge and harm perceptions of e-cigarettes, as well as a reduced self-reported likelihood of e-cigarette use in the year following gameplay [43].

While many of the research studies focused on digital health interventions are health behavior theory-driven [19] and have shown intervention effectiveness [43, 44], reporting of design approaches or inclusion of stakeholders in the design process (e.g., participatory research) is rather inconsistent [20, 45]. Recent reviews on design methodologies in digital health interventions identified a need for more transparency in the reporting of usability, acceptability, feasibility, and effectiveness when evaluating digital health technologies and the extent to which users were involved in the design process [20, 45]. Regardless of which digital technology is employed, the scientific literature often lacks a detailed discussion of human-computer interaction or design principles, especially for health interventions for the pediatric population.

Design Frameworks

Multiple existing frameworks can guide the design of health technologies, including the sociotechnical-pedagogical usability [46], Technology Acceptance Model (TAM), and the UCD [47, 48]. The *sociotechnical-pedagogical* framework is a design framework [49] that integrates the technological, social, and pedagogical dimensions of learner experience to evaluate and improve technology-mediated learning systems to support educational goals, social presence, and collaboration with various learning activities and assessments for learners [46]. Although the sociotechnical-pedagogical usability framework has been successfully used with college students and adult learners [50, 51], it has not been tailored to the unique needs of pediatric populations. For example, this approach does not account for the developmental stages of children or the pivotal roles of parents, teachers, and other adults in supporting young people. It also does not account for the broader social-ecological environment (i.e., school, community, neighborhood, and culture) in which the technology is being deployed [52, 53].

Another design framework is Davis's *Technology Acceptance Model (TAM)* [54]. This framework explores determinants of technology adoption, such as perceived ease of use, usefulness, and acceptance of digital environments [54]. However, when applied to children, determinants also include fun and playability as unique factors that influence children's adoption of technologies [54, 55], which is lacking in the TAM.

UCD is a common design framework that considers the needs of stakeholders and end users throughout all stages of the development cycle to enhance the final product's usability [20, 47] and acceptability [3, 47, 48]. Because of UCD's broad design approach, it can be applied to all types of digital health interventions, including those that address substance use prevention and treatment [3, 7, 20, 22]. However, the application of the UCD approach in studies of digital health interventions for children and youth is rare, and many lack descriptions of the UCD process and evaluation (e.g., usability, acceptability, feasibility, and effectiveness) [3, 45, 56]. Specifically, studies utilizing UCD approaches fail to address the gap in creating age-appropriate interventions with the desired level of interactivity and engagement for the involved pediatric group which are reported as essential determinants for children's adoption of technologies [3, 7, 22]. For example, MiSARA—a smartphone app that focused on reducing escalation in adolescent substance use—used an iterative participatory approach to design and develop the technology [22]. The study evaluated the application for acceptability and feasibility. While the participants rated the app as easy to use and preferred it over web-based interventions, they consistently desired greater interactivity, a recreational approach, and overall engagement with the technology [22]. This suggests that although studies are employing participatory methods and UCD principles, they fail to adapt these approaches effectively to design youth-friendly interventions [7]. These findings are problematic for researchers when trying to replicate or unpack the UCD characteristics of digital health interventions in similar contexts [57].

The Child-Computer Interaction (CCI) is an arm [58] within the Human-Computer Interaction (HCI) discipline that describes design processes that focus on children and youth. It includes several UCD frameworks adapted for children technology design and development [55, 59, 60], such as CARSS (Context, Activities, Roles, Stakeholders, Skills; a participatory learner-centered design approach involving children as child-learner) [61], BRIDGE (treats children as active and authentic stakeholders within meaningful communities of practice) [62], and BLUEBELLS (a balance of child-centered design and expert design) [63]. These frameworks highlight that children's interactions with technology significantly diverge from adults', prioritizing fun and playability over straightforward usability [55, 64].

While each of these frameworks (socio-technical, TAM, UCD) has important components that focus on children and youth as the primary users of technologies, they have not been used within the context of digital *health* interventions for substance use prevention and treatment for children and youth. Research suggests that a teenager's involvement in substance use is vulnerable to several influences, including social-ecological factors and developmental factors [15]. Existing design frameworks lack evidence of addressing the social ecology and child development factors of the pediatric group and do not present a comprehensive approach for designing digital health technologies with children and youth. We, therefore, propose a novel comprehensive framework to develop effective, engaging, and acceptable digital health interventions to address substance use in children and youth—*The Child-Centered Design Framework for Digital Health Interventions*. A particularly novel aspect of CCD is its deliberate inclusion of the social-ecological dimension, which considers the physical and relational context of the child, the culture surrounding the child and technology implementation. We describe the CCD Framework in detail in the following sections.

The Child-Centered Design (CCD) Framework for Digital Health Interventions

The CCD Framework expands on the conventional user-centered design approach by integrating well-established theories from multiple disciplines that are highly relevant to children, technology, and substance use—including social and behavioral science, product design, child development, and education. Building upon our own experience of designing digital technologies to address substance use in children and adolescents and the existing literature [4, 19, 39, 43, 44, 59], the CCD Framework includes five dimensions that should be addressed in the design process: 1) Social-Ecology, 2) Child Development, 3) Theories and Frameworks, 4) Interaction Design Methods, Tools, and Techniques for Children, and 5) Evaluation (Fig. 1). We describe each of these Dimensions in detail below.

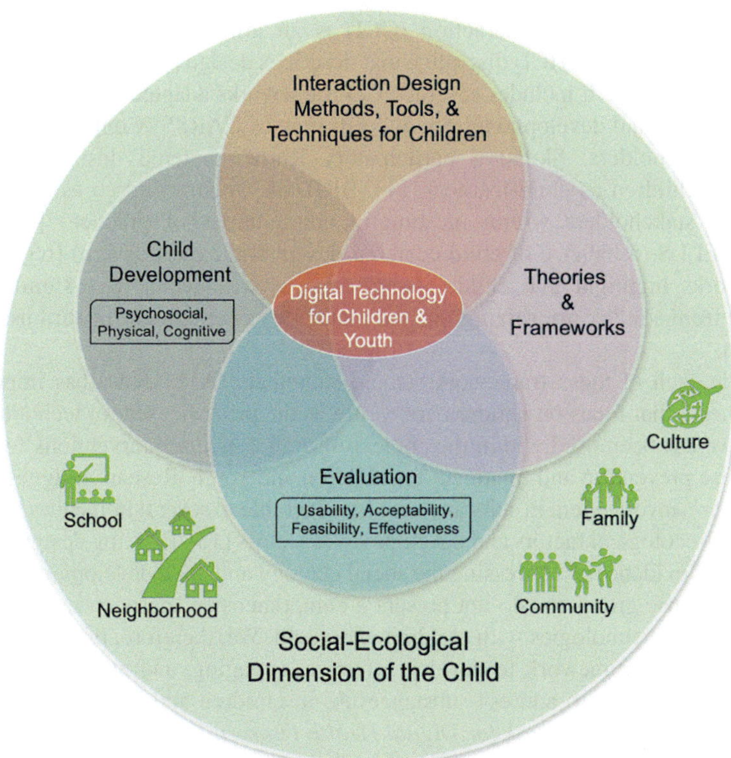

Fig. 1 Child-Centered Design (CCD) Framework for children and youth

Social-Ecological Dimension

The social-ecological dimension in the CCD framework recognizes the importance of family, school, neighborhood, community, and cultural contexts within which the child grows up, as well as the interactions between the child and their surrounding environment [65, 66]. Literature on digital health technologies recommends incorporating the social-ecological dimension into the design of digital health systems to address the lived social experiences of children and youth effectively [57, 67]. For example, Tachtler et al. [57] used the lens of the social-ecological model of resilience to conduct co-design workshops with migrant youth to design mental health apps. Research suggests that incorporating the social-ecological dimension into the design of digital health systems allows for a better representation of the diversity of relations and roles in the social ecology surrounding the child's life [67]. It recognizes the factors influencing the culture, values, and policy regulations shaping the development of the individual over time, thereby allowing designers to reflect on the complex interplay of social relations and technology interactions impacting the technology design and usage [57, 67].

Child Development Dimension

The child development dimension considers the child's cognitive, physical, and psychosocial abilities and limitations. Several theories have been proposed to improve the understanding of the intricacies of child development, including Jean Piaget's theory of cognitive development [68], Erik Erikson's theory of psychosocial development [69], and sociocultural theories by Lev Vygotsky [70]. These theories can guide the development of age-appropriate support and interventions within digital technologies.

For instance, according to Piaget's theory of cognitive development, children ages 2–7 years have not yet developed abilities to understand abstract concepts or consider things from another person's perspective [71]. Thus, to best account for the child's comprehension and literacy levels, narrative content must be straightforward, linear, and presented from the child's perspective. From a usability perspective, a younger child is typically less able to navigate a complex user interface than an older child with more advanced motor skills and cognitive abilities [72] (Fig. 2). For example, Gestwicki [73] notes that children under the age of eight typically cannot control their eye muscles to read for more than 15–20 min. They need to refocus their eyes periodically after engaging with long texts. This suggests that effective design strategies should incorporate a mix of text, audio, and video to maintain engagement and minimize the number of screens to limit the time required to interact with a technology [74].

Regarding psychosocial development, younger children are beginning to form friendships and learn social skills but continue to look to adults to reinforce healthy

Fig. 2 Example of a contrast scenario of a child attempting to understand and interact with an advanced digital technology and an adolescent displaying disinterest while swiping through a simple digital technology. (Image Source: Generated with chat.openai.com)

choices and praise [75]. Thus, designers may consider including features that allow adults to navigate content alongside their children. On the other hand, older children, especially adolescents, are developing more complex and nuanced cognitive and psychosocial abilities, including those related to abstract thinking and intense peer relationships [75]. Research has shown that the opinions of peers can significantly impact an adolescent's behaviors related to health, including substance use [43]. In this context, Lev Vygotsky's social development theory explains cognitive development in children through social interaction, thereby suggesting that learning is a social process and happens primarily through interaction with more knowledgeable others (i.e., teachers, parents, and peers) [70]. As such, a digital health intervention might include peer support groups or forums where individuals can interact with peers who may have experience managing similar health conditions [76]. Additionally, narrative content within digital technologies might depict peers as teachers or coaches in the content [43, 44].

Theories and Frameworks

The theories and frameworks dimension highlights health behavior and learning theories that explain the underlying mechanisms through which a digital intervention may influence a change in an individual's behavior related to health outcomes [77]. Recent reviews [78] have reported that social cognitive theory, motivational interviewing, and family interaction theory are the most common behavior change theories used in web, computer, mobile, or tablet-based health systems that address adolescent substance use [4, 19]. Social cognitive theory posits that people learn from one another through observation, imitation, and modeling [79]. From this theoretical perspective, a digital health system might include social features that allow children to observe and emulate health-promoting behaviors from peers or characters within the system. Another example of a guiding theory for digital health systems is the theory of planned behavior [80], which suggests that intentions are the primary predictor of behavior. From this perspective, a digital health system might include skill practice for children and youth to plan for and enact specific behaviors through goal setting and tracking progress.

The theories and framework dimension also includes the application of learning theories. Learning theories can help designers and developers understand how behaviors are acquired, how they can be altered, and how social learning influences healthy decision-making. A recent review [81] identified commonly reported learning theories, including problem-based learning, Bandura's social learning theory [82], and Mayer's cognitive theory of multimedia learning [83] in intervention trials of health professions' digital education. For instance, Bandura's social learning theory [82] suggests that learning occurs through the observation and adoption of others' behaviors within a social context. This theory emphasizes the role of self-efficacy (the belief in one's own ability to succeed) and cultural context in shaping behaviors [82]. From this theoretical perspective, a digital health intervention might include social features, such as online support groups or features to enhance self-efficacy, such as reward elements [84]. Mayer's cognitive theory of multimedia learning [83] emphasizes that the way information is presented can enhance understanding by aligning with cognitive processes, or if poorly executed, can hinder learning by overwhelming or confusing the learner. Digital health systems [85] have incorporated multimedia learning theory to design interventions' screen layout, information architecture, and videos or break intervention content into chunks, or modules, to reduce cognitive load on individuals [85–87].

These theories inform the design of the system's content and features and ensure that it aligns with the psychological and behavioral processes that influence health behaviors. The goal with the theories and frameworks dimension is not to suggest which theories to use, rather determining what theory (ies) best explains the health behavior that the digital intervention is trying to target and its determinants so that those theories can be employed in the design process. By leveraging theories and frameworks, designers and developers can create more effective and engaging digital health systems grounded in evidence-informed behavior change principles.

Interaction Design Methods, Tools, and Techniques

This dimension includes age-appropriate and skill-sensitive design approaches that allow children to contribute meaningfully to the design process. Child-Computer Interaction (CCI), a multidisciplinary research area within HCI, examines children's activities, behaviors, concerns, and abilities as they engage with digital technologies with the participation of others, primarily adults, in environments that the adults partially (or fully) control and regulate [55, 58, 60]. Within CCI, children can have four distinct roles in the technology design research process: users, testers, informants, and design partners [88]. For instance, children interact with a final product as users. As testers, they test the prototypes and provide design recommendations. As informants, they strictly offer information, and as design partners, they take on the role of equal stakeholders throughout the technology's design process [58, 88]. Examples of role-specific tools and techniques developed within CCI are outlined in Table 1.

Researchers need to be aware of several special considerations when using interaction design methods for technologies that target substance use in children and youth. For instance, youth have legal rights to confidentiality when seeking information or services related to substance use. Substance use is also linked with potential legal consequences and overarching safety concerns [6, 60]. As a result, youth may be reticent to report their perceptions, attitudes, or experiences accurately during the design of digital technologies for substance use [6]. Recent reviews of research involving children and youth in the technology design process documented that most studies did not report on informed consent or ethical approvals received to conduct the study [59, 96]. In this context, Markopoulos et al. [97] recommend clearly justifying children's participation in the research or design processes, transparently communicating their contributions, and how their contributions influence the eventual designs. Similarly, Read et al. [98] presented a tool of checklists (CHECk1 and CHECk2) to critically consider the reasons for involving children before and after a design activity, such as what platforms or technologies are we planning to use and why, and which children will we design with and why?

Table 1 Role-specific tools and techniques for children and adolescents involved in technology design

Role	Examples of tools and techniques
User	• Outdoor Play Observation Scheme [89] (OPOS—a user observation technique) • Observational System of Motor Skills [90] (OSMOS—a video coding technique where a grid is used to analyze coordinative skills of children, e.g., balance)
Informant	• This-or-that [91] (an interviewing tool to identify what children like or dislike after interacting with a technology) • Big Paper Technique [92] (a paper prototyping technique)
Tester	• Smileyometer [93] (a survey instrument to gather opinions about technology from children) • Again-again Table [93] (a table-based tool requiring an understanding of only 1 question—"do you want to do this again?" and 3 responses—"yes," "no," or "maybe" for each activity)
Design partner	• Bags-of-stuff [94] (a prototyping technique in which children and adults use big bags filled with art supplies to create low-fidelity prototypes) • DisCo [95] (a computer-based design tool for children and adults to collaborate as design partners)

Evaluation Dimension

The Evaluation dimension incorporates an examination of the usability, acceptability, feasibility, and effectiveness of digital health systems for children and youth [99]. *Usability* is a measure of how well an individual interacts with technology in a specific context to achieve a defined goal [100]. Basic measures of usability include task completion rate (percentage of users who were able to complete the task in a study), the time a task requires to be completed by the user, number of errors, and users' subjective satisfaction [100]. In the context of digital health systems, usability can be measured using multiple methods, including qualitative methods (e.g., interviews, observations) and quantitative methods (questionnaires, time taken to complete a task, task completion rate) [101]. Research suggests that while quantitative measures are more commonly used to evaluate the usability of digital health systems, qualitative methods such as think-aloud, interviews, and observations enrich the quantitative evaluations by pinpointing the problems that need to be addressed [101].

Acceptability is defined as a user's willingness to use a system [5]. Acceptability of digital health systems is highly contextual, subject to change, and shaped by current social and cultural norms [5]. A recent review demonstrated that the acceptability of digital health systems can be measured at different stages of the development lifecycle of a system either by using qualitative data via observations, interviews or focus groups or quantitative measures, such as standardized surveys (e.g., SUS) [102].

The *feasibility* of a digital health system refers to assessing whether the system works as intended in a given context or within a specific environment [17]. This implies measuring the extent of an individual's engagement and adherence to the study. Studies on health systems with pediatric populations have reported collecting data pertaining to recruitment success (total number of enrolled participants) and attrition rates (decrease in the number of participants contributing to the study) to measure the feasibility of digital health interventions [17, 103, 104]. Studies have also developed and validated questionnaires to measure the feasibility of health systems, such as the four-item questionnaire called Feasibility of Intervention Measure (FIM) [105].

Effectiveness refers to evaluating whether the digital health system achieves the intended results in a specific context [17, 106]. According to WHO guidelines to conduct digital health evaluations, effectiveness can be assessed in terms of outputs (improvements in performance and user adoption), outcomes (changes that emerge as a result of inputs and processes in a digital health system), and impacts (medium- to long-term effects produced by an intervention) [17]. From this perspective, research suggests adopting a hybrid approach to evaluate the effectiveness of digital health systems and interventions, that is, examining the effects of both the intervention and the implementation processes in a real-world setting [17, 107].

While each of these constructs is critical to evaluate digital interventions, it is important to note the lack of available age-appropriate measures within each construct to evaluate interventions specifically with a pediatric population. The application of the CCD Framework can provide a baseline structure that researchers and practitioners can use to inform the development or selection of age-appropriate measures and/or the modification of existing measures that evaluate the usability, feasibility, acceptability, and effectiveness of digital interventions for children and youth. For example, applying the CCD framework may inform the use of the Smileyometer [93] (a picture-based usability scale from the Interaction Design Methods, Tools, and Techniques dimension) to measure acceptability or usability among children.

Future Research Needs

We developed the CCD Framework to encourage researchers and designers to consider the Social-Ecology, Child Development, Theories and Frameworks, Interaction Design Methods, Tools, and Techniques for Children, and Evaluation dimensions when developing digital health technologies for substance use prevention and treatment among children and youth. Further, we encourage the use of existing CCD frameworks and methods within the CCI discipline, such as CARSS [61] and BRIDGE [62] and consider approaches on how to adapt these frameworks to health interventions, including substance use prevention and treatment. We also encourage future researchers and designers to validate usability, acceptability, feasibility, and effectiveness measures developed with the CCD Framework in the context of health interventions for children and youth.

As technology expands to include biometrics, artificial intelligence (AI), and machine learning [108, 109], newer CCD approaches will need to be explored and our framework expanded. In this context, a hypothetical example of the design-oriented implication of the CCD Framework would be using AI and machine learning to create adaptive digital health behavior change interventions for personalized preventive healthcare for children and youth, e.g., using narrative-centered behavior change environments that simulate real-world problem scenarios to help youth develop strategies for managing risky situations involving substance use [108, 109]. Another example could be using continuous customization of machine learning models to design adaptive health technologies and personalized interventions for substance use in children and youth [109].

Researchers and practitioners may need to examine ways to expand the CCD Framework when working with multiple dimensions that may be time-consuming. One way to address this is by explaining the appropriateness of health interventions utilizing advanced technologies [110] for children and youth by involving human experts, such as child development experts who are well informed on how the intervention should be designed for children with varying skills and abilities [110]. Further, to explore the potential of the CCD Framework adequately, it is essential to work across the multidisciplinary divisions of social and behavioral science, product design, child development, and education. In this vein, digital healthcare interventions should embrace more participatory approaches, bringing in experts from all dimensions of the CCD Framework for both the development of interventions and transfer of knowledge [20, 45].

Summary and Conclusion

Youth substance use is a leading public health problem that digital health technologies may help prevent and address [7, 8]. Although multiple studies of relevant digital interventions exist, reporting of their UCD outcomes for children and youth is rare, and many studies lack descriptions of the UCD process and evaluation (e.g., usability, acceptability, feasibility, and effectiveness). We highlight a novel framework that can help guide the design of digital health technologies that address substance use in children and adolescents. The CCD Framework prioritizes the unique perspectives and physical, cognitive, and psychological developmental stages of children and adolescents, thus bridging a gap in digital technology design. We encourage future researchers to critique, utilize, and refine this framework to establish a unified set of principles, heuristics, and best practices for child-centered design.

Competing Interests This study has not received any funding.

Kanu Priya Singh *is funded by the* Connecticut Department of Public Health.

Deepa Camenga *has received research grants from the* National Institutes of Health, Elevance Foundation, and Connecticut Department of Public Health.

Kammarauche Aneni *has received research grants from the* National Institutes of Health, National Institute on Drug Abuse, Center for Technology and Behavioral Health, Doris Duke Foundation, and the Yale Center for Clinical Investigation.

Kimberly Hieftje *has received research grants from the* National Institutes of Health and Connecticut Department of Public Health.

The authors have no conflicts of interest to declare that are relevant to the content of this chapter.

References

1. Adolescent Substance Use: America's #1 Public Health Problem. National Center on Addiction and Substance Abuse at Columbia University https://eric.ed.gov/?id=ED521379 (2011).
2. Hollis, C. et al. Annual Research Review: Digital health interventions for children and young people with mental health problems – a systematic and meta-review. J. Child Psychol. Psychiatry **58**, 474–503 (2017).
3. Liverpool, S. et al. Engaging Children and Young People in Digital Mental Health Interventions: Systematic Review of Modes of Delivery, Facilitators, and Barriers. J. Med. Internet Res. **22**, e16317 (2020).
4. Monarque, M., Sabetti, J. & Ferrari, M. Digital interventions for substance use disorders in young people: rapid review. Subst. Abuse Treat. Prev. Policy **18**, 1–29 (2023).
5. Perski, O. & Short, C. E. Acceptability of digital health interventions: embracing the complexity. Transl. Behav. Med. **11**, 1473–1480 (2021).

6. Ollier, J. et al. Can digital health researchers make a difference during the pandemic? Results of the single-arm, chatbot-led Elena+: Care for COVID-19 interventional study. Front. Public Health **11**, 1185702 (2023).
7. Hawke, L. D. et al. What makes mental health and substance use services youth friendly? A scoping review of literature. BMC Health Serv. Res. **19**, (2019).
8. Yarnell, S., Li, L., MacGrory, B., Trevisan, L. & Kirwin, P. Substance Use Disorders in Later Life: A Review and Synthesis of the Literature of an Emerging Public Health Concern. Am. J. Geriatr. Psychiatry **28**, 226–236 (2020).
9. Richter, L. Rethinking Substance Use Prevention: An Earlier and Broader Approach. https://drugfree.org/reports/rethinking-substance-use-prevention-an-earlier-and-broader-approach/ (2024).
10. Oyefeso, A., Ghodse, H., Clancy, C., Corkery, J. & Goldfinch, R. Drug abuse-related mortality: a study of teenage addicts over a 20-year period. Soc. Psychiatry Psychiatr. Epidemiol. **34**, 437–441 (1999).
11. Salas-Wright, C. P. et al. Trends in Substance Use Prevention Program Participation Among Adolescents in the U.S. J. Adolesc. Health **65**, 426–429 (2019).
12. 2022 NSDUH Detailed Tables | CBHSQ Data. SAMHSA https://www.samhsa.gov/data/report/2022-nsduh-detailed-tables (2023).
13. Hall, W. D. et al. Why young people's substance use matters for global health. Lancet Psychiatry **3**, 265–279 (2016).
14. Gray, K. M. & Squeglia, L. M. Research Review: What have we learned about adolescent substance use? J. Child Psychol. Psychiatry **59**, 618–627 (2018).
15. Griffin, K. W. & Botvin, G. J. Evidence-Based Interventions for Preventing Substance Use Disorders in Adolescents. Child Adolesc. Psychiatr. Clin. N. Am. **19**, 505–526 (2010).
16. Brown, C. A., Belfield, C. R. & Field, S. J. Cost effectiveness of continuing professional development in health care: a critical review of the evidence. BMJ **324**, 652–655 (2002).
17. World Health Organization. Monitoring and Evaluating Digital Health Interventions: A Practical Guide to Conducting Research and Assessment. (World Health Organization, Geneva, 2016).
18. Stockings, E. et al. Prevention, early intervention, harm reduction, and treatment of substance use in young people. Lancet Psychiatry **3**, 280–296 (2016).
19. Aneni, K. et al. A Systematic Review and Meta-Analysis Evaluating the Efficacy of Stand-Alone Digital Interventions to Prevent Substance Use Among Adolescents | Current Addiction Reports. Springer **10**, 378–395 (2023).
20. Duffy, A., Christie, G. J. & Moreno, S. The Challenges Toward Real-world Implementation of Digital Health Design Approaches: Narrative Review. JMIR Hum. Factors **9**, e35693 (2022).
21. Lipari, R. N. & Park-Lee, E. Key Substance Use and Mental Health Indicators in the United States: Results from the 2018 National Survey on Drug Use and Health. https://www.samhsa.gov/data/ (2019).
22. Coughlin, L. N. et al. Developing an Adaptive Mobile Intervention to Address Risky Substance Use Among Adolescents and Emerging Adults: Usability Study. JMIR MHealth UHealth **9**, e24424 (2021).
23. Davies, E. L., Lonsdale, A. J., Hennelly, S. E., Winstock, A. R. & Foxcroft, D. R. Personalized Digital Interventions Showed no Impact on Risky Drinking in Young Adults: A Pilot Randomized Controlled Trial. Alcohol Alcohol **52**, 671–676 (2017).
24. Sinadinovic, K. et al. Guided web-based treatment program for reducing cannabis use: a randomized controlled trial. Addict. Sci. Clin. Pract. **15**, 9 (2020).
25. Walukevich-Dienst, K., Lewis, E. M., Neighbors, C., Green, J. C. & Buckner, J. D. Online personalized feedback intervention reduces cannabis-related problems among college students with high problem distress. Exp. Clin. Psychopharmacol. **29**, 14–22 (2021).
26. Vargas-Martínez, A. M. et al. Measuring the effects on quality of life and alcohol consumption of a program to reduce binge drinking in Spanish adolescents. Drug Alcohol Depend. **205**, 107597 (2019).

27. Vogel, E. A. et al. The Put It Out Project (POP) Facebook Intervention for Young Sexual and Gender Minority Smokers: Outcomes of a Pilot, Randomized, Controlled Trial. Nicotine Tob. Res. **22**, 1614–1621 (2020).
28. Schwinn, T. M., Schinke, S. P., Keller, B. & Hopkins, J. Two- and three-year follow-up from a gender-specific, web-based drug abuse prevention program for adolescent girls. Addict. Behav. **93**, 86–92 (2019).
29. Marsch, L. A. et al. Evaluating the Effectiveness of a Web-Based Program (POP4Teens) to Prevent Prescription Opioid Misuse Among Adolescents: Randomized Controlled Trial. JMIR Public Health Surveill. **7**, e18487 (2021).
30. Teesson, M. et al. Combined prevention for substance use, depression, and anxiety in adolescence: a cluster-randomised controlled trial of a digital online intervention. Lancet Digit. Health **2**, e74–e84 (2020).
31. Carrà, G. et al. Impact of a Mobile E-Health Intervention on Binge Drinking in Young People: The Digital–Alcohol Risk Alertness Notifying Network for Adolescents and Young Adults Project. J. Adolesc. Health **58**, 520–526 (2016).
32. Carrà, G. et al. Risk Estimation Modeling and Feasibility Testing for A Mobile Ehealth Intervention for Binge Drinking among Young People: The D-ARIANNA (Digital-Alcohol Risk Alertness Notifying Network for Adolescents and Young Adults) Project. Subst. Abuse **36**, 445–452 (2015).
33. Kazemi, D. M. et al. Real-time demonstration of a mHealth app designed to reduce college students hazardous drinking. Psychol. Serv. **16**, 255–259 (2019).
34. Ferreira-Brito, F. et al. Game-based interventions for neuropsychological assessment, training and rehabilitation: Which game-elements to use? A systematic review. J. Biomed. Inform. **98**, 103287 (2019).
35. Baranowski, T., Buday, R., Thompson, D. I. & Baranowski, J. Playing for Real: Video Games and Stories for Health-Related Behavior Change. Am. J. Prev. Med. **34**, 74–82.e10 (2008).
36. Andrew, L. et al. Serious games for health promotion in adolescents – a systematic scoping review. Educ. Inf. Technol. **28**, 5519–5550 (2023).
37. Hieftje, K., Edelman, E. J., Camenga, D. R. & Fiellin, L. E. Electronic Media–Based Health Interventions Promoting Behavior Change in Youth: A Systematic Review. JAMA Pediatr. **167**, 574–580 (2013).
38. Metcalf, M., Rossie, K., Stokes, K., Tallman, C. & Tanner, B. Virtual Reality Cue Refusal Video Game for Alcohol and Cigarette Recovery Support: Summative Study. JMIR Serious Games **6**, e9231 (2018).
39. Tsamitros, N., Sebold, M., Gutwinski, S. & Beck, A. Virtual Reality-Based Treatment Approaches in the Field of Substance Use Disorder. Springer Link **8**, 399–407 (2021).
40. Javaid, M. & Haleem, A. Virtual reality applications toward medical field. Clin. Epidemiol. Glob. Health **8**, 600–605 (2020).
41. Bozgeyikli, L. L., Bozgeyikli, E., Katkoori, S., Raij, A. & Alqasemi, R. Effects of Virtual Reality Properties on User Experience of Individuals with Autism. ACM Trans. Access. Comput. **11**, 1–27 (2018).
42. Huang, K.-T. et al. Augmented Versus Virtual Reality in Education: An Exploratory Study Examining Science Knowledge Retention When Using Augmented Reality/Virtual Reality Mobile Applications. Cyberpsychology Behav. Soc. Netw. **22**, 105–110 (2019).
43. Weser, V. U. et al. Evaluation of a virtual reality E-cigarette prevention game for adolescents. Addict. Behav. **122**, 107027 (2021).
44. Weser, V. U. et al. A quasi-experimental test of a virtual reality game prototype for adolescent E-Cigarette prevention. Addict. Behav. **112**, 106639 (2021).
45. Vandekerckhove, P., de Mul, M., Bramer, W. M. & de Bont, A. A. Generative Participatory Design Methodology to Develop Electronic Health Interventions: Systematic Literature Review. J. Med. Internet Res. **22**, e13780 (2020).

46. Jahnke, I., Schmidt, M., Pham, M. & Singh, K. Sociotechnical-Pedagogical Usability for Designing and Evaluating Learner Experience in Technology-Enhanced Environments. Learn. User Exp. Res. 127–145 (2020).
47. ISO 9241-210:2019(en), Ergonomics of human-system interaction — Part 210: Human-centred design for interactive systems. International Standard Organization https://www.iso.org/obp/ui/#iso:std:iso:9241:-210:ed-2:v1:en.
48. ISO 9241-110:2020(en), Ergonomics of human-system interaction — Part 110: Interaction principles. International Standard Organization https://www.iso.org/obp/ui/#iso:std:iso:9241:-110:ed-2:v1:en.
49. Jonassen, D. H. Instructional Design as Design Problem Solving: An Iterative Process. Educ. Technol. **48**, 21–26 (2008).
50. Li, S., Singh, K., Riedel, N., Yu, F. & Jahnke, I. Digital learning experience design and research of a self-paced online course for risk-based inspection of food imports. Food Control **135**, 108698 (2022).
51. Pham, M., Singh, K. & Jahnke, I. Socio-technical-pedagogical usability of online courses for older adult learners. Interact. Learn. Environ. **31**, 2855–2871 (2023).
52. Earnshaw, V. A. et al. Disclosure, Stigma, and Social Support Among Young People Receiving Treatment for Substance Use Disorders and Their Caregivers: a Qualitative Analysis. Int. J. Ment. Health Addict. **17**, 1535–1549 (2019).
53. Wisdom, J. P., Cavaleri, M., Gogel, L. & Nacht, M. Barriers and facilitators to adolescent drug treatment: Youth, family, and staff reports. Addict. Res. Theory **19**, 179–188 (2011).
54. Davis, F. D. Perceived Usefulness, Perceived Ease of Use, and User Acceptance of Information Technology. MIS Q. **13**, 319–340 (1989).
55. Read, J. C. & Bekker, M. M. The Nature of Child Computer Interaction. in (2011). https://doi.org/10.14236/ewic/HCI2011.43.
56. Søgaard Neilsen, A. & Wilson, R. L. Combining e-mental health intervention development with human computer interaction (HCI) design to enhance technology-facilitated recovery for people with depression and/or anxiety conditions: An integrative literature review. Int. J. Ment. Health Nurs. **28**, 22–39 (2019).
57. Tachtler, F., Talhouk, R., Michel, T., Slovak, P. & Fitzpatrick, G. Unaccompanied Migrant Youth and Mental Health Technologies: A Social-Ecological Approach to Understanding and Designing. in Proceedings of the 2021 CHI Conference on Human Factors in Computing Systems 1–19 (ACM, Yokohama Japan, 2021). https://doi.org/10.1145/3411764.3445470.
58. Tsvyatkova, D. & Storni, C. A review of selected methods, techniques and tools in Child–Computer Interaction (CCI) developed/adapted to support children's involvement in technology development. Int. J. Child-Comput. Interact. **22**, 100148 (2019).
59. Lehnert, F. K. et al. Child–Computer Interaction: From a systematic review towards an integrated understanding of interaction design methods for children. Int. J. Child-Comput. Interact. **32**, 100398 (2022).
60. Hourcade, J. P. Child-Computer Interaction. (Self-Publication, 2015).
61. Good, J. & Robertson, J. CARSS: A Framework for Learner-Centred Design with Children. Int. J. Artif. Intell. Educ. IOS Press **16**, 381–413 (2006).
62. Iversen, O. S. & Brodersen, C. Building a BRIDGE between children and users: a socio-cultural approach to child–computer interaction. Cogn. Technol. Work **10**, 83–93 (2008).
63. Kelly, S. R., Mazzone, E., Horton, M. & Read, J. C. Bluebells: a design method for child-centred product development. in Proceedings of the 4th Nordic conference on Human-computer interaction: changing roles 361–368 (ACM, Oslo Norway, 2006). https://doi.org/10.1145/1182475.1182513.
64. Wang, E. Y. et al. Predicting pediatric healthcare provider use of virtual reality using a technology acceptance model. JAMIA Open **6**, ooad076 (2023).
65. Bronfenbrenner, U. Environments in developmental perspective: Theoretical and operational models. in Measuring environment across the life span: Emerging methods and concepts

3–28 (American Psychological Association, Washington, DC, US, 1999). https://doi.org/10.1037/10317-001.
66. Earls, F. & Carlson, M. The Social Ecology of Child Health and Well-Being. Annu. Rev. Public Health **22**, 143–166 (2001).
67. Murnane, E. L., Walker, T. G., Tench, B., Voida, S. & Snyder, J. Personal Informatics in Interpersonal Contexts: Towards the Design of Technology that Supports the Social Ecologies of Long-Term Mental Health Management. Proc. ACM Hum.-Comput. Interact. **2**, 1–27 (2018).
68. Piaget, J. Piaget's theory of cognitive development. in Childhood cognitive development: The essential readings (Blackwell, Malden, MA, 2000).
69. Erikson, E. H. Childhood and Society. (W.W. Norton & Company).
70. Vygotsky, L. S. Mind in Society: Development of Higher Psychological Processes. (Harvard University Press, 1978). https://doi.org/10.2307/j.ctvjf9vz4.
71. Piaget, J. The construction of reality in the child. in The growth of logical thinking from childhood to adolescence (Basic Books, New York, 1954). https://doi.org/10.4324/9781315009650.
72. Druin, A. Cooperative inquiry: developing new technologies for children with children. in Proceedings of the SIGCHI conference on Human factors in computing systems the CHI is the limit - CHI '99 592–599 (ACM Press, Pittsburgh, Pennsylvania, United States, 1999). https://doi.org/10.1145/302979.303166.
73. Gestwicki, C. Developmentally Appropriate Practice: Curriculum and Development in Early Education. (Cengage Learning, Belmont, Calif, 2013).
74. Harbeck, J. D. & Sherman, T. M. Seven Principles for Designing Developmentally Appropriate Web Sites for Young Children. Educ. Technol. **39**, 39–44 (1999).
75. Brown, B. B. & Larson, J. Peer Relationships in Adolescence. in Handbook of Adolescent Psychology, Vol. 2: Contextual influences on adolescent development (eds. Lerner, R. M. & Steinberg, L.) 74–104 (Wiley, Hoboken, NJ, US, 2009). https://doi.org/10.1002/9780470479193.adlpsy002004.
76. Plagerson, S. Integrating mental health and social development in theory and practice. Health Policy Plan. **30**, 163–170 (2015).
77. Duncan, L. R., Hieftje, K. D., Culyba, S. & Fiellin, L. E. Game playbooks: tools to guide multidisciplinary teams in developing videogame-based behavior change interventions. Transl. Behav. Med. **4**, 108–116 (2014).
78. Michie, S. & Prestwich, A. Are interventions theory-based? Development of a theory coding scheme. Health Psychol. **29**, 1–8 (2010).
79. Bandura, A. Social Foundations of Thought and Action: A Social Cognitive Theory. (Prentice-Hall, 1986).
80. Ajzen, I. The theory of planned behavior. Organ. Behav. Hum. Decis. Process. **50**, 179–211 (1991).
81. Bajpai, S., Semwal, M., Bajpai, R., Car, J. & Ho, A. H. Y. Health Professions' Digital Education: Review of Learning Theories in Randomized Controlled Trials by the Digital Health Education Collaboration. J. Med. Internet Res. **21**, e12912 (2019).
82. Bandura, A. Social Learning Theory. (General Learning Press, 1971).
83. Mayer, R. E. The Cambridge Handbook of Multimedia Learning. (Cambridge University Press, 2014).
84. Handani, S. W. & Kusuma, T. P. P. Comparing Learning Media Applications Using Gamification Theory. in 2018 3rd International Conference on Information Technology, Information System and Electrical Engineering (ICITISEE) 37–42 (2018). https://doi.org/10.1109/ICITISEE.2018.8720994.
85. Aronson, I. D., Marsch, L. A. & Acosta, M. C. Using findings in multimedia learning to inform technology-based behavioral health interventions. Transl. Behav. Med. **3**, 234–243 (2013).
86. Aronson, I. D. & Bania, T. C. Race and Emotion in Computer-Based HIV Prevention Videos for Emergency Department Patients. AIDS Educ. Prev. **23**, 91–104 (2011).

87. Bickel, W. K., Marsch, L. A., Buchhalter, A. R. & Badger, G. J. Computerized Behavior Therapy for Opioid-Dependent Outpatients: A Randomized Controlled Trial. Exp. Clin. Psychopharmacol. **16**, 132–143 (2008).
88. Druin, A. The role of children in the design of new technology. Behav. Inf. Technol. **21**, 1–25 (2002).
89. Ma, Y., Bekker, T., Ren, X., Hu, J. & Vos, S. Effects of playful audio augmentation on teenagers' motivations in cooperative physical play. in Proceedings of the 17th ACM Conference on Interaction Design and Children 43–54 (ACM, Trondheim Norway, 2018). https://doi.org/10.1145/3202185.3202729.
90. Landry, P. et al. Design strategy to stimulate a diversity of motor skills for an exergame addressed to children. in Proceedings of the 12th International Conference on Interaction Design and Children 84–91 (ACM, New York New York USA, 2013). https://doi.org/10.1145/2485760.2485781.
91. Zaman, B. & Abeele, V. V. Laddering with young children in User eXperience evaluations: theoretical groundings and a practical case. in Proceedings of the 9th International Conference on Interaction Design and Children 156–165 (ACM, Barcelona Spain, 2010). https://doi.org/10.1145/1810543.1810561.
92. Woodward, J. et al. Using Co-Design to Examine How Children Conceptualize Intelligent Interfaces. in Proceedings of the 2018 CHI Conference on Human Factors in Computing Systems 1–14 (ACM, Montreal QC Canada, 2018). https://doi.org/10.1145/3173574.3174149.
93. Read, J. C. Validating the Fun Toolkit: an instrument for measuring children's opinions of technology. Cogn. Technol. Work **10**, 119–128 (2008).
94. Guha, M. L., Druin, A. & Fails, J. A. Cooperative Inquiry revisited: Reflections of the past and guidelines for the future of intergenerational co-design. Int. J. Child-Comput. Interact. **1**, 14–23 (2013).
95. Walsh, G. et al. DisCo: a co-design online tool for asynchronous distributed child and adult design partners. in Proceedings of the 11th International Conference on Interaction Design and Children 11–19 (Association for Computing Machinery, New York, NY, USA, 2012). https://doi.org/10.1145/2307096.2307099.
96. Van Mechelen, M., Baykal, G. E., Dindler, C., Eriksson, E. & Iversen, O. S. 18 Years of ethics in child-computer interaction research: a systematic literature review. in Proceedings of the Interaction Design and Children Conference 161–183 (ACM, London United Kingdom, 2020). https://doi.org/10.1145/3392063.3394407.
97. Markopoulos, P., Read, J. C. & Giannakos, M. Design of Digital Technologies for Children. in Handbook of Human Factors and Ergonomics (eds. Salvendy, G. & Karwowski, W.) 1287–1304 (Wiley, 2021). https://doi.org/10.1002/9781119636113.ch49.
98. Read, J. C. et al. CHECk: a tool to inform and encourage ethical practice in participatory design with children. in CHI '13 Extended Abstracts on Human Factors in Computing Systems 187–192 (ACM, Paris France, 2013). https://doi.org/10.1145/2468356.2468391.
99. Gilbey, D., Morgan, H., Lin, A. & Perry, Y. Effectiveness, Acceptability, and Feasibility of Digital Health Interventions for LGBTIQ+ Young People: Systematic Review. J. Med. Internet Res. **22**, e20158 (2020).
100. Preece, J. et al. Human-Computer Interaction. (Addison-Wesley, 1994).
101. Maramba, I., Chatterjee, A. & Newman, C. Methods of usability testing in the development of eHealth applications: A scoping review. Int. J. Med. Inf. **126**, 95–104 (2019).
102. Nadal, C., Sas, C. & Doherty, G. Technology Acceptance in Mobile Health: Scoping Review of Definitions, Models, and Measurement. J. Med. Internet Res. **22**, e17256 (2020).
103. de Lima, A. L. S. et al. Feasibility of large-scale deployment of multiple wearable sensors in Parkinson's disease. PLOS ONE **12**, e0189161 (2017).
104. Davies, E. H. et al. Demonstrating the feasibility of digital health to support pediatric patients in South Africa. Epilepsia Open **6**, 653–662 (2021).
105. Weiner, B. J. et al. Psychometric assessment of three newly developed implementation outcome measures. Implement. Sci. **12**, 108 (2017).

106. Haynes, B. Can it work? Does it work? Is it worth it?: The testing of healthcare interventions is evolving. BMJ **319**, 652–653 (1999).
107. Curran, G. M., Bauer, M., Mittman, B., Pyne, J. M. & Stetler, C. Effectiveness-implementation Hybrid Designs: Combining Elements of Clinical Effectiveness and Implementation Research to Enhance Public Health Impact. Med. Care **50**, 217 (2012).
108. Ozer, E. M. et al. Fostering Engagement in Health Behavior Change: Iterative Development of an Interactive Narrative Environment to Enhance Adolescent Preventive Health Services. J. Adolesc. Health **67**, S34–S44 (2020).
109. Rowe, J. P. & Lester, J. C. Artificial Intelligence for Personalized Preventive Adolescent Healthcare. J. Adolesc. Health **67**, S52–S58 (2020).
110. Kaimara, P., Oikonomou, A. & Deliyannis, I. Could virtual reality applications pose real risks to children and adolescents? A systematic review of ethical issues and concerns. Virtual Real. **26**, 697–735 (2022).

Technology and Innovation in Pediatric Surgery: The Past, The Cutting Edge, and Future Possibilities

Bianca Fischer, Nadeen Alturki, and Keith Kuenzler

Introduction to Minimally Invasive Surgery

Technology allowed surgery to evolve. Without innovation and engineering, surgeons' concepts for improved instrumentation would have never been accomplished. In fact, the practice of surgery at its outset used little technology. However, now, for most of us, it is becoming increasingly difficult to remember a time when traditional "open" surgery was the norm.

Open surgery is performed with large incisions, granting the surgeon access to the desired body cavity or the ability to deliver the target organ up to the surface. Unsurprisingly, long incisions that cut through the muscles of the chest or abdomen result in significant postoperative pain, great potential for wound complications, protracted recovery times, and extended hospital stays. In addition, as wound tension increases with the length of an incision, wound dehiscence and hernias are more common with open incisions [1]. Minimally invasive surgery (MIS) spares patients from these large incisions and their possible adverse consequences. MIS is an umbrella term used to encompass surgical techniques, which utilize "keyhole" incisions of just a few millimeters in diameter. This allows for the insertion of long, thin instruments towards the target anatomy. The terms used for the MIS technique are laparoscopy within the abdomen (Fig. 1) and thoracoscopy within the chest (Fig. 2).

B. Fischer · N. Alturki
Department of Surgery, Hackensack University Medical Center, Hackensack, NJ, USA
e-mail: bianca.fischer@hmhn.org; nadeen.alturki@hmhn.org

K. Kuenzler (✉)
Department of Surgery, Hackensack University Medical Center, Hackensack, NJ, USA

Joseph M Sanzari Children's Hospital, Hackensack, NJ, USA

New York University Langone Medical Center, New York, NY, USA
e-mail: keith.kuenzler@nyulangone.org

J. L. Wells (ed.), *A Clinical Lens on Pediatric Engineering*,
https://doi.org/10.1007/978-3-031-84655-7_11

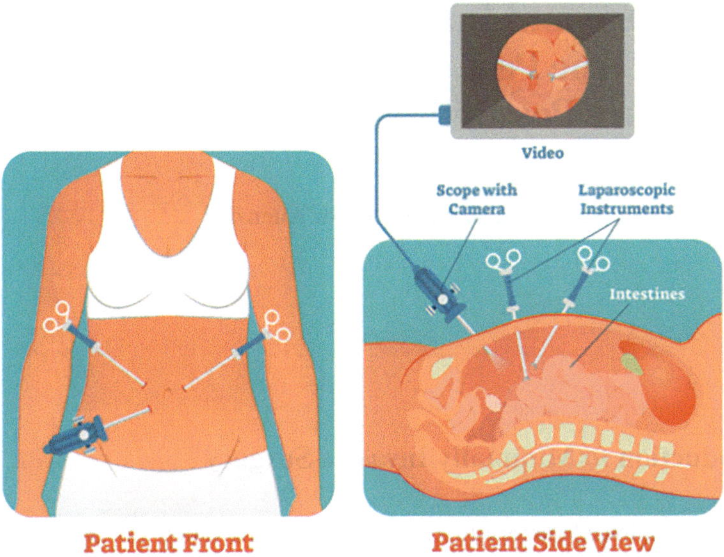

Fig. 1 Laparoscopy. (Courtesy VectorMine/iStockphoto.com)

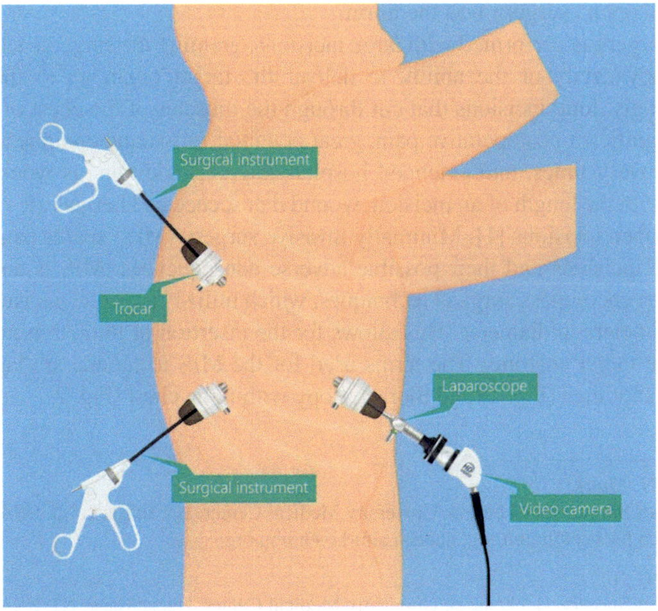

Fig. 2 Thoracoscopy. (Courtesy Pepermpron/Shutterstock.com)

In order to operate using the MIS platform, surgeons must create a visible working space within the desired body cavity. This is executed by insufflating carbon dioxide into the cavity to create an operative field. Additionally, a telescope fitted with a camera and a fiber optic light delivery system are required. The small incisions of MIS permit high-definition magnified visualization and precise dissection while resulting in decreased tissue trauma, physiologic stress, wound dehiscence, hernia occurrence, postoperative pain and healing time [1].

Engineering and technological advancements enabled each component of MIS to progress. The application of MIS to children and, then, infants provides even more complex challenges owing to the geometrical and physiological limits of this patient population. The constant transformation in this field created an essentiality for innovators and engineers to foster new technology into practice. Collaborative efforts across engineering and surgery led to the production of minimally invasive instruments sized for children. This, in turn, expanded the application of MIS from more elementary procedures in adults to the most complex procedures in children.

Early Development and Application of Pediatric MIS

The origins of MIS date back to the 1930s, when laparoscopy started as a diagnostic procedure. Hans Christian Jacobaeus, an internist from Sweden, first used laparoscopy to evaluate intra-abdominal fluid [2]. He worked with German physician Ludolph Brauer to adopt it in a therapeutic approach and treated patients with tuberculosis utilizing thoracoscopy as a means of freeing the lung from scar tissue to prevent lung collapse [3]. Another key contributor in the history of laparoscopic surgery was Heinz Kalk, who performed laparoscopic biopsies of the liver and spleen. Kalk designed a device to gain safe access to the thorax or abdominal cavity. It contained a spring that retracted its sharp point after entry through the abdominal wall to avoid injury to vital structures. He also invented a thirty-degree telescope, which greatly improved the visibility of the surgical field by facilitating viewing at different angles, particularly required around fixed structures [4]. Gynecologists were the first specialists to perform increased therapeutic procedures using laparoscopy [5]. In the 1970s, Kurt Semm, a German gynecologist, progressed from simple diagnostics to advanced cutting, suturing, and knot tying by developing laparoscopic scissors, needle holders, clamps and retractors with long handles for tissue manipulation from outside the body. These instruments were placed through fixed ports, a concept popularized by Semm that remains in use today. He not only performed multiple gynecologic procedures but also performed the first laparoscopic appendectomy in 1980 [6]. Five years later, Erich Mühe, a general surgeon in Germany, performed the first laparoscopic cholecystectomy [5, 7]. Appendectomy and cholecystectomy remain two of the most commonly performed MIS procedures. One of the critical steps to these advancements in the 1980s was the development of solid-state camera technology, responsible for the first wave of electronic digital cameras and portable video systems [8]. This gave surgeons the optical resolution for the anatomic detail necessary to perform increasingly complex operations. Unlike the original telescope, surgeons could now stand upright to view the video monitor and perform operations using both hands, a major improvement over keeping one eye locked onto the objective of the laparoscope held by one hand [5].

Notably, all of these advances occurred in the adult patient. Early on, pediatric surgeons resisted the adoption of these techniques in children, largely due to technological shortcomings including 10 mm diameter telescopes and instruments, as well as insufflators that deliver large amounts of carbon dioxide, which carry the potential to create dangerously high intra-abdominal pressures in small children. Over time, instrument manufacturers offered miniature devices and reliable insufflators safe for the use in small children. This allowed pediatric surgeons to begin their laparoscopic journey in the early 1990s. They followed the same route as the adult surgeons by beginning with laparoscopic appendectomies and cholecystectomies [1]. Other early barriers that limited MIS were high cost due to new technology, specialized equipment, and additional training of staff. It is well-recognized that there is a large up-front cost for hospitals introducing the new technology. However, once established, MIS is demonstrated to generate an overall cost advantage [9]. Another initial criticism was the prolonged length of minimally invasive operations. Expectedly, as MIS training increased and learning curves improved, there was notable reduction in operative times [1].

In pediatric surgery, until the late 1990s and early 2000s, instruments were large and simply not designed for infants and children. Therefore, for most "traditionally" trained pediatric surgeons, procedures were limited to diagnostic maneuvers (to examine or biopsy) or simple "excision" operations (appendectomy, cholecystectomy). The development and downsizing of instruments permitted the performance of precise, complex operations and led to the expansion of pediatric MIS. Specifically, the creation of ports, telescopes and instruments sized at 5 mm and 3 mm made MIS techniques applicable to many more pediatric operations (Fig. 3).

As one example, a pyloromyotomy is a relatively common infant procedure for a disease called pyloric stenosis. The pylorus is a complex channel of muscle fibers at the stomach outlet that regulates the emptying of gastric contents into the small intestine for digestion. In infants with pyloric stenosis, the muscle hypertrophies for

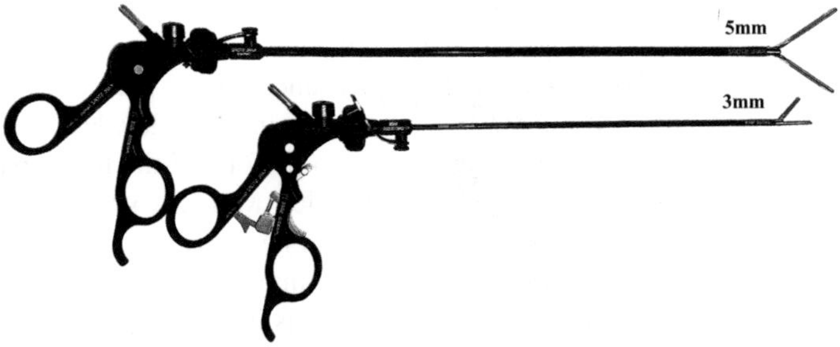

Fig. 3 Comparison of 5 mm and 3 mm laparoscopic bowel graspers

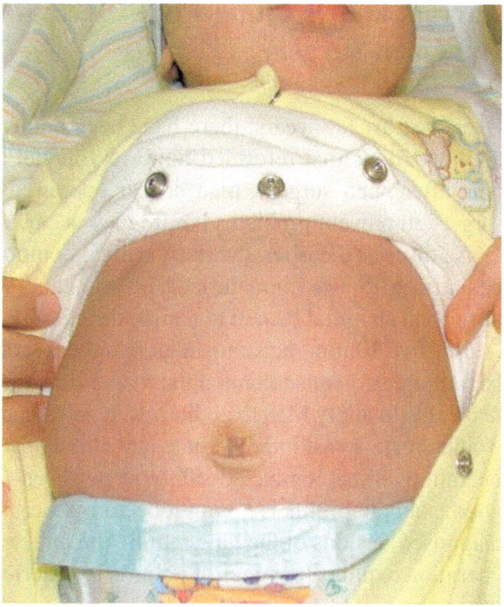

Fig. 4 Infant 2 weeks after laparoscopic pyloromyotomy

unknown reasons, resulting in a 2–8 week old vomiting baby whose stomach cannot empty. The corrective operation requires surgeons to incise the pylorus without causing injury to the inner lining of the stomach. This was previously performed using a 3–4 cm right upper quadrant open incision, which cuts through the rectus abdominus muscle and leaves a scar that may grow with the patient. Now with laparoscopy, 3 mm incisions achieve the same goal but with minimal pain and no long-term visible scar (Fig. 4).

Unsurprisingly, pediatric surgeons began to broaden the scope of MIS to thoracic operations. A standard open thoracotomy is quite painful given the division of large muscles of the chest, such as latissimus dorsi, serratus anterior, and intercostals. So, whilst a patient after laparotomy may attempt to reduce pain by limiting movement, a child after thoracotomy feels pain with each breath. Needless to say, this can compromise normal deep breathing and coughing essential to the postoperative recovery process, which increases the risk for pneumonia, infection and other related complications. Open thoracotomy often requires long hospital stays, epidural infusion of pain medication, and opiate pain management. Additionally, in infants, the surgical closure of a thoracotomy incision carries risks of future chest wall deformities, scoliosis and shoulder girdle instability. This is postulated to be due to tight approximation of adjacent ribs creating a unilateral, tensile component on the lateral chest wall, which causes asymmetry as the child grows. Hence, for children requiring thoracic surgery, there is a particular advantage to the MIS approach. Not only does it reduce postoperative pain significantly, but thoracoscopy also does not create these mechanical distortions due to utilizing small incisions between the ribs [10].

Current and "Cutting-Edge" Techniques in MIS and Open Surgery

Over the past few years, there have been numerous "game-changing" refinements and inventions that reformed how we conduct complex MIS in infants, eliminate pain after open surgery, plan cancer operations using 3D printing, perform fetal surgery to prevent problems before they occur, and even avoid surgery altogether.

As discussed earlier, the most critical component allowing the establishment of pediatric MIS was miniaturized instruments. Where 10 mm diameter tools may have been an initial benefit to adults, the cumulative tissue trauma and scarring from four or five 10 mm incisions in an infant exceeded that of an open operation. In the beginning, surgeons tolerated the long 5 mm and 10 mm vessel sealers and staplers to be able to apply MIS to pediatrics. It was clear; however, 3 mm short instruments would create a vast improvement and allow the most delicate and complex thoracic operations to be conducted safely in newborns and babies under 6 months old. A good example of this is a lung lobectomy. Certain congenital (and often prenatally diagnosed) lesions of the lung require the affected lobe to be entirely removed. This operation requires branches of pulmonary arteries, veins, and bronchi to the affected lobe to be sealed and divided. The surgeon must trust the engineering and function of the MIS instruments to execute these maneuvers without fail. Massive bleeding or airway leak can be catastrophic while working through keyhole incisions without immediate open access to the chest cavity. Fortunately, in the past decade, reliable 3 mm diameter 20 cm long vessel sealers (reduced from 5 mm, 36 cm) and 5 mm diameter tissue staplers (decreased from 10–12 mm diameter) have been developed to revolutionize thoracoscopic lung lobectomy. Children often are able to go home the day after surgery with almost no scarring visible in a few weeks.

For the engineers, the conundrum was how to safely deliver the proper energy in such a small instrument. Historically, the control of blood vessels and bleeding was by suturing. Other tools to control bleeding were then introduced such as metal clips, stapling devices and electrocautery. The first electric modality introduced used monopolar electric current, which passes through tissue via the use of two electrodes. Energy is widely dispersed from an active smaller electrode on the cautery device to a larger return electrode, a grounding pad, which is placed on the patient's skin. Monopolar instruments are commonly used for hemostatic division of tissue. However, limitations include injury to tissues from thermal spread by inadvertent capacitive coupling or damage to the insulation shaft. Capacitive coupling is a phenomenon in which a surrounding conductor, such as another instrument, may absorb stray current and damage nearby tissue. The advent of bipolar electrocautery overcame some of the limitations seen with monopolar electrocautery. Bipolar instruments contain an active electrode and a return electrode at the tip of the instrument. Rather than passing through the patient to the grounding pad, the current is distributed through the tissue between the jaws at the tip of the instrument. Hence, bipolar instruments resulted in less potential damage to surrounding tissue and less risk of capacitive coupling [11]. To further refine this, bipolar vessel "sealers" were then created to essentially bond the walls of the target vessel together. There are now small sealers that range in thermal spread of 1–2 mm. Traditional staplers were not only of large diameter, necessitating longer incisions, but often could not fit into an infant's chest to be opened and deployed due to the length of the functioning end. Thinner and shorter staplers now allow the division of bronchi and other tissues in the chest and abdomen to be performed in a minimally invasively manner even in newborns (Fig. 5).

As previously described, open thoracotomy is an extremely painful incision. Although MIS techniques have allowed surgeons to avoid many thoracotomies, there are situations in which open surgery is unavoidable, or chest wall reconstruction is required. In these cases, pain control can be a challenge and traditionally was accomplished with intravenous and oral opiate medications. Particularly in children and adolescents, there are concerns for tolerance and abuse. Cryoanalgesia is an innovative modality that greatly improves postoperative pain after thoracic surgery. This refers to the use of cold temperature to disrupt peripheral nerve function. A cryoprobe uses high-pressure carbon dioxide or nitrous oxide to rapidly cool nerve axons to between −50 and −70 °C. In open thoracic or thoracoscopic surgery, the probe may be applied directly onto intercostal nerves and results in temporary disruption of the nerve fibers. The end result is regional numbness up to 3 months until the nerves regenerate—reducing and possibly eliminating the need for intravenous or oral analgesia. The most common clinical application in pediatric surgery is the correction of pectus excavatum. In this deformity, there is gradual and sometimes severe inward growth of the costal cartilage which results in a depressed sternum in an adolescent. The reconstructive surgery involves inserting a curved metal bar into the chest forcibly and instantaneously bracing the sternum into a more normal configuration. Cryoablation is now routinely utilized to freeze several intercostal nerves bilaterally. Cryoablation during surgery has significantly improved pain control for patients and shortened the length of hospital stay [12, 13]. In our practice, most patients who previously remained in the hospital for 5–6 days before being discharged on several weeks of oral opiates, with cryoanalgesia, are now discharged the day after surgery with over-the-counter acetaminophen or ibuprofen.

New technology has also improved preoperative preparation in children with solid tumors. When planning a tumor resection operation, the exact location of the tumor and its relationship to nearby vital structures is key to a successful surgical

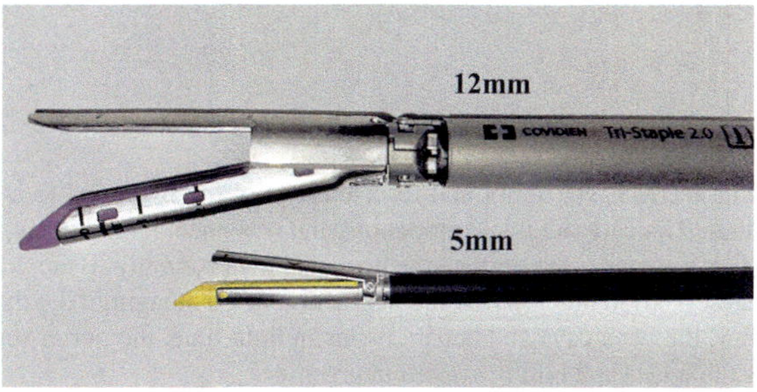

Fig. 5 A 12 mm stapler compared to a 5 mm surgical stapler

outcome. Most commonly, surgeons view two-dimensional (2D) cross-sectional imaging preoperatively. Even in-person interpretation reviewing images with a radiologist leaves the surgeon to mentally reconstruct images and carry these into the three-dimensional (3D) reality of the operation. As an emerging technology, 3D printing began to have a role in tumor resection planning. 2D medical images from a computed tomography scan (CT) or magnetic resonance imaging (MRI) can be converted into a 3D printable construct, which allows physicians to visually see the relation of the tumor with important structures (Fig. 6). 3D printing can also be used to educate students and patients' family members to help to visualize the tumor anatomy and describe the proposed surgery with its risks and benefits [14].

The availability of miniaturized instruments also allowed the expansion of one of the newest advancements in the surgical field, fetal surgery. At present, it is only offered at select institutions and for very limited indications. Largely due to the developments in technology, diagnosis of fetal anomalies now occurs earlier in pregnancy. Fetal surgery developed as physicians identified this window of opportunity for possible intervention while still in utero. Originally, it was only deemed necessary if the fetus had a lethal disease. As maternal safety during fetal surgery improved, non-lethal fetal disease was evaluated for surgical repair. Operating on a fetus while in utero in a minimally invasive manner is termed ***fetoscopic***. One of the diseases approved for fetoscopy is myelomeningocele (MMC), which is a defect of the spine, spinal cord, and spinal canal where a sac containing parts of the spinal cord and nerves protrudes at the affected vertebral level. If left untreated, the child can develop severe neurologic deficits from intrauterine trauma. Fetoscopic surgery entails general anesthesia for the mother where the uterus is exposed after an abdominal incision. With the use of ultrasonography, the fetus and placenta are visualized to plan port placement. Carbon dioxide is infused through a port similar to laparoscopic procedures to expand the working field. The fetus is anesthetized

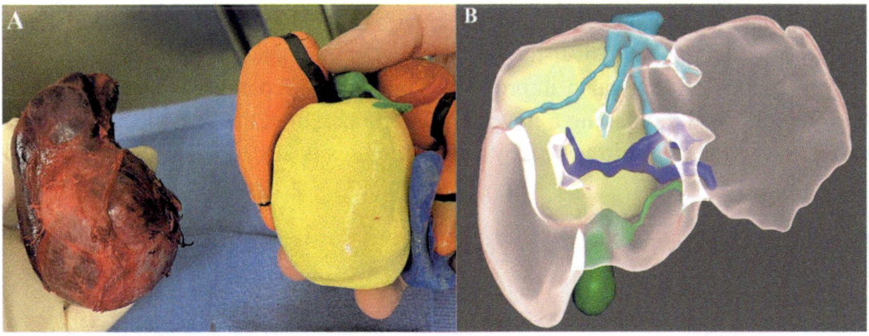

Fig. 6 (**a**) Liver tumor and resected lobe of the liver next to a 3D printed model, which demonstrated a nearly identical model of what was encountered in surgery. The model shows the normal liver in orange, the tumor in yellow, and the proximity to the vena cava (which was protected and preserved) in blue. (**b**) 3D reconstruction of CT imaging. The liver is translucent, the tumor in yellow, the vena cava and hepatic veins in light blue, the portal venous system in dark blue, and the gallbladder and biliary system in green

and the spinal cord defect is identified and closed with sutures or with a patch. The outcomes of MMC repaired prenatally versus postnatally were compared and found that prenatal repair of MMC led to improved mental and motor function [15].

Simple advancements in devices and their use have even led to the avoidance of surgery altogether. One such example of this is the evolution in the treatment of gastroschisis [16, 17]. Gastroschisis is a congenital anomaly of the anterior abdominal wall resulting in eviscerated abdominal contents. Historically, patients were subjected to immediate sutured closure. The immediate repair of gastroschisis poses potentially devastating risks to the neonate, including increased intra-abdominal hypertension, hypoventilation, reduced cardiac output, bowel ischemia and poor wound healing. Only the defects that were unable to be sutured primarily underwent delayed closure, where in the interim, protruding abdominal contents were covered with transparent handmade devices, called ***silos***. Surgeons would fashion an array of synthetic materials and often fix or suture it to the edge of the abdominal wall defect to create a custom silo. Over time, surgeons realized that delayed closure is applicable to the entire disease entity. In 2004, the technique of sutureless closure was described in a series of patients [18]. Silos cover and protect the bowel, reduce swelling, while also allowing direct visualization of the bowel. A silo is placed and the protruding abdominal contents are gradually reduced back into the abdominal cavity and the neonate's abdomen slowly expands to accommodate it (Fig. 7). The remaining defect is then allowed to close spontaneously (Fig. 8). The invention of preformed silos, a flexible inert silicone bag fitted with a spring-loaded ring, led to undelayed coverage of eviscerated organs at birth and the induction of immediate treatment (Fig. 9). Risks with operative management were found to be unnecessary, preformed silos were readily available off the shelf and in fact led to a cosmetic closure [16–18].

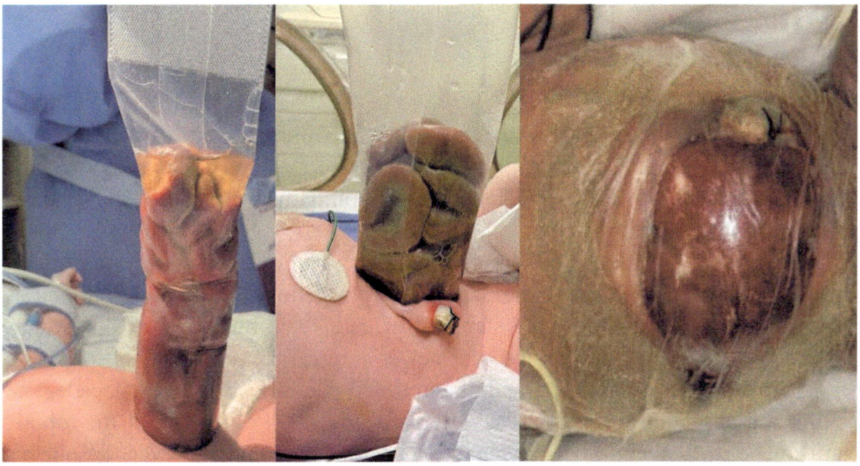

Fig. 7 The transparent bag covers the eviscerated bowel and is gradually reduced

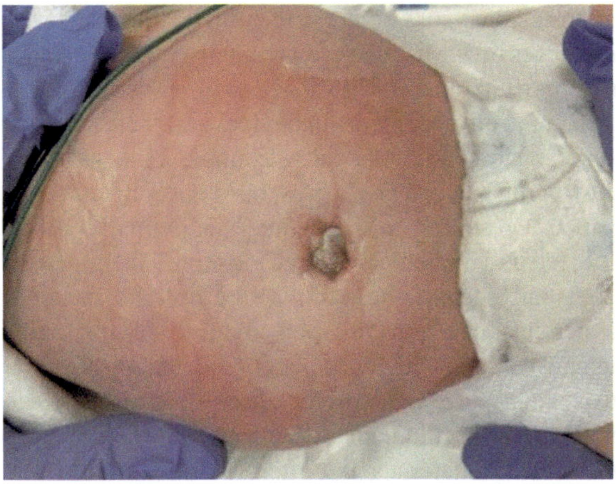

Fig. 8 Neonate's abdomen 6 weeks after treatment completion with sutureless closure for gastroschisis using silo

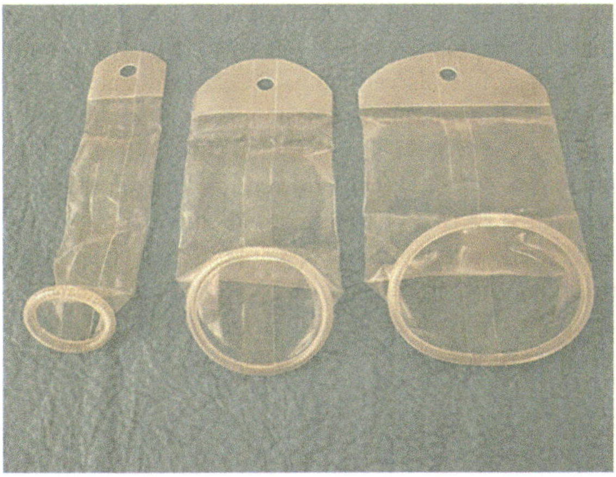

Fig. 9 Spring-loaded silicone silo bags of various sizes based on defect diameter, from left to right sizes include 4 cm, 7.5 cm, and 10 cm

Future Possibilities

Innovation within the pediatric surgery field, both within surgery or with adjuncts around surgery, led to the development of highly sophisticated and individualized equipment geared towards the pediatric population. However, there is still endless potential for evolution.

Decades ago, traditional open surgery was revolutionized with the advent of laparoscopic surgery. Conventional laparoscopy was novel and unfamiliar, and as we have described, it is now the standard of care for a multitude of operations. Currently, there is the introduction of a newer surgical technique, robotic surgery. Similar to laparoscopy, its use in adults preceded pediatrics by over a decade, and its first use in pediatric surgery was described in 2001. Proponents of the robotic platform boast a plethora of advantages over laparoscopy; however, despite its advertised benefits, pediatric robotic-assisted surgery remains in its infancy [19].

Robotic surgery mirrors MIS benefits seen in laparoscopy, such as small cosmetic incisions, decreased postoperative pain and opiate use, decreased recovery times and shorter length of stays. It offers additional advantages as well, through its unique design. The robotic endoscope has three-dimensional visualization capable of magnifying images 10 to 15-fold, surpassing the depth perception and magnification capable with loupes, which are magnifying glasses worn during open surgery, or traditional laparoscopes. Robotic instruments mimic the movement of the human wrist, facilitating seven degrees of freedom rather than the four degrees of freedom in laparoscopy. There is also the ability to utilize multiple robotic arms, forgoing the need for surgical assistance. Furthermore, arms can lock into place, which allows for reliable and indefinite retraction, removing the risk of human muscle fatigue and potential loss of control of an instrument. The robotic platform features tremor filtration and motion scaling, where larger motions at the surgeon console can be scaled to translate into smaller motions of the robotic arms up to 7:1. These features allow surgeons to become more capable of fine accurate controlled movements, particularly attractive and advantageous with a smaller body habitus, such as in the pediatric patient. Single-site robotic surgery has furthered the field where one 2.5 cm site contains one endoscope and three interchangeable instruments to perform the surgery. The robot is operated by the surgeon via a physically separate ergonomically designed console. This particular feature may open the possible avenue of telesurgery and telementoring. Theoretically, this could increase patient access to highly specialized surgeons and even contain global humanitarian potential [19–21].

Nevertheless, its slow adoption in the pediatric surgery realm is attributable to multiple barriers, more than likely to its construction with the adult patient in mind. The only FDA-approved robotic system for use in children is the da Vinci Surgical System, which only has 5 mm and 8 mm instruments and robotic endoscopes are 8.5 mm or 12 mm in size [20]. Robotic sizing is one large prohibitive reason for the uptake of this platform in pediatric surgery. The robot is 6 feet tall, larger than many pediatric patients, and it could potentially limit access to the patient by the anesthesiologist or surgical assistant. Another issue is the recommendation for robotic ports to be spaced 8 cm apart to avoid robotic arm collisions, space many pediatric patients do not have. The trocars should also be inserted to the level of its remote center which is 5.61 cm deep, which may be the entire width of a neonatal hemithorax or hemiabdomen. However, size is not the only issue. An additional large institutional drawback to robotics is cost, not only of the new technology but also staff and operational training, increased operative times as surgeons learn the new technique and annual maintenance [19–21]. Therefore, not only is the robotic design not geared towards pediatrics (yet), but also the cost of re-training an entire field into robotic surgery may be paramount and take a few generations of surgeons for adoption at its full potential. As with laparoscopic adoption, the arrival and uptake of a new technique is clearly underway; we only require the gear to bring it to fruition.

Ultimately, there is an abundance of possibility to address the surgical needs of children in a meaningful manner. History informs the future, and it clearly demonstrates the importance of the identification of these needs by surgeons and the creative talent of engineers to move toward a common goal of benefiting the pediatric population.

Competing Interests The authors have no conflict of interest to declare relevant to the content of this chapter.

References

1. Blinman T, Ponsky T. Pediatric minimally invasive surgery: laparoscopy and thoracoscopy in infants and children. *Pediatrics*. 2012;130(3):539–549. https://doi.org/10.1542/peds.2011-2812
2. Litynski GS. Laparoscopy--the early attempts: spotlighting Georg Kelling and Hans Christian Jacobaeus. *JSLS*. 1997;1(1):83–85.
3. Jacobaeus HC. The Cauterization of Adhesions in Artificial Pneumothorax Treatment of Pulmonary Tuberculosis under Thoracoscopic Control. *Proc R Soc Med*. 1923;16(Electro Ther Sect):45–62.
4. Litynski GS. Laparoscopy between the world wars: the barriers to trans-atlantic exchange. Spotlighting Heinz Kalk and John C. Ruddock. *JSLS*. 1997;1(2):185–188.
5. Nakayama KD. The minimally invasive operations that transformed surgery. Published online 2017.
6. Litynski GS. Kurt Semm and an automatic insufflator. *JSLS*. 1998;2(2):197–200.
7. Litynski GS. Endoscopic surgery: the history, the pioneers. *World J Surg*. 1999;23(8):745–753. https://doi.org/10.1007/s002689900576
8. Fossum ER. Camera-on-a-chip: Technology Transfer from Saturn to your Cell Phone. *Technol Innov*. 2013;15(3):197–209. https://doi.org/10.3727/194982413X13790020921744
9. Fox D, Morrato E, Campagna EJ, et al. Outcomes of laparoscopic versus open fundoplication in children's hospitals: 2005–2008. *Pediatrics*. 2011;127(5):872–880. https://doi.org/10.1542/peds.2010-1198
10. Lawal TA, Gosemann JH, Kuebler JF, Glüer S, Ure BM. Thoracoscopy versus thoracotomy improves midterm musculoskeletal status and cosmesis in infants and children. *Ann Thorac Surg*. 2009;87(1):224–228. https://doi.org/10.1016/j.athoracsur.2008.08.069
11. El-Sayed MM, Saridogan E. Principles and safe use of electrosurgery in minimally invasive surgery. *Gynecol Pelvic Med*. 2021;4:6. https://doi.org/10.21037/gpm-2020-pfd-10
12. Talsma J, Kusakavitch M, Lee D, et al. Forgotten Branch of the Intercostal Nerve: Implication for Cryoablation Nerve Block for Pectus Excavatum Repair. *J Pediatr Surg*. 2023;58(12):2435–2440. https://doi.org/10.1016/j.jpedsurg.2023.05.006
13. Eldredge RS, McMahon L. Intercostal nerve cryoablation therapy for the repair of pectus excavatum: a systematic review. *Front Surg*. 2023;10:1235120. https://doi.org/10.3389/fsurg.2023.1235120
14. Safhi AY. Three-Dimensional (3D) Printing in Cancer Therapy and Diagnostics: Current Status and Future Perspectives. *Pharm Basel Switz*. 2022;15(6):678. https://doi.org/10.3390/ph15060678
15. Yamashiro KJ, Farmer DL. Fetal myelomeningocele repair: a narrative review of the history, current controversies and future directions. *Transl Pediatr*. 2021;10(5):1497–1505. https://doi.org/10.21037/tp-20-87
16. Dekonenko C, Fraser JD. Approaches for Closing Gastroschisis. *Adv Pediatr*. 2020;67:123–129. https://doi.org/10.1016/j.yapd.2020.03.005
17. Marven S, Owen A. Contemporary postnatal surgical management strategies for congenital abdominal wall defects. *Semin Pediatr Surg*. 2008;17(4):222–235. https://doi.org/10.1053/j.sempedsurg.2008.07.002
18. Sandler A, Lawrence J, Meehan J, Phearman L, Soper R. A "plastic" sutureless abdominal wall closure in gastroschisis. *J Pediatr Surg*. 2004;39(5):738–741. https://doi.org/10.1016/j.jpedsurg.2004.01.040

19. Jacobson JC, Pandya SR. Pediatric robotic surgery: An overview. *Semin Pediatr Surg*. 2023;32(1):151255. https://doi.org/10.1016/j.sempedsurg.2023.151255
20. Denning NL, Kallis MP, Prince JM. Pediatric Robotic Surgery. *Surg Clin North Am*. 2020;100(2):431–443. https://doi.org/10.1016/j.suc.2019.12.004
21. Mei H, Tang S. Robotic-assisted surgery in the pediatric surgeons' world: Current situation and future prospectives. *Front Pediatr*. 2023;11:1120831. https://doi.org/10.3389/fped.2023.1120831

Pediatric Engineering: Pediatric Orthopedics

Andrew Chou and Karim Masrouha

Origins and the Innovation Gap

The origin of the word "orthopedics" comes from the Greek language. It is a combination of *orthos* (to straighten) and *pedos* (a child). As such, the specialty started out as an endeavor to help correct deformities and fix fractures in children. The treatment options for children with fractures and limb or spine deformities, for many years, included traction and casting with Plaster of Paris, which continues to be used to this day. The development of metallic implants for use in orthopedic injuries did not occur until the late 1800s, and stainless-steel implants that did not have corrosive properties were not readily available until the 1920s [1]. This later innovation, along with the development of anesthesia and aseptic techniques, which were popularized in the 1840s and 1870s, respectively, allowed for the development of a subspecialty within orthopedic surgery dedicated to the treatment of children [2, 3].

Being a discipline focused on the surgical management of musculoskeletal conditions, which frequently involves the use of implants for stabilization, orthopedic surgery is an obvious choice for health technology innovation and medical device development. From minimally invasive procedures such as arthroscopy to spinal pedicle screws and intelligent joint replacements, it is no surprise that orthopedic implants represent 20% of all devices approved by the U.S. Food and Drug Administration (FDA) annually [4]. While surgical implants have been tailored to

A. Chou
Department of Orthopaedic Surgery, KK Women's and Children's Hospital, Singapore, Singapore
e-mail: andrew.chou.c.c@singhealth.com.sg

K. Masrouha (✉)
Department of Orthopedic Surgery, NYU Langone Health, New York, NY, USA
e-mail: karim.masrouha@nyulangone.org

J. L. Wells (ed.), *A Clinical Lens on Pediatric Engineering*,
https://doi.org/10.1007/978-3-031-84655-7_12

the unique anatomy and physiology of children, from earlier implants such as rush rods and Luque wires to more modern pediatric-specific proximal and distal femur plates and pedicle screws, pediatric orthopedics overall lags in innovation compared to the other orthopedic subspecialties such as joint replacement surgery, spine surgery, and sports medicine. The FDA previously estimated that the pediatric medical device market lagged behind their adult counterparts by 5–10 years and FDA surveys of numerous physician groups show 66% of respondents felt there was a need for additional devices that grow with children [4]. More often than not, adult implants and devices are used off-label or repurposed in children. While technically legal, this results in suboptimal results and dissatisfied surgeons [4–6]. In fact, the 2005 American Academy of Orthopaedic Surgeons (AAOS) survey on Pediatric Device Development found that liability was a major concern for surgeons using devices off-label, with the majority of respondents feeling that the orthopedic implant industry was only "somewhat" meeting the demands of the pediatric population [6].

The causes of this innovation gap and the lack of pediatric-specific implants for many decades are multifactorial. One likelihood is that many pediatric deformities and injuries can frequently be treated by non-surgical means. Due to the remodeling potential and high rate of bony union in skeletally immature patients, surgical reduction (realignment) and fixation are not always necessary. The field is still exploring which fractures need to be reduced and fixed and which can be treated by casting alone. There are currently ongoing randomized controlled trials (RCTs) investigating this topic in at least two types of fractures [7].

The logistical challenges that may be exacerbating the innovation gap also include:

1. **Limited market.** Children's musculoskeletal conditions are a comparatively smaller market with lower disease burden compared to their adult counterparts. Moreover, while congenital conditions are rare, they are often complex and difficult to treat, with suboptimal outcomes. Consequently, implant companies are disincentivized financially to invest in research and development of pediatric-specific solutions, while startups may find the pediatric population too small or challenging to address in the first place [4, 6, 8, 9].
2. **Lower reimbursement.** Treatments for pediatric conditions often have lower reimbursement rates, which again leads to financial disincentives for companies looking to enter the pediatric orthopedic space [6, 10]. In fact, high research and development costs with poor prospects of profitability were previously cited as major impediments to pediatric innovation at an "Industry for Kids" summit at the AAOS meeting in 2012 [4, 6].
3. **Challenging, complex regulatory environment.** Due to the invasiveness of surgical implants, the majority of pediatric orthopedic innovations are considered Class III medical devices and must undergo the highest level of regulation by the FDA [6]. Furthermore, the mean expenditure for Class III medical devices requiring clinical trials has been estimated to be USD $54 million and can easily quadruple in certain cases [4]. Pathak et al. found that only two of 124 Class III

medical devices approved by the FDA from 2016 to 2021 were dedicated to pediatric patients, and 23 had dual pediatric/adult use [11]. While pediatric devices may be eligible for Humanitarian Device Exemption (HDE), the HDE approval pathway lacks transparency and the complex, lengthy, and costly regulatory process of bringing orthopedic implants to market can be inhibitory for innovators [4–6, 9]. In fact, the consensus of 13 separate orthopedic manufacturers at the 2012 AAOS "Industry for Kids" summit was that premarket approval, the standard approval pathway for pediatric devices, was too expensive for small-market pediatric devices [6]. Furthermore, while devices can be used off-label once approved for a specific indication, the FDA restricts the collection of any research data for devices used off-label except under an investigational device exemption (IDE) protocol, which is subject to their strict scrutiny [6, 9]. In the instance of seeking regulatory approval for vertebral body tethering for adolescent idiopathic scoliosis, this required surgeon-sponsored IDE studies, necessitating significant time, funding, and exposure to liability to comply with FDA regulations [9].

4. **Difficulty conducting research.** As children are fundamentally a vulnerable population, conducting clinical research for pediatric devices faces greater barriers compared to their adult counterparts [6]. Not all pediatric hospitals have the research capacity and infrastructure in place to conduct clinical trials for pediatric devices, thereby limiting the total number of eligible sites. Additionally, given the ethical concerns for testing novel devices in this patient population, the need for longitudinal follow-up data in growing children, and unique challenges of recruiting patients with rare and/or congenital conditions, conducting the necessary clinical trials in pediatrics can be a financial, ethical, and logistical nightmare [4, 6, 9].

Closing the Gap

In a bid to close this innovation gap, Congress passed the Pediatric Medical Safety and Improvement Act in 2007, culminating in the genesis of the FDA's Pediatric Device Consortium, which aims to support medical device development in children via funding and regulatory support [4, 6]. Public-private partnerships have also aimed to develop an ecosystem for pediatric innovation, such as the System of Hospitals for Innovation in Pediatrics—Medical Device (SHIP-MD) strategic framework launched in 2021, which brings together organizations such as the FDA's Center for Devices and Radiological Health, National Institutes of Health (NIH), American Academy of Pediatrics (AAP), and the medical device industry. By focusing on policy reforms such as increasing breakthrough device designation, building up pediatric clinical trial networks, and forming a pediatric-specific device review team, the SHIP-MD initiative hopes to reduce process risk and accelerate the development of pediatric devices [4].

Likewise, the Pediatric Orthopedic Society of North America (POSNA) and Scoliosis Research Society (SRS) Task Force for Pediatric Medical Devices in 2014 made numerous recommendations to facilitate regulatory approval for appropriate medical devices in children, including developing industry-clinician collaborations, adapting and improving existing 510k and HDE pathways for pediatric use, enhancing post-market surveillance, and expanding research data to include those from overseas and existing multicenter study groups/registries [6]. A recent study of Class III medical devices with pediatric orthopedic indications showed that all were approved via the HDE pathway [4]. In addition to having a less stringent regulatory process, devices approved via the HDE pathway can be labeled as Humanitarian Use Device (HUD) if they treat patients with rare conditions affecting <8000 individuals in the United States annually, after which they can be sold for a profit if the condition both affects pediatric patients and is FDA approved for use in children. In contrast, devices with HDE approval for adults cannot be sold for profit.

Children Are Not Small Adults

Treating children comes with unique and challenging considerations:

1. **Size matters.** Children are not small adults and the anatomy and physiology of their musculoskeletal system reflect that. Given that the FDA defines pediatric patients as those <21 years old, in comparison to adult implants, pediatric implants require a wider range of sizes that accommodate the changing morphology and size of children's bones [4, 8].
2. **Growth plates and remodeling.** Due to the presence of the growth plate, or physis, in pediatric bones, surgical implants and treatment modalities often need to avoid or spare this structure in order to preserve a child's growth potential. While certain technologies have been able to capitalize on the physis as part of their solution, such as 8-plates for guided growth and transphyseal screws, which allow for proximal femur fixation without growth plate tethering, it can prove challenging for implants used in the trauma setting, as avoiding the physis can come at the cost of construct stability or screw purchase [4, 8]. Continued growth may also result in the incorporation of implants within the bone and implant migration, which may result in difficult removal of hardware and secondary deformity (Fig. 1). In the spine, fixation at an early age may require sequential lengthening procedures, resulting in increased cost and morbidity. This also creates challenges for nonbiologic reconstructions in the setting of bone loss from infection or tumors, discussed in the next section.
3. **Complex care considerations.** Beyond their anatomic and physiological differences, children fundamentally differ from adults in terms of their activity, perioperative care, and considerations. Pediatric patients are often at the height of their physiological prime and engage in sports and physical activities at a greater

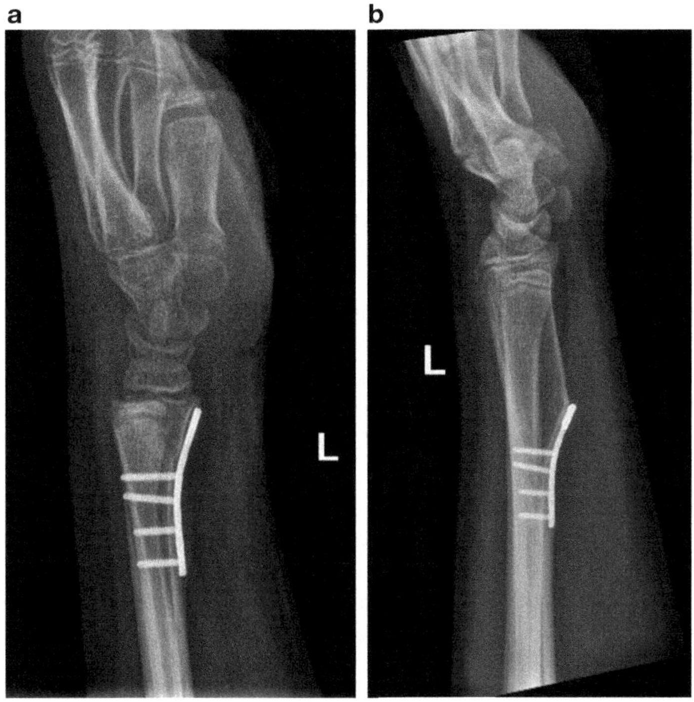

Fig. 1 (**a**) Lateral radiograph of an 11-year-old male who underwent open reduction and internal fixation of a distal radius fracture. (**b**) Follow-up imaging at 15 years of age demonstrates a healed fracture with secondary deformity (Courtesy of NYU Langone Health)

frequency and intensity than their adult counterparts, which means physiotherapy and rehabilitation may not be required. Conversely, adhering to strict postoperative weight-bearing or activity restrictions can be problematic in children. Pediatric patients are also innately a vulnerable population with complex social dynamics, so treating a child means treating their whole family/caregivers. As with any vulnerable population, involving patients in their own care while also addressing the concerns and stresses of caregivers can be a challenge in and of itself but essential in providing the most appropriate care [4, 6, 8].

Condition-Specific Implants

Despite the challenges, there have been success stories in pediatric orthopedic innovation. These include implants and devices for spine and lower extremity deformity, fracture fixation, and reconstruction after tumor resection.

Spine Deformity

While orthopedics is a primarily surgical specialty, the non-operative aspects of care cannot be ignored and have continued to evolve with time. The history of bracing for the treatment of adolescent idiopathic scoliosis reaches as far back as the fifth century by the ancient Greeks, although the Boston, Rigo Cheneau, and Milwaukee braces look nothing like Hippocrates' hippocratic ladder or bench [12]. Improvements in materials, appearance, and functionality have helped improve patient satisfaction and compliance with use [13].

Historically, most spine deformity was corrected with a combination of traction, bracing, and casting. It was not until the late 1800s that surgeons attempted to surgically correct deformity and fuse the spine [14]. There were attempts at both instrumented fusion as well as non-instrumented fusion, which relied heavily on casting. These techniques were developed by Hadra, Hibbs, and Lange [15–17]. Initially, many of the deformities treated resulted from infectious diseases such as tuberculosis or poliomyelitis and, once refined, were attempted on idiopathic scoliosis and spondylolisthesis. In the mid-twentieth century, Harrington developed a method of fixation and fusion using rods and hooks on each end of the construct [18, 19]. This method had initial success but resulted in long-term problems due to the reliance on a long construct involving only the posterior elements of the spine. This technique was refined in the late 1970s by Luque, who used sublaminar wires for segmental spinal fixation; however, there were still issues with the long construct, as well as neurological complications. This was followed by the development of an anterior approach to correct the deformities, which was initially successful; however, it was fraught with complications such as hardware failure and pseudarthrosis. This led to the most recent developments in spine fusion, first introduced by Cotrel and Dubousset, with the use of two rods connected to laminar hooks or transpedicular screws [20]. These screws were more stable, crossed all columns of the spine, and had better load-bearing properties. The screws are linked to pre-bent rods and so later advancements included polyaxial screws to allow for more reliable placement of the rods.

For early-onset scoliosis, growing rods have been used to allow for correction and stabilization using pedicle screws and rods, while also accommodating growth in younger children (Fig. 2). More recently, a magnetic growing rod system has been used; however, the complications related to loss of lengthening, hardware failure, and metallosis have hindered its widespread use [21]. Another example of

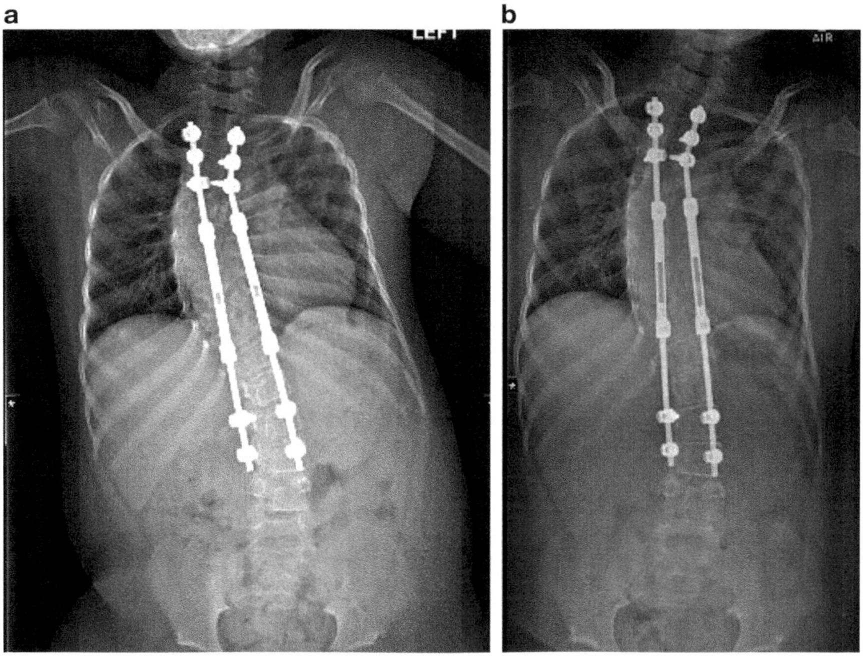

Fig. 2 (**a**) Antero-posterior (AP) radiograph of the spine following insertion of the growing rods and (**b**) AP radiograph of the spine after lengthening the rods (Courtesy of NYU Langone Health)

recent innovation is the Vertical Expandable Prosthetic Titanium Rib (VEPTR) (Fig. 3), which was developed by the late pediatric orthopedic surgeon Dr. Robert M. Campbell Jr., for the treatment of thoracic insufficiency syndrome, a rare disease often associated with congenital scoliosis, which affects fewer than 5000 children in the United States annually. The device was FDA-approved via HDE in 2004 after a 14-year clinical trial [22].

Surgical navigation has been applied in several areas of orthopedic surgery, but it is most frequently used in spine surgery. Navigation is performed with a software, which combines cross-sectional imaging performed intraoperatively, fiducials, and a camera to allow for accurate and safe placement of transpedicular screws [23] (Fig. 4).

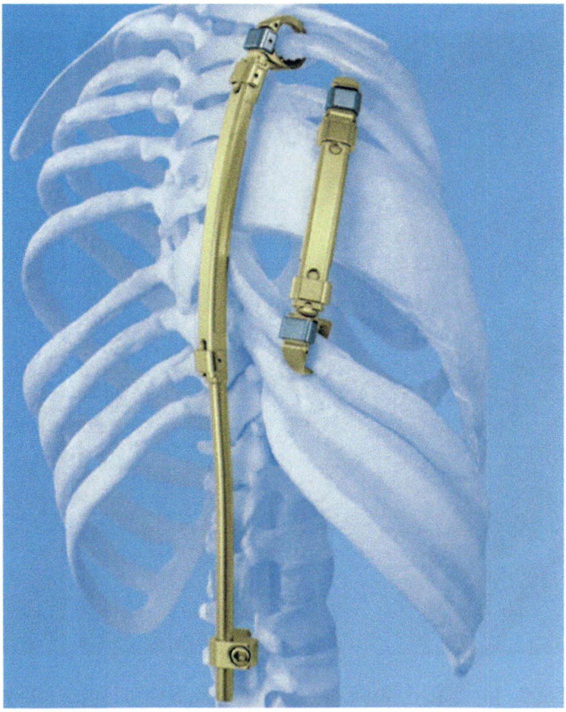

Fig. 3 Digital rendering of the Vertical Expandable Prosthetic Titanium Rib (VEPTR). (Courtesy of Synthes®, with permission)

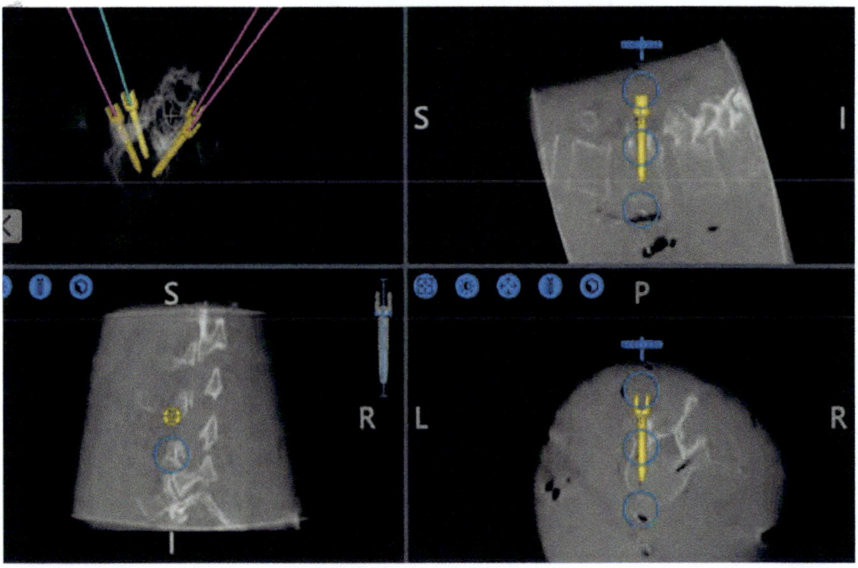

Fig. 4 Intraoperative image demonstrating the use of navigation software for accurate screw insertion (Courtesy of NYU Langone Health)

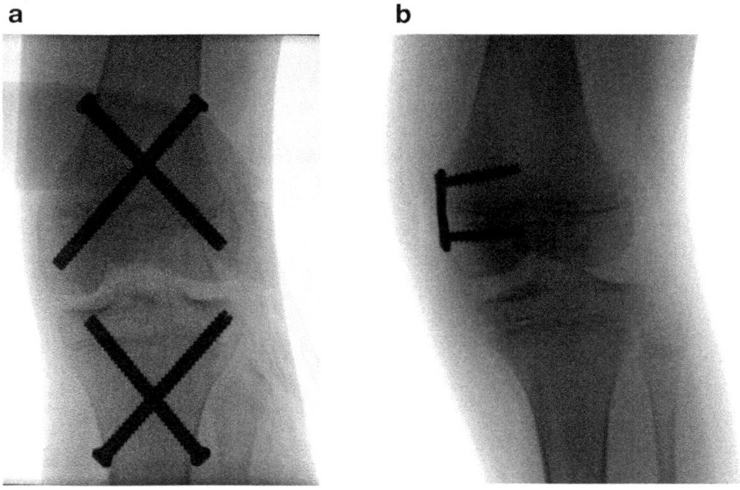

Fig. 5 (**a**) AP radiograph of the knee demonstrating the transphyseal screw epiphysiodesis, and (**b**) AP radiograph of the knee demonstrating the 8-plate for hemiepiphysiodesis (Courtesy of NYU Langone Health)

Lower Extremity Growth Modulation and Angular Correction

The treatment of angular deformity and limb length differences is an important aspect of pediatric orthopedics. Many of the treatment options make use of the physis. Historically, angular deformity was treated with metallic staples; however, these implants were fraught with hardware complications, including loosening and breakage, particularly in obese patients. This changed with the introduction of 8-plates and the use of transphyseal cannulated screws (Fig. 5) [24].

There are early reports demonstrating the use of a modified technique, using half of a hemiepiphysiodesis plate on either side of the growth plate, connected with a strong suture in 45-degree inclines on either side of the bone to allow for rotational correction. These are still in development; however, the early results demonstrate proof of concept [25].

Deformities, however, are not always two-dimensional. Adolescent Blount's disease, for example, includes angular deformities in two planes, a torsional deformity, and a limb length difference. A simple angular correction in the coronal plane would not be sufficient to correct the entirety of the deformity. Therefore, circular frames, which were designed and first used by Dr. Gavriil Ilizarov, were particularly useful in addressing this problem. The challenge was that the frames were cumbersome to apply, the instructions for gradual correction were difficult for the patients to understand, and there was a high risk for pin site complications as these frames had to stay on for many weeks. Over the past two decades, the use of the original Ilizarov frames has been widely replaced with hexapod frames, initially developed by Dr. Charles Taylor. These frames allowed for nine degrees of freedom and

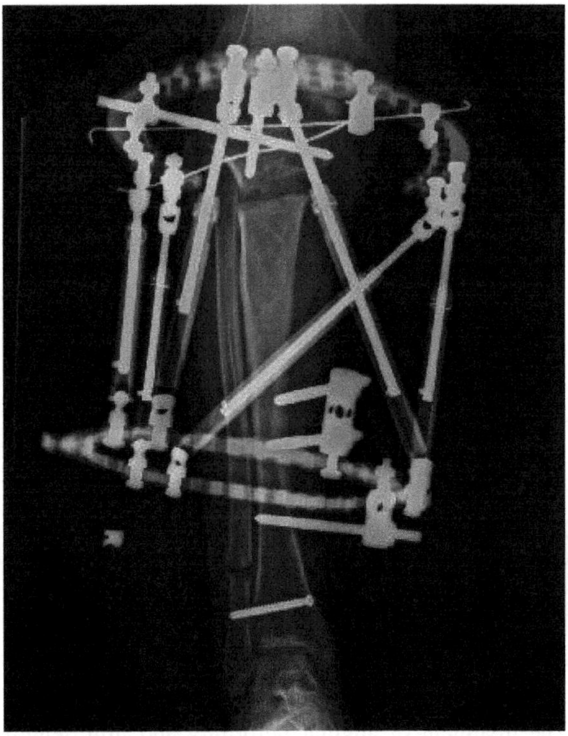

Fig. 6 AP radiograph of the tibia following tibia and fibula osteotomies and application of a Taylor Spatial Frame (Courtesy of NYU Langone Health)

revolutionized the correction of complex congenital or post-traumatic/post-infectious deformities, nonunions, and limb length differences (Fig. 6). The frames available today come with personalized software-generated treatment plans that can be modified as needed during the treatment phase. Newer frames have motorized struts that allow for automated correction of the deformity [26].

More recently, lengthening of long bones can be achieved using magnetic intramedullary devices. This obviates the need for a bulky frame, removes the risk of pin site infection, and simplifies the lengthening process for patients [27] (Fig. 7).

Proximal and distal femur anatomic locking plates with different sizes and offsets to accommodate for infants, young children, and adolescents have simplified and standardized the surgical approaches to deformity correction around the hip and distal femur.

Casting for Clubfoot or Fractures

The Ponseti method for treating structural clubfoot was revolutionary in that not only were treatment outcomes improved, but management with serial casting was vastly superior to surgical reconstruction [28]. Lastly, while immobilization with

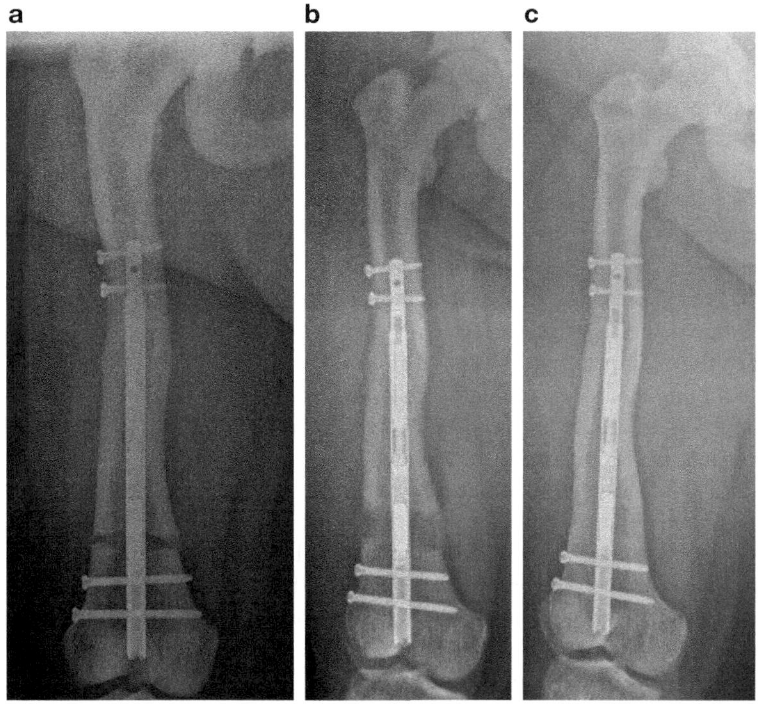

Fig. 7 (**a**) AP radiograph of the femur following femoral osteotomy and insertion of a magnetic lengthening nail to correct a limb length difference. (**b**) AP radiograph of the femur following lengthening. Note the regenerate bone formed at the lengthening site. (**c**) AP radiograph following consolidation of the regenerate (Courtesy of NYU Langone Health)

Plaster of Paris or fiberglass casts has remained integral to fracture management for decades, pioneering startups such as Cast21 and ActivArmor have made headlines for incorporating patient-specific designs, improved comfort and hygiene, and new technologies like 3D printing and waterproof isocyanate to improve patient satisfaction and outcomes with casting [29, 30].

Reconstruction Following Tumor Resection

Bone tumors can be divided into benign and malignant. The vast majority of bone tumors in children are benign; however, primary bone sarcomas do occur and their local control includes wide surgical resection. The most common locations include the distal femur, proximal tibia, and proximal humerus. The challenges of reconstruction following the resection of malignant bone tumors in the extremities include the sacrifice of the physis, which impacts the growth of the limb, the small canal of the residual bone, which would be needed to accommodate the stem of a prosthesis, and difficulty having younger children follow postoperative restrictions, which may result in early failure or wear of the implant. Extendible implants have been developed over the years [31]. The earlier versions required a small open procedure to allow for lengthening, though the newer versions include a coil, which can be turned using an external magnet. These implants were very promising for patients who had an expected limb length difference of 5–8 cm, since small differences could be treated with cessation of growth on the contralateral side or even a shoe lift. However, the medium- and long-term results showed that these implants were fraught with complications, and few remain on the market. One implant had a 100% failure rate at 10 years, with failure defined as a need for revision of the implant [32]. This is in contrast to the standard endoprosthetic implants, which have a 10-year survival of approximately 66% [33].

When the anticipated limb length difference is very large and there are no good options for biologic reconstruction with structural allograft or autograft materials, a rotationplasty procedure has been advocated since it allows for a functional below-knee amputation as opposed to an above-knee amputation or hip disarticulation procedure [34]. Despite this procedure having a very high success rate, and excellent functional outcomes, it can be difficult for the parents of these children to accept the idea and the appearance of the limb following the surgery. It is important to introduce parents to the procedure early on following the diagnosis and to meet regularly with them throughout the duration of neoadjuvant chemotherapy to answer questions and potentially allow them to meet other patients who have had this procedure done in order for it to be more acceptable. Alternate options, which are not as popular in North America, include radiation and/or cryotherapy of the resected specimen, and reimplantation into the patient [35]. Of course, this is a temporary option since the implanted autograft is essentially a dead bone with no growth potential.

Newer extendible implants that address the challenges of the older ones may preclude the need for this procedure, by allowing for a stable, custom-made implant that is extendible, durable, and which can potentially be exchanged for a standard implant at the end of growth.

Bone Void Fillers

When benign bone tumors in children develop a pathologic fracture or are large enough that patients experience pain from an impending fracture, they can be treated by intralesional curettage and grafting. The curettage portion of the procedure is to remove the tumor tissue and the grafting portion is to create both a void filler and to provide a scaffold into which the patient's native bone will grow. As such, these fillers need to be osteoconductive, which means that bone is able to grow onto them [36]. An example of this includes cancellous allograft or synthetic grafts, which include calcium sulfate, calcium phosphate, and silicas (Fig. 8). Adding demineralized bone matrix may allow for the material to become osteoinductive, which means that it recruits the cells needed to build bone. The ideal graft material would be

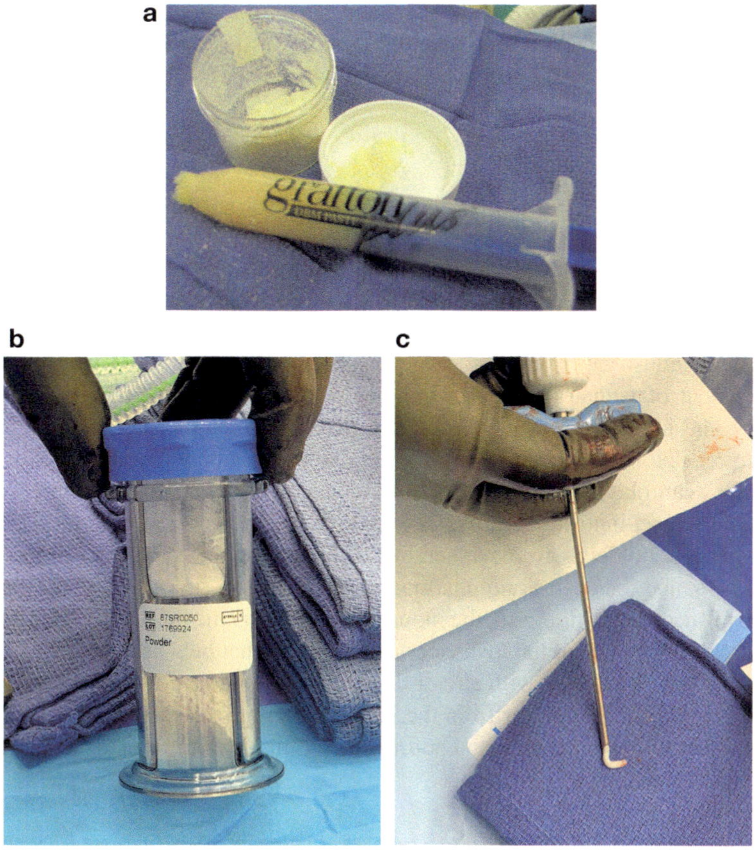

Fig. 8 (**a**) Clinical image of cancellous bone graft material and demineralized bone matrix paste. (**b**) Clinical imaging of synthetic calcium sulfate–calcium phosphate graft as a powder prior to mixing and (**c**) clinical image of the same graft as an injectable paste after mixing (Courtesy of NYU Langone Health)

structurally strong, obviating the need for the use of metallic implants to support the bone as it heals, would be both osteoconductive and osteoinductive, and would inhibit rather than stimulate tumor growth. This last point is particularly important since many of the benign, yet aggressive, bone tumors in children have a high recurrence rate, frequently exceeding 20%.

Digital Technology

With the exponential growth in digital technologies such as artificial intelligence (AI), augmented/virtual reality (AR/VR), and 3D printing, the explosive growth in digital health has also spread to pediatric orthopedics. This includes applications in

AI-assisted diagnosis of developmental dysplasia of the hips, prediction models for contralateral slipped capital femoral epiphysis, VR-assisted cast-removal to reduce anxiety in patients, and 3D printing of patient-specific casts [37].

Conclusion

Although providing financial support and reducing regulatory barriers are beneficial to incentivizing the industry to support innovation in pediatric orthopedics, it is unlikely sufficient to catch up with the rest of the thriving orthopedic implant and innovation industry. Ultimately, clinicians need to advocate and champion the development of pediatric-specific solutions for children. By using their voice to improve the landscape via increasing compensation, revising complex and challenging regulatory pathways, and improving industry collaborations, surgeons and physicians can play their part in driving innovation for children's musculoskeletal health. Likewise, innovators looking to enter the pediatric orthopedic market must align the stakeholders—from vulnerable children and their parents, surgeons, to researchers, and the device and pharmaceutical industry—in order to have the best chance of success. By doing so, the industry can develop the most appropriate, practical, user-friendly, and reliable devices and implants that help provide the best possible outcomes in patients with any of the varied orthopedic conditions that impact pediatric patients. Though far from the lowest hanging fruit, untapped opportunity awaits those able to anticipate and address the complex innovation environment in pediatric orthopedics.

Competing Interests The authors have no conflicts of interest to declare that are relevant to the content of this chapter.

References

1. Prasad K, Bazaka O, Chua M, Rochford M, Fedrick L, Spoor J, et al. Metallic Biomaterials: Current Challenges and Opportunities. Materials (Basel). 2017;10(8).
2. Newsom SW. Pioneers in infection control-Joseph Lister. J Hosp Infect. 2003;55(4):246–53.
3. Zimmer M. History of anaesthesia: early forms of local anaesthesia. Eur J Anaesthesiol. 2014;31(1):1–12.
4. Grote C, Dubin J, Otsuka N. Mind the Gap: Innovation in Pediatric Orthopaedics Is Lagging: AAOS Now; 2023 [12/20/2023]. Available from: https://www.aaos.org/aaosnow/2023/dec/clinical/clinical01/.
5. Espinoza J, Shah P, Nagendra G, Bar-Cohen Y, Richmond F. Pediatric Medical Device Development and Regulation: Current State, Barriers, and Opportunities. Pediatrics. 2022;149(5).
6. Sheha ED, Hammouri Q, Snyder BD, Campbell RM, Jr., Vitale MG, Devices PSTFfPM. Off-label use of pediatric orthopaedic devices: important issues for the future. J Bone Joint Surg Am. 2014;96(3):e21.

7. About the Study – COMET. Available from: https://www.aaos.org/aaosnow/2023/dec/clinical/clinical01/.
8. Barbella M. Pediatric Orthopedics Need a Minor Adjustment." Orthopedic Design & Technology. 2021. Available from: https://www.odtmag.com/issues/2021-05-01/view_features/pediatric-orthopedics-need-a-minor-adjustment/.
9. Larson AN. Innovation With Ethics in Pediatric Orthopaedics. J Pediatr Orthop. 2022;42(Suppl 1):S39–S43.
10. Espinoza JC. The Scarcity of Approved Pediatric High-Risk Medical Devices. JAMA Netw Open. 2021;4(6):e2112760.
11. Pathak K, Narang C, Hwang TJ, Espinoza JC, Bourgeois FT. High-risk Therapeutic Devices Approved by the US Food and Drug Administration for Use in Children and Adolescents From 2016 to 2021. JAMA Pediatr. 2023;177(1):98–100.
12. Gartsman GM, Brinker MR, Khan M, Karahan M. Self-assessment of general health status in patients with five common shoulder conditions. J Shoulder Elbow Surg. 1998;7(3):228–37.
13. Bashir AZ, Dinkel DM, Pipinos, II, Johanning JM, Myers SA. Patient Compliance With Wearing Lower Limb Assistive Devices: A Scoping Review. J Manipulative Physiol Ther. 2022;45(2):114–26.
14. Mohan AL, Das K. History of surgery for the correction of spinal deformity. Neurosurg Focus. 2003;14(1):e1.
15. Hadra BE. The classic: Wiring of the vertebrae as a means of immobilization in fracture and Potts' disease. Berthold E. Hadra. Med Times and Register, Vol22, May 23, 1891. Clin Orthop Relat Res. 1975(112):4–8.
16. Hibbs RA. A report of fifty-nine cases of scoliosis treated by the fusion operation. By Russell A. Hibbs, 1924. Clin Orthop Relat Res. 1988(229):4–19.
17. Lange F. Support for the spondylitic spine by means of buried steel bars, attaced to the vertebrae. Journal of Bone and Joint Surgery. 1910;2(2):344–61.
18. Hall JE. Spinal surgery before and after Paul Harrington. Spine (Phila Pa 1976). 1998;23(12):1356–61.
19. Harrington PR. Treatment of scoliosis. Correction and internal fixation by spine instrumentation. J Bone Joint Surg Am. 1962;44-A:591–610.
20. Cotrel Y, Dubousset J. [A new technic for segmental spinal osteosynthesis using the posterior approach]. Rev Chir Orthop Reparatrice Appar Mot. 1984;70(6):489–94.
21. Tsirikos AI, Roberts SB. Magnetic Controlled Growth Rods in the Treatment of Scoliosis: Safety, Efficacy and Patient Selection. Med Devices (Auckl). 2020;13:75–85.
22. Waldhausen JH, Redding GJ, Song KM. Vertical expandable prosthetic titanium rib for thoracic insufficiency syndrome: a new method to treat an old problem. J Pediatr Surg. 2007;42(1):76–80.
23. Wilson JP, Jr., Fontenot L, Stewart C, Kumbhare D, Guthikonda B, Hoang S. Image-Guided Navigation in Spine Surgery: From Historical Developments to Future Perspectives. J Clin Med. 2024;13(7).
24. Keny S. Trends in Guided Growth for Correction of Angular Deformities around the Knee: Past, Present, and Future. Journal of Clinical Orthopaedics. 2023;8(2):57–61.
25. Paley D, Shannon C. Rotational Guided Growth: A Preliminary Study of Its Use in Children. Children (Basel). 2022;10(1).
26. Paloski M, Taylor BC, Iobst C, Pugh KJ. Pediatric and adolescent applications of the Taylor Spatial Frame. Orthopedics. 2012;35(6):518–27.
27. Bafor A, Iobst CA. What's New in Limb Lengthening and Deformity Correction. J Bone Joint Surg Am. 2022;104(16):1419–25.
28. Rastogi A, Agarwal A. Long-term outcomes of the Ponseti method for treatment of clubfoot: a systematic review. Int Orthop. 2021;45(10):2599–608.
29. Smith J, DePhillipo, N, David, S, Nicolay, K, Wentz, S, Steckler, A, Westberg C. 3D-printed short arm casts: reliability, validity, feasibility compared with conventional waterproof fiberglass casts. Journal of 3D Printing in Medicine. 2022;6(4):175–85.

30. Stefanescu K, Timlin CL, Moy AS, Zapotoczny G. Reduced Isocyanate Release Using a Waterproof, Resin-Based Cast Alternative Relative to Fiberglass Casts. Toxics. 2023;11(12).
31. Schinhan M, Tiefenboeck T, Funovics P, Sevelda F, Kotz R, Windhager R. Extendible Prostheses for Children After Resection of Primary Malignant Bone Tumor: Twenty-seven Years of Experience. J Bone Joint Surg Am. 2015;97(19):1585–91.
32. Masrouha K, Abboud M, Saab R, Muwakkit SA, Khoury N, Haidar R, et al. Long-term follow-up of children treated with the Repiphysis expandable prosthesis for lower extremity bone sarcoma. J Pediatr Orthop B. 2022;31(2):e258–e63.
33. Moses Li ML, Wong KC, Chiu WK, Kumta SM. Intermediate-term results and risk factors analysis of tumor endoprosthesis in paediatric patients after the resection of lower extremity bone sarcoma. J Orthop Surg (Hong Kong). 2022;30(3):10225536221132403.
34. Sakkers R, van Wijk I. Amputation and rotationplasty in children with limb deficiencies: current concepts. J Child Orthop. 2016;10(6):619–26.
35. Hamed Kassem Abdelaal A, Yamamoto N, Hayashi K, Takeuchi A, Miwa S, Tsuchiya H. Epiphyseal Sparing and Reconstruction by Frozen Bone Autograft after Malignant Bone Tumor Resection in Children. Sarcoma. 2015;2015:892141.
36. Gross RH. The use of bone grafts and bone graft substitutes in pediatric orthopaedics: an overview. J Pediatr Orthop. 2012;32(1):100–5.
37. Luo S, Deng L, Chen Y, Zhou W, Canavese F, Li L. Revolutionizing pediatric orthopedics: GPT-4, a groundbreaking innovation or just a fleeting trend? Int J Surg. 2023;109(11):3694–7.

Interface of Engineering with Autism and Neurodiversity

Wendy J. Ross

Introduction

Without realizing it, the first form of engineering most people remember using is the built-in sharpener on the box of 64 Crayola crayons. Without it, they could not have continued to color once the wax hit the paper. That simple piece of engineering made children more independent with coloring. And guess what? It benefited adults, too. This brings home the point (no pun intended) that acts of engineering developed for some tend to benefit everyone. Independence is contingent on altering elements in the surroundings, rather than the people. Therefore, the acts of engineering need not just focus on the hand holding the crayon but the accessibility of the crayon itself. Society benefits from multiple types of minds and perspectives. Society also derives an advantage from biomedical engineering designed to support those minds in accessing their worlds.

Engineering is "the branch of science and technology concerned with the design, building, and use of engines, machines, and structures…the work done by, or the occupation of, an engineer." It is also "the action of working artfully to bring something about." [1] Engineering can be something as complicated as a car or as simple as a walking stick, or a built-in crayon sharpener. Biomedical engineering unites engineering design and biological science with the aim of developing solutions for people to meet their healthcare goals. A critical component of biomedical engineering is patient-centered design. Specifically, for children, having a pediatric lens improves the quality of life both in the present, and in the future, because it lays a foundation which has implications for long-term health and well-being.

Over the years, biomedical engineering has addressed the needs of many. In the disability community, with the advent of tools like wheelchairs, hearing aids,

W. J. Ross (✉)
Jefferson Center for Autism and Neurodiversity, Jefferson Health, Philadelphia, PA, USA
e-mail: Wendy.Ross@Jefferson.edu

J. L. Wells (ed.), *A Clinical Lens on Pediatric Engineering*,
https://doi.org/10.1007/978-3-031-84655-7_13

cochlear implants, glasses, and more, the field has provided access to those with ambulatory, visual, and hearing deficits. Physical disabilities were initially at the forefront of the disability movement because they present externally, unlike developmental or neurological differences like autism, dyslexia, and ADHD. Those with minds that think or process their worlds differently are often invisible, making them harder to identify and to serve. The Americans with Disabilities Act (ADA) does protect those with invisible disabilities, but there are fewer specific guidelines. Engineering plays an expanding role in improving the lives of those with neurodiversity and will contribute to clearer standards for future policies for supporting this population.

About Dyslexia

Dyslexia is a disability in the acquisition of reading skills. It affects 20% of the population and represents 80–90% of those with learning disabilities [2–4]. It was first reported in the 1800s with a description of a child with average intelligence, hearing, and vision, who was found to struggle with reading despite the ability to do higher math problems [5, 6]. Children with dyslexia struggle primarily with learning to read, further impacting their lives by comparatively reducing their exposure to vocabulary and ideas expressed in books. Furthermore, when children do not develop reading skills as expected, they tend to compensate with difficult behaviors, compounding the false theory that they are not trying to learn to read.

A better understanding of dyslexia occurred after cases of individuals who lost their ability to read after a traumatic brain injury were identified. These injuries indicated that there was something in brain structure influencing the ability to read. This brain-based etiology led to the use of biomedical tools to look inside the brain in order to understand developmental dyslexia better.

Dyslexia Engineering and Research

In 1987, the National Institute of Child Health and Human Development, a branch of the National Institutes of Health (NIH), established centers to investigate dyslexia. They looked both at the profiles of children with dyslexia and the structures of their brains. When magnetic resonance imaging (MRI) was created, magnets and radio waves were used to see details of the brain anatomy that may influence activity. An MRI can show structure but not function. Positron emission tomography (PET) scans can show function by injecting a radioactive compound into the bloodstream to measure blood flow to different parts of the brain with the assumption that increased blood flow is an indicator of increased activity. Drawbacks to PET scans include the complexity of the equipment to make the radioactive compound and the invasiveness of the injection. A better option became the functional MRI (fMRI), which allows scientists to observe, non-invasively, how the brain works. The fMRI, like the PET scan, operates under the assumption that cognitive efforts like reading require increased metabolic demands and, therefore, increased blood flow. Increased blood flow can be illustrated by the increased magnetic elements of the blood, seen by fMRI. The elevated level of oxygen in this blood has a stronger magnetic signal, which is picked up by the fMRI.

Functional MRI is more patient-friendly than PET scans; however, MRIs can also be challenging for young children as they require the child to stay still. Some children even require anesthesia for the completion of this test. In the fMRI, the patient needs to be awake so they can be asked to look at a periscope to see tasks outside of the scanner. The ability to rhyme is a critical component of reading. An individual might be shown two different series of letters and asked if they rhyme. Dyslexic brains show different functional structural images than non-dyslexic brains on this task. Essentially, the fMRI creates a map of what happens in the brain when it reads. Mapping the neural pathways in good readers illuminates the different brain activation in dyslexic readers. Although the fMRI is used primarily in research, the findings impact intervention by demonstrating that adjusting teaching instruction and amount of repetition can improve blood flow to the parts of the brain responsible for reading. In other words, fMRI can identify the differences in dyslexic brains at baseline and, furthermore, evaluate the effects of specific interventions for instruction in reading. This put to rest the idea that children with difficulty reading were just being difficult [2].

Diagnosis of Dyslexia

Although functional MRIs are used in research, they have not taken on a role in the practical diagnosis of dyslexia. Psychologists diagnose dyslexia through a series of tests looking at cognition and performance in tasks related to reading.

Interventions for Dyslexia

The 1993 National Council on Disability's report to the President and Congress of the United States estimated that up to 75% of children with learning disabilities could remain in typical class if supplied with the appropriate assistive technology [7]. It also estimated that appropriate assistive technology could lower the need for other school-related services.

According to the federal special education law (Individuals with Disabilities Education Act (IDEA), P.L. 101-476), public schools are obligated to provide a free and appropriate public education in the least restrictive environment to all learners. Once a learning disability is identified, an individualized educational plan is written with goals and benchmarks, specially designed instruction and accommodations for each student. Part of this process can include a SETT evaluation. SETT stands for Student, Environment, Tasks, and Tools that facilitate participation in the curriculum. Often, these are technological products of engineering.

Technology benefits typical learners as well. During the pandemic, when school buildings were closed, all students needed equipment like laptops or iPads and internet services in order to access their education. Collateral technology like headphones was also needed to enable these students to attend their online learning in a home where other family members might have been simultaneously engaged in remote learning or employment.

There are a myriad of technologies used to promote learning. They can include everything from mainstream calculators and computers to augmentative assistive communication devices for those who cannot speak (further addressed in the autism section). Many technology-based reading programs exist for children with dyslexia. Audiobooks can expose students to literature without them having to read words, filling the gap in vocabulary and ideas that often develop in nonreaders. Technology that links the text to the audio can cement reading skills by combining the auditory component with highlighted print [7–9].

Programs to promote reading skills can provide text, record the student reading, and reward the reader with video prizes. Reading is comprised of both decoding, the literal steps of sounding out the words, and comprehension, understanding the text. Electronic programs can also promote reading comprehension with supplemental questions and visual cues [10].

Individuals with dyslexia frequently also have difficulty writing. Recording devices can reduce the need to take notes in classes. Dictation software additionally supports these individuals and has become mainstream even in adulthood. Medical professionals, for example, frequently dictate their patient notes, which can be faster than typing them into the electronic medical record. Artificial intelligence (AI) can enhance a student's ability to generate written assignments; however, it provokes ethical questions as to ownership of the work product.

About ADHD

Attention-deficit/hyperactivity disorder (ADHD) is described by the Diagnostic and Statistical Manual of Mental Disorders (DSM-5), the most comprehensive resource to inform clinicians and researchers about mental health challenges, as a persistent pattern of inattention and/or hyperactivity-impulsivity that interferes with functioning and learning [11]. Inattention refers to having difficulty sustaining focus and attending to a desired activity, topic, or situation. To meet the criteria, the person should have normal sensory systems like hearing and vision and no medical reason for inattention, like seizures or thyroid disorders. There should be no other primary mental health conditions like anxiety to account for inattention and distractibility. Other psychosocial contributors or distracting life situations should also be ruled out at the outset. Lack of attention should not be linked to defiance or a deficit in comprehension. Of note, individuals with ADHD can hyperfocus on preferred tasks [12].

Organization can be a challenge in ADHD. Steps of executive functioning like gathering materials necessary for a task, allowing enough time for it, beginning it, and taking it to completion, can all be negatively impacted.

Hyperactivity describes a person who is moving around constantly, especially in situations when it is not appropriate. Fidgeting, tapping and excessive talking are all symptoms.

Impulsivity describes a person who may act without thinking or struggle with self-control. An impulsive person may interrupt others or make important decisions without considering long-term consequences. They can struggle with delaying gratification.

ADHD can be primarily inattentive, primarily hyperactive, or combined.

ADHD Engineering and Research

Neuroimaging is an important part of research done on ADHD. Magnetic resonance imaging (MRI) and diffusion tensor imaging studies have shown differences in volume, surface-based measures (cortical thickness, surface area, and gyrification) and white matter integrity in different cerebral regions in children and adults with ADHD compared to population norms [13, 14]. Differences in the basal ganglia, prefrontal structures, and the corpus callosum have been the most consistently reported findings. The variation in brain structure supports the concept that ADHD is grounded in brain-based differences as opposed to being a willful pattern of behavior. Hemodynamic (functional MRI, functional near-infrared spectroscopy, positron emission tomography, single-photon emission computed tomography) and magnetoencephalography measurements have also shown differences in neural activity during the execution of neuropsychological tasks and during rest in widespread regions of the brain. Importantly, multimodal studies combining structural and functional methods have shown an intercorrelation between structural and functional abnormalities in ADHD. Differences in neuroimaging have clarified the etiology of ADHD, decreasing negative perceptions of affected individuals [15].

ADHD Diagnosis

There are no biomedical tools that diagnose ADHD independent of clinician input. There are some tools that are implemented as contributors to professional opinion. Assessments like the Test of Variables of Attention (T.O.V.A.) may be helpful in giving the diagnosing clinician additional information, but they should not be used in isolation to diagnose ADHD. The T.O.V.A. calculates response time variability (consistency), response time (speed), commissions (impulsivity), and omissions (focus and vigilance). These calculations are then compared to a large age- and gender-matched normative sample, as well as to a sample population of individuals independently diagnosed with ADHD [16].

ADHD Intervention

Medical treatment for ADHD includes medications like stimulants. Behavioral therapy is a first-line intervention in younger patients. Biomedical interventions can include tools like neurofeedback. The concept of neurofeedback for ADHD is based on evidence that some people with ADHD show differences in brain activity, especially in the frontal lobe. This part of the brain affects concentration and organization. When used for ADHD, a device reads a child's brain wave activity and sends feedback via a game-like challenge. Supporters claim this process can train the brain to behave differently and reduce ADHD symptoms. Over time, the hypothesis is that having a child adjust his brain waves through practice and repetition of the neurofeedback game will help him concentrate for longer periods of time.

During neurofeedback, electrodes are placed on a child's head or in a headband to monitor his brain activity. The child watches a video and listens for a prompt to adjust something in it. He focuses his mind on the task, which sends a message through the electrodes on his scalp to the computer. This occurs without a mouse, joystick or keyboard. As the computer senses electrical activity in the child's brain, it sends him feedback by making the adjustment. This feedback helps the child figure out how to change the way he is paying attention. Evidence for the effectiveness of neurofeedback is inconclusive.

One of the core disruptors of ADHD includes executive functioning deficits. Educational interventions include programming to enhance these skills. Time management is often a high-priority skill that can be enhanced by posting schedules, agendas, and graphic organizers. Tools that support this skill exist on Google Classroom integrated with Google Calendar. By organizing the workflow of assignments and providing students with access to a schedule created in Google Calendar, teachers can help keep students on track and organized.

Many schools have online learning management systems, which can promote teacher-student collaboration and communication, help parents follow child progress, provide quizzes and tests, and provide mobile access to promote convenience.

Electronic educational platforms can help students summarize literature, organize essays, and proofread. Some applications specifically target overall executive functioning skills. Some children with ADHD have trouble attending during class. Tools like Smart Pens can help bridge gaps by recording a class, and linking audio recordings to notes. They replay the audio when notes are tapped. They save, organize, and play back the notes on laptops. The smart pens also enable the sharing of notes and audio recordings.

As mentioned earlier, the advent of artificial intelligence (AI) has become controversial in the educational space. Students can ask AI to complete assignments for them. Some consider it a reasonable tool while others see it as a path for cheating. AI can also provide erroneous information. Over time, the impact of AI as a friend or foe of education will further unfold.

About Autism

Autism is a neurodevelopmental condition defined by differences in three areas: (1) social skills, (2) communication, and (3) behavior. The prevalence of autism has been rising over the past two decades. Viewpoints of autism have also changed considerably during this time. While for some, autism is a true disorder or disability, for others it is considered a difference or a neurodiversity [17–19].

Autism presents as a range. Some with autism have no form of communication or social skills and engage in sensory-seeking or avoidant behaviors that can interfere with their ability to interact with the world around them. On the other end of the spectrum, autistic individuals might have language but struggle with applying it socially. Even when socially interested, social cues are missed. They may exhibit unexpected behaviors or focus on a specific area of interest. This portion of the chapter will describe various levels of autism and then use these descriptions to help explain the intersection of autism with engineering.

Autism Components

Sensory and Behavioral

The sensory systems of autistic individuals can be under or over-sensitive to their environments, or even both, depending on the stimuli. For example, some might be overly sensitive to noises, even while tending to make loud noises themselves. Textures, colors, and temperatures of food can be avoided, sought, or simply restricted. Clothing textures can also be challenging. Identifying body position in the spatial environment can be a struggle with the seeking of movement or confined spaces to feel secure. Temple Grandin, a well-known author and renowned author on autism and animal behavior, channeled her lived experience of seeking tight spaces into the engineering of a squeeze machine for herself, a device that compressed her body and calmed her. She used the concept of this personal device to develop humane strategies for channeling animals in the meat industry. Subsequently, weighted and/or compression garments have impacted the fashion of children with autism as a way of helping to improve their sensory experience and leverage it to increase their ability to self-regulate and engage in the world around them [20].

Some sensory-seeking, repetitive, or self-stimulatory behaviors exhibited by children with autism are used to help calm or activate themselves. This can present as repetitive movements or seeking either tight or wide open spaces. Spinning, jumping, flapping, flicking light switches, and opening and closing doors are examples of repetitive behaviors. Avoidant behaviors can occur as well and some children with autism might have a tendency to elope. Unexpected behaviors can present as atypical play where a child might not use a toy functionally as it was designed; but, instead, explore it from different angles or take it apart in an effort to figure out how it works. They might exhibit atypical visual regard, looking at items from different angles. Perseverative behaviors can also present as a specific interest upon which they might have a tendency to monologue [21].

Communication

Language is composed of parts. One is receptive language, or the understanding of language. Another is expressive language: how language is used to get needs met or describe interest in something. Expressive language is further divided into a subcategory of the ability to make speech or sounds orally. Furthermore, expressive language can involve nonverbal interactions. Some messages are not delivered through spoken words but use nonverbal strategies like the use of eye gaze or gestures like pointing.

Communication in autistic children is assessed by examining these various components. Some children with autism have stronger receptive skills than expressive skills. They understand more than they can speak. Often the intelligence of these children is underestimated, as adults tend to assume that when people do not speak, they do not understand. Some autistic children have better expressive than receptive language. They may have memorized or simply repeated language, even scripting from television shows they have seen. Because they have spoken language, it is often assumed that they understand more than they might truly comprehend. An example of this is when children echo or repeat the last thing they have heard or repeat an entire phrase by rote. Some children may use memorized language but apply it functionally. For example, they may have a scripted phrase that they use appropriately to get a need met even if they have not independently spontaneously combined the words. Language gets more complicated when we consider that while those who can speak may be able to use spoken language to get their needs met or to convey interest, they may not be able to use the language socially. These might be individuals who use language to monologue or talk about areas of interest but struggle to have a reciprocal conversation.

Communication is a critical element of autism. When children cannot communicate their needs, it can lead to anxiety and negative behaviors secondary to frustration. When children cannot speak or communicate, their behavior becomes their form of communication.

Social Skills

Social skills can range from seeming to be dissociated from surrounding people to being simply not able to read a social cue or situation. Social skills also involve a degree of being able to take the perspective of others. Reciprocal play and conversations are different than expected in autistic children.

Autism and Research

One of the connections found between autism and engineering involves the phenotype of an individual drawn to science-based cognitive pursuits and employment. Data indicated that fathers and grandfathers of children with autism were found more than twice as often in the field of engineering, compared with fathers and grandfathers of other children [22, 23].

While some on the autism spectrum consider it a variation, others have severe features which can be disabling. As the numbers of those impacted with autism have increased, a better understanding of its cause has been a focus of many researchers. Understanding the genetic subtypes of autism can potentially inform prognosis, medical management, and assessment of familial recurrence risk. Some genetic profiles have already been identified. Other efforts to elucidate the genetic underpinnings of autism include sequencing the entire genome. It is also critical, then, to match the genotypes to the phenotypes in large samples, necessitating the creation of a methodology for storing and sharing "big data" securely while allowing for in-depth analysis by multiple users around the globe [24].

Structural magnetic resonance imaging (MRI) is another methodology that has been used to elucidate the etiology of autism. Patients with Autism Spectrum Disorder (ASD), especially under the age of 6 years, were observed to have increased whole brain volume. While other findings can be inconsistent, consistent findings differentiating autistic brains included increased volume in the frontal and temporal lobes, increased cortical thickness in the frontal lobe, increased surface area and cortical gyrification, and increased cerebrospinal fluid volume, as well as reduced cerebellum volume and reduced corpus callosum volume. Brain structures correlate to behavioral manifestations. For example, increased frontal lobes can contribute to social and executive functioning deficits. The corpus callosum is a bundle of nerves that connects the left and right sides of the brain. If it is missing or partially missing, the sides do not communicate well, resulting in a range of challenges from deficits in intelligence to autism or even seizures. Although, at this time, this technology has not been used clinically as a diagnostic tool or to guide therapy, it may one day provide support in these endeavors [25, 26].

Functional MRI uses specialized pulse sequences to localize metabolic correlates of neural activity linked to relevant neurocognitive processes. Unlike positron emission tomography (PET) and single-photon emission computed tomography

(SPECT), fMRI does not rely on radiotracers and is noninvasive. fMRI has provided insights into the pathophysiological mechanisms of autism by showing dysfunctional activation in key areas of verbal and non-verbal communication, social interaction, and executive function. Understanding the function of the brain can help provide areas for intervention [27].

While the structural and functional roles of components of the brain implicated in autism are beginning to be clarified, engineering has begun to provide insight into varying phenotypes of autism by providing methodologies for operationalizing measurement of one of the features of autism, atypical social-emotional reciprocity. It has been hard to recreate social situations as early experiments used computer modeling, which does not capture the true bidirectional nature of social interactions in natural settings. Recently, an advance was made in this arena when an automated facial analysis system was used to examine the interpersonal coordination of positive affect between children with ASD and others [28]. This novel method of describing emotional reciprocity has research and future clinical applications.

Machine-based learning is starting to be used to identify those with autism in medical records. This can help identify comorbidities, effective interventions, and also social factors that impact outcomes.

Autism Diagnosis

Autism diagnosis involves an examination of social skills, language, and behavior in children. The standard of pediatric care is that pediatric primary care providers do a Modified Checklist for Autism in Toddlers (MCHAT) at 18 and 24 months to screen patients and identify those in need of further diagnostic referral [29]. Specialists like developmental pediatricians, neurologists, or psychiatrists tend to give a definitive diagnosis using tools like the Autism Diagnostic Observation Schedule, which requires extensive training and often a prolonged period of time to implement and generate a report [30]. A shortage of providers and the length of time that full evaluations consume can be rate-limiting factors in diagnosis.

During the Covid-19 pandemic, telehealth assessments for autism gained popularity. While children with overt signs of autism could be identified this way, it is not clear that children with more subtle presentation could be identified. To assist in autism diagnosis, eye-tracking measurement of visual engagement was evaluated relative to expert clinical diagnosis at specialty autism assessment centers. In children aged 16–30 months, eye-tracking measurement of social visual engagement was predictive of expert clinical diagnosis. It was found to correlate with measures of social disability, verbal ability, and nonverbal cognitive ability. With delays in evaluation often lasting a year or more, this advanced technology has the capacity to enhance diagnosis and, thus, also impact the trajectory of the future by providing intervention earlier. Eye-tracking assessments were also found to demonstrate equitable measures regardless of race, gender, or language [31].

The importance of early childhood identification and intervention cannot be overstated. Minds have increased neuroplasticity at younger ages, making earlier interventions have increased potency. Neuroplasticity refers to the ability of the mind to rewire itself in response to stimuli [32]. The brains of children are still in a period of growth developmentally, enhancing their flexibility to reconfigure and, therefore, acquire abilities in different ways. As a child's brain is still actively developing, there is more neuroplasticity than in an adult brain, meaning that children are more able to learn novel skills and to adapt to new environments. Each progressive gain builds on the next, amplifying its effect over time. Moreover, learning and using new skills increases the possibility of automatizing them to a point where they become a foundation for further building, generalization and transference of skills [33].

Autism Interventions

Behavioral Support

Entire industries around creating devices and spaces to help optimize sensory experiences of children with autism have evolved. These can include anything from the engineering of clothing, blankets that incorporate varying levels of weight, play spaces, and sensory gyms. Lines of furniture that involve movement or aspects that shield the individual from their surroundings can be supportive. Other engineered measures involve portable items like noise canceling or reduction headphones to reduce ambient noises that can be overwhelming. It can also include handheld items like fidgets that can have various textures, or movement or visual components. Entire immersive experiences like Snoezelen rooms have also been developed (Fig. 1). These rooms have highly intense light-based and visually stimulating mobile aspects. In addition, they can record user preferences, identifying individuals and resetting to their preferences upon their entry to the room. Sensory-friendly spaces are critical. Some autistic children will elope from places that overwhelm them. If they do elope, technology like GPS tracking devices exist that can help lead to their safe return.

When individuals enter a facility, ways of measuring the use and efficacy of the spaces in supporting them have been accomplished with technology like heat mapping. Heat maps are graphical representations of information. For example, technology can be used to track the movement of individuals in a room and create a graph showing how many people gravitate to one space and for how long [34]. If autistic individuals cannot provide verbal feedback to their environments, devices to measure biometric impact can provide an understanding of the response. Of course, things like inter-heart rate variability, skin temperature changes, and alterations in blood pressure (BP) must be correlated to emotional states to make this truly clinically impactful. This is an area still in its infancy. A complicating factor in the use

Fig. 1 A Snoezelen room at Shrub Oak International School, Westchester, NY. Photo by Caitlin Sweetapple

of biometric measures is that many of them necessitate that the participant use a wearable like a monitor or wrist band or ring, which can trigger a confounding sensory experience. Newer technology involving the use of radio waves will look at non-wearable means of measuring biometric feedback. The potential safety implication of radio wave feedback will have to be weighed against any benefits.

Communication Intervention

Engineering can intercede on many levels in communication. When children do not have the motor components of speech, they might be able to use a low tech solution like physical cards to express their needs, picture exchange communication system (PECS) [35]. These cards might have icons or photographs of desired items. They may also simply have *yes* and *no* on them. Yes and no questions can be easier than having to generate an open-ended response. If children can access communication with the cards, they may eventually be able to use augmentative and alternative communication (AAC) devices (Fig. 2). An AAC is a tablet or laptop that helps someone with a speech or language impairment to communicate. These devices have the icons on a screen, or multiple screens. The child must be able to navigate these screens and touch the desired words or phrases, which triggers the machine to speak for them (Fig. 3). When children lack the fine motor skills to point to items, but still have the cognitive ability to express themselves, machines that use eye gaze

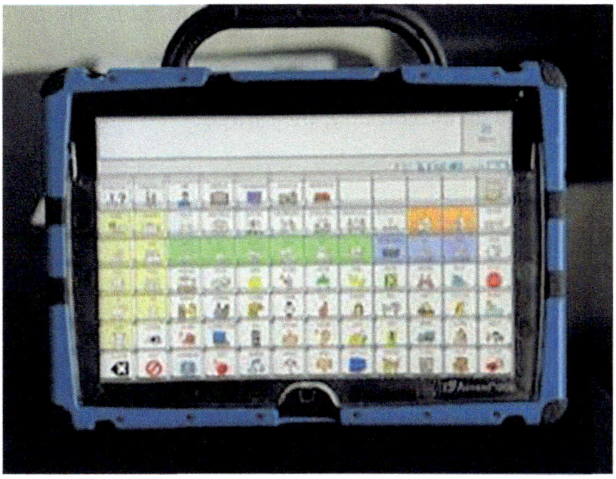

Fig. 2 An augmentative and alternative communication (AAC) device. Photo by Lindsay Carfley

Fig. 3 A child using an AAC device. Photo by Lindsay Carfley

instead of touch can be used for communication. Parents sometimes fear that the use of a device indicates that spoken language development will be aborted. Current research indicates that the use of a communication device can accelerate spoken language (unless there are true oral motor deficits) by more quickly assisting the child in understanding the value of communication and increasing their motivation to acquire this skill [36, 37].

Finally, when an individual with autism has the desire to communicate but simply cannot speak verbally, they might also use a keyboard to type responses.

Social Skills, Play, and Learning

Play is considered the work of children. It is not a frivolous pursuit but rather the way children experience and learn from the world around them. It enables them to be in a creative and safe space to practice new skills that enhance and integrate communication, social, fine and gross motor skills. Play is actually considered a human right by the United Nations. Frequently, autistic children do not engage with toys as expected or as their designers intended. This is not necessarily negative, as the experience of having more than one way of engaging in play can support overall development. New toys have been developed that can stimulate, guide, and increase play skills. Immersive play experiences involving virtual reality can help children explore new worlds, increase opportunities to explore vocational or skill-building tasks, and practice them in a safe space. Social learning toys and dolls that can help children learn and practice social skills have been an emerging area of popular play items [38].

Moving Toward Independence

Technology has facilitated the development of independent living, vocational, and employment skills. Individual educational plans always have a section for transition to adulthood and schools need to plan for this programming. Providing exposure to multiple types of employment through virtual reality has opened up possibilities that may have otherwise been missed in terms of finding a vocational interest for which an individual has an aptitude. It has also contributed to independent living strategies. Parents who had not initially imagined their child living independently are now starting to conceptualize a living situation where tools like Alexa and technological support can create opportunities not previously available. Self-driving cars might be a method of transportation safer than public transportation for some children as they transition into adulthood. The advent of cell phones, especially with accommodative apps and GPS, has helped autistic individuals communicate and be located. Apple has a selection of accessible apps for those with cognitive challenges that can also support organizational skills, reduce distraction on your phone, allow the individual to type instead of speak to communicate, and reduce background noise while streaming content.

Autism, Technology, and Health

Autistic individuals are more prone to seizures over the course of their lifetime. Some seizures can be subtle. The development of the electroencephalogram (EEG) can help determine the existence as well as the type and frequency of seizures.

Sleep studies can help clarify and, therefore, address sleeping needs. The importance of sleep is critical. It is not just a time of rest but a time where learning is consolidated in the brain. Sometimes children with autism demonstrate decreased sleep requirements while their caregivers continue to need to rest. Illuminating further the architecture of their sleep can individually help children as well as support the identification of strategies for the entire family system to rest [39].

Healthcare "passports" are tools developed to gather and efficiently convey information that can make a doctor's visit possible. They consist of a series of questions about patient preferences and information. For example, the person can let the provider know how they best communicate and other preferences, like needing to be told before a clinician touches or examines them. Knowing and accommodating these preferences can make the visit more feasible. Some of these healthcare "passports" are installed on apps that the individual holds on a device. Others are completed online and imported directly to the electronic medical record. At our hospital, one question asks what in their surroundings provides comfort. For instance, if a patient likes rivers, images of rivers can then be streamed on a smart board in the examination room to decrease anxiety while waiting for the visit [40–42].

In hospital settings, some of the examination chairs are able to check weights, decreasing the need for an additional transition point to a scale and then to the room. Ambient listening technology in the examination rooms is starting to be used to annotate the encounter so that the clinician does not have to stop to take notes along the way and can, therefore, focus more on the patient's comfort and expedite the visit.

Conclusion

Biomedical engineering has hugely impacted many aspects of life for children with neurodiversity like those with Dyslexia, ADHD, and Autism. Aspects of neurodiversity that underlie these conditions, such as communication deficits, sensory challenges, and differences in learning, can affect children with other diagnoses and impact those children considered neurotypical as well. Technological interventions for neurodiverse children often benefit everyone.

From enabling research that has elucidated the nature of these conditions to facilitating diagnosis and intervention, biomedical engineering has significantly improved the lives and futures of others. Specifically, these biomedical interventions have been designed to create more functional matches between individuals and their environments, targeting each as a nexus for change. In so many ways, the positive impact of biomedical engineering reaches far beyond its designated beneficiaries, and its influence is continuing to build.

We are all the beneficiaries of these brilliant advances and welcome their partnership.

Competing Interests The author has no conflicts of interest to declare that are relevant to the content of this chapter.

References

1. Oxford University Press. Engineering. n.d. 29 January 2024. <https://www.oxfordlearnersdictionaries.com/definition/english/engineering>.
2. Shaywitz, SE, JE Shaywitz and BA Shaywitz. "Dyslexia in the 21st century." Curr Opin Psychiatry 34.2 (2021): 80–86.
3. Shaywitz, SE, JR Gruen and BA Shaywitz. "Management of dyslexia, its rationale, and underlying neurobiology." Pediatr Clin North Am 54.3 (2007): 609–23.
4. Shaywitz, S. "Dyslexia." NEJM 338.5 (1998–01): 307–12.
5. Kirby, P. "Dyslexia debated, then and now: a historical perspective on the dyslexia debate." Oxf Rev Educ 46.4 (2020): 472–486.
6. Anderson, PL and R Meier-Hedde. "Early case reports of dyslexia in the United States and Europe." J Learn Disabil 34.1 (2001): 9–21.
7. National Council on Disability. Annual Report to the President and the Congress of the United States. Washington, DC, 1993.
8. Calculator, S and CM Jorgenson. "Integrating AAC instruction into regular education settings: Expounding on best practices." Augmentative and Alternative Communication 7 (1991): 204–214.
9. Allen, AA, et al. "The effectiveness of aided augmented input techniques for persons with developmental disabilities: a systematic review." Augment Altern Commun 33.3 (2017): 149–159.
10. Zabala, JS. Setting the Stage For Success: Building Success through Effective Selection and Use of Assistive Technology Systems. 2000. 16 April 2024. <https://www.ldonline.org/ld-topics/assistive-technology/setting-stage-success-building-success-through-effective-selection>.
11. American Psychiatric Association. Diagnostic and Statistical Manual of Mental Disorders, 5th edition. Arlington, VA: American Psychiatric Association, 2013.
12. Austerman, J. "ADHD and behavioral disorders: Assessment, management, and an update from DSM-5." Cleve Clin J Med 82 (2015).
13. Pereira-Sanchez, V and FX Castellanos. "Neuroimaging in attention-deficit/hyperactivity disorder." Curr Opin Psychiatry 34.2 (2021): 105–111.
14. Samea, F, et al. "Brain alterations in children/adolescents with ADHD revisited: A neuroimaging meta-analysis of 96 structural and functional studies." Neurosci Biobehav Rev 100 (2019): 1–8.
15. Albajara, SA, T Villemonteix and I Massat. "Structural and functional neuroimaging in attention-deficit/hyperactivity disorder." Dev Med Child Neurol 61.4 (2019): 399–405.
16. Forbes, GB. "Clinical utility of the Test of Variables of Attention (TOVA) in the diagnosis of attention-deficit/hyperactivity disorder." J Clin Psychol 54.4 (1998): 461–76.
17. Maenner, MJ, Z Warren and AR Williams. "Prevalence and Characteristics of Autism Spectrum Disorder Among Children Aged 8 Years — Autism and Developmental Disabilities Monitoring Network, 11 Sites, United States." MMWR Surveill Summ 72 (2023).
18. Rosen, NE, C Lord and FR Volkmar. "The Diagnosis of Autism: From Kanner to DSM-III to DSM-5 and Beyond." J Autism Dev Disord 51.12 (2021): 4253–4270.

19. Baskin, JH, M Sperber and BH Price. "Asperger syndrome revisited." Rev Neurol Dis 3.1 (2006): 1–7.
20. Grandin, T and M Scariano. Emergence: labeled autistic. Novato, CA: Arena Press, 1986.
21. Koegel, R, et al. "Using Perseverative Interests to Improve Interactions Between Adolescents with Autism and their Typical Peers in School Settings." J Posit Behav Interv 14.3 (2012): 133–141.
22. Baron-Cohen, S, et al. "Is There a Link between Engineering and Autism?" Autism 1.1 (1997): 101–109.
23. Baron-Cohen, S. "Autism and the technical mind: children of scientists and engineers may inherit genes that not only confer intellectual talents but also predispose them to autism." Sci Am 307.5 (2012): 72–5.
24. Yuen RKC et al. "Whole genome sequencing resource identifies 18 new candidate genes for autism spectrum disorder." Nat Neurosci 20.4 (2017): 602–611.
25. Pagnozzi, AM, et al. "A systematic review of structural MRI biomarkers in autism spectrum disorder: A machine learning perspective." Int J Dev Neurosci 71 (2018): 68–82.
26. Faizo, NL. "A narrative review of MRI changes correlated to signs and symptoms of autism." Medicine 101.34 (2022): e30059.
27. Dichter, GS. "Functional magnetic resonance imaging of autism spectrum disorders." Dialogues Clin Neurosci 14.3 (2012): 319–51.
28. Sariyanidi, E, et al. "Comparison of Human Experts and AI in Predicting Autism from Facial Behavior." CEUR Workshop Proc 3359 (2023): 48–57.
29. Wieckowski, AT, et al. "Sensitivity and Specificity of the Modified Checklist for Autism in Toddlers (Original and Revised): A Systematic Review and Meta-analysis." JAMA Pediatr 177.4 (2023): 373–383.
30. Randall, M, et al. "Diagnostic tests for autism spectrum disorder (ASD) in preschool children." Cochrane Database Syst Rev 7.7 (2018): CD009044.
31. Jones, W, et al. "Eye-Tracking-Based Measurement of Social Visual Engagement Compared With Expert Clinical Diagnosis of Autism." JAMA 330.9 (2023): 854–865.
32. Innocenti, GM. "Defining neuroplasticity." Handb Clin Neurol 184 (2022): 3–18.
33. ADA.gov. Introduction to the Americans with Disabilities Act. n.d. 16 Apr 2024. https://www.ada.gov/topics/intro-to-ada/#top.
34. Smith, AD. "Spatial navigation in autism spectrum disorders: a critical review." Front Psychol 6.31 (2015).
35. Charlop-Christy, MH, et al. "Using the picture exchange communication system (PECS) with children with autism: assessment of PECS acquisition, speech, social-communicative behavior, and problem behavior." J Appl Behav Anal 35.3 (2002): 213–31.
36. Barton-Hulsey, A, S Phinney and S Collins. "Augmentative and Alternative Communication Supports for Language and Literacy in Preschool: Considerations for Down Syndrome and Autism Spectrum Disorder." Semin Speech Lang 42.4 (2021): 345–362.
37. Brignell, A, et al. "Communication interventions for autism spectrum disorder in minimally verbal children." Cochrane Database Syst Rev 11.11 (2018): CD012324.
38. Paldam, E, et al. "A robot or a dumper truck? Facilitating play-based social learning across neurotypes." Autism Dev Lang Impair 7 (2022).
39. Devnani, PA and AU Hegde. "Autism and sleep disorders." J Pediatr Neurosci 10.4 (2015): 304–7.
40. Dharampuriya, PR and SL Abend. "Roadmap for Creating Effective Communication Tools to Improve Health Equity for Persons With Intellectual and Developmental Disabilities." Front Health Serv 2 (2022).
41. Heifetz, M and Y Lunsky. "Implementation and evaluation of health passport communication tools in emergency departments." Res Dev Disabil 72 (2018): 23–32.
42. Soh, TY Tuan, et al. "Adoption of a Digital Patient Health Passport as Part of a Primary Healthcare Service Delivery: Systematic Review." J Pers Med 12.11 (2022): 1814.

Index

J. L. Wells (ed.), *A Clinical Lens on Pediatric Engineering*,
https://doi.org/10.1007/978-3-031-84655-7

9783031846540